# Village Technology Handbook

Volunteers in Technical Assistance
1815 North Lynn Street
Arlington, Virginia 22209 USA

**Village Technology Handbook**



(This is the third edition of a manual first published in 1963, with the support of the U. S. Agency for International Development, and revised in 1970, which has gone through eight major printings.)

Manufactured in the United States of America.

Set in *Times Roman* type on an IBM personal computer, a gift to VITA from International Business Machines Corporation, using WordPerfect software donated by WordPerfect Corporation.

Published by: Volunteers in Technical Assistance
1815 North Lynn Street, Suite 200
Arlington, Virginia 22209 USA

10 9 8 7 6 5 4 3 2 1

**Library of Congress Cataloging-in-Publication Data**

Village technology handbook.

Bibliography: p. 413
1. Building--Amateurs' manuals. 2. Do-it-yourself work. 3. Home economics, Rural--Handbooks, manuals, etc. I. Volunteers in Technical Assistance.
TH148.V64 1988 620'.41734 88-5700
ISBN 0-86619-275-1

# Village Technology Handbook

# Table of Contents

AGRICULTURE

## FOOD PROCESSING AND PRESERVATION

## CONSTRUCTION

# Foreword

The Village Technology Handbook has been an important tool for development workers and do-it-yourselfers for 25 years. First published in 1963 under the auspices of the U.S. Agency for International Development, the *Handbook* has gone through eight major printings. Versions in French and Spanish, as well as English, are on shelves in bookstores, on desks in government offices and local organizations, in school libraries and technical centers, and in the field kits of village workers around the world. The technologies it contains, like the chain and washer pump, the evaporative food cooler, and the hay box cooker, have been built for technology fairs and demonstration centers throughout the developing world–and more importantly, have been adopted and adapted by people everywhere.

Because the *Handbook* has been a faithful friend for so long, this revision was approached with care. As even the best of friendships needs an occasional reassessment, our question was how to update the book without damaging its fundamental utility–to avoid throwing the baby out with the bath water.

We began by circulating sections of the book to VITA Volunteers with expertise in the various technical areas. We asked them to take a good hard look at what was presented and let us know what should be revised, updated, discarded, replaced. The volunteers' replies affirmed what tens of thousands of users around the world have recognized over the years, that the basic material was sound. Where they suggested changes, additions, and deletions, we have done our best to oblige.

Concurrently, we reviewed the comments that many of those users have sent to us over the years. Comments on what worked, what caused trouble, and what would be nice to have included. With so much going on in the development of small-scale, village technologies, the latter category was extensive. But because so much of the original book is still very applicable today, we opted to make the additions and changes selectively. We made the decision to add to this volume where it seemed most feasible, and to begin to compile a companion volume that will cover a selection of those other technologies.

Since the *Handbook* is primarily intended for "do-it-yourselfers" in villages and rural regions, most space still is allocated to the development of water resources and to agriculture. And rather than simply replacing everything and starting over, this new edition reorganizes some sections, updates several of the original articles, and includes a number of new ones on frequently requested topics. The new articles cover energy efficient stoves, the use of wind power to pump water, stabilized earth construction, a novel ceramics kiln, small-scale candle and paper production, high yield gardening, oral rehydration therapy, and malaria control. An all-new reference section is also provided.

VITA is committed to assisting sustainable growth: that is, to progress, based on expressed needs, that increases self reliance. Access to clearly presented technical information is a key to such growth. VITA searches out, develops, and disseminates techniques and devices that contribute to self suffiency. *The Village Technology Handbook* is one such VITA effort to support sustainable growth with easy to read technical information for the communities of the world.

VITA Volunteers are similarly committed to helping VITA help others, and many of them were involved in this project, reviewing material in their technical fields. VITA wishes to thank Robert M. Ross and David C. Neubert for reviewing the sections on agriculture; Phil D. Weinert, Charles G. Burney, Walter Lawrence, and Steven Schaefer, water resources and purification; Malcolm C. Bourne and Norman M. Spain, food processing and preservation; Dwight R. Brown and William Perenchio, construction; Charles D. Spangler, sanitation; Jeff Wartluft, Mark Hadley, Marietta Ellis, Gerald Kinsman, and Peter Zweig, home improvement; Dwight Brown and Victor Palmeri, crafts and village industries; and Grant Rykken, communications.

Most especially, we would like to thank VITA Volunteer engineer and literacy specialist Len Doak, who was coaxed out of retirement and away from the fishing docks to coordinate the revision, sort out the comments, and pull the new pieces together.

VITA staff who were involved included Suzanne Brooks, administrative support and graphics; Julie Berman, administrative support; Margaret Crouch, editorial; and Maria Garth, typesetting.

And finally, this effort has given all of us a new respect for Dan Johnson, one of VITA's "founding fathers" and currently a member of the Board of Directors, who devoted a year of his life to putting the original *Handbook* together a quarter of a century ago. That so much of that work has stood the test of time is due in no small measure to the care with which he and the other VITA Volunteers who worked with him approached their task.

**–VITA Publications**
January 1988

# Notes on Using the Handbook

## INTRODUCTION

The *Village Technology Handbook* contains eight major subject sections, each containing several articles. The articles cover both the broad topic areas such as agriculture, as well as specific agricultural projects such as building a scraper.

If you are planning an entirely new project you would benefit by reading the entire section through. If you are planning a specific project (such as building a wind-driven water pump) only that article need be read.

The skills needed for each of the projects described vary considerably, but none of the projects requires more than the usual construction and trade skills such as carpentry, welding, or farming that are generally found in most modest sized villages.

When the materials suggested in the *Handbook* are not available, it may be possible to substitute other materials. Be careful to make any changes in dimensions made necessary by such substitutions.

If you need translations of articles from the *Handbook*, we ask that you let us know. The book itself has been translated into English, French, and Spanish, and some individual articles may be available in other languages.

The articles in the *Handbook* came from many sources. Your comments and suggestions for changes, difficulties with any of the projects described, or ideas for new articles are welcome. Those kinds of comments were a very important element in preparing this revised edition, and we expect to rely on them in the future as well. Please send your comments so that we may continue to share.

## SUMMARY OF THE HANDBOOK BY SECTION

### Section 1. Water

Water resources are so vital that extensive coverage is provided. Much of this material is from the original, but it has been reorganized and updated. The sequence of articles begins with principles of hydrology that explain where underground water is likely to be found. This is followed by articles on types of wells and how to make well drilling tools and how to drill or dig the wells.

Next come articles on practical methods to lift water from wells and to transport it. Articles on several pumps and water piping occur here. A new article on wind-driven pumps is in this section. A number of charts and tables help in the calculation of pipe size and water flow.

Water storage and purification are the topics of the next series of articles. This section is unchanged from the earlier edition, but several new references are listed.

## Section 2. Health and Sanitation

Next to pure water, sanitation is one of the most critical health needs of any society. This section begins with two brief articles on the principles for disposal of human waste. These are followed by details of how to build various types of latrines. Also included is an article on bilharziasis (schistosomiasis) and a new articles on malaria control and oral rehydration therapy.

## Section 3. Agriculture

Seven topics are covered, beginning with earth moving devices to level fields and build irrigation ditches. This is followed by directions for an irrigation system based on concrete tile, including how to make the tile in the field. A variety of material on raising poultry is included, and a new article on small, high yield gardens has been added.

## Section 4. Food Processing and Preservation

The articles in this section describe storage and handling of different types of food, evaporative coolers and other cold storage technologies, and a variety of other storage and processing systems and devices. The section has been revised and updated and new references have been added.

## Section 5. Construction

Much of this section deals with construction of buildings and walls using concrete or bamboo. A new article on stabilized earth construction has been added, and instructions for making glues to use in construction are also included.

## Section 6. Home Improvements

Washing clothes, cooking, making soap, and making bedding are covered here. An important new addition is an article on the construction of an energy efficient cookstove developed in West Africa. The stove has shown more than double the fuel efficiency of the traditional open fire.

## Section 7. Crafts and Village Industry

Traditional crafts that lend themselves to development as small businesses are discussed in this section--pottery, hand papermaking, and candle making. Ceramic kilns described include an alternative kiln design fueled by waste motor oil.

## Section 8. Communications

This section remains unchanged from the original on the premise that while changes in communications could actually fill volumes on their own, there are many places in developing areas where the simple technologies presented here are still quite useful. Simple writing instruments and silk screen printing are discussed. The skills and materials described should be available in most rural villages.

# SOURCES OF ADDITIONAL INFORMATION

Each article in the *Handbook* concludes with one or more source references. These and other sources of information have been compiled into the new expanded Reference section at the back of the book. VITA publications that are listed may be ordered directly from VITA Publications, Post Office Box 12028, Arlington, Virginia 22204 USA.

You may also request technical assistance from VITA Volunteer experts by writing to VITA, 1815 North Lynn Street, Suite 200, Arlington, Virginia 22209 USA.

# About VITA

Volunteers in Technical Assistance (VITA) is a private, nonprofit, international development organization. It makes available to individuals and groups in developing countries a variety of information and technical resources aimed at fostering self sufficiency–needs assessment and program development support; by-mail and on-site consulting services; information systems training; and management of long-term field projects.

Throughout its history, VITA has concentrated on practical and workable technologies for development. It has collected, organized, tested, synthesized, and disseminated information on these technologies to more than 70,000 requesters and hundreds of organizations in the developing countries. As the information revolution dawned, VITA found itself in a leadership position in the effort to bring the benefits of that revolution to those in the Third World who are traditionally passed over in the development process.

Perhaps of greatest significance is VITA's emphasis on technologies that are commercially viable. These have the potential of creating new wealth through adding value to local materials, thereby creating jobs and increasing income as well as strengthening the private sector. We have increasingly translated our experiences in information management to the implementation of projects in the field. This evolution from information to implementation to create jobs, businesses, and new wealth is what VITA is really about. It provides missing links without creating dependency.

VITA places special emphasis on the areas of agriculture and food processing, renewable energy applications, water supply and sanitation, housing and construction, and small business development. VITA's activities are facilitated by the active involvement of thousands of VITA Volunteer technical experts from around the world, and by its documentation center containing specialized technical material of interest to people in developing countries.

VITA currently publishes over 150 technical manuals, papers, and bulletins, many available in French and Spanish as well as English. Manuals deal with construction or implementation details for such specific topics as windmills, reforestation, water wheels, and rabbit raising. In addition, VITA Technical Bulletins present plans and case studies of specific technologies to encourage further experimentation and testing. The technical papers–"Understanding Technology"–offer general introductions to the applications and necessary resources for technologies or technical systems. Included in the series are topics that range from composting to

Stirling engines, from sanitation at the community level to tropical root crops. Publications catalogues are available upon request.

*VITA News* is a quarterly magazine that provides an important communications link among far-flung organizations involved in technology transfer and adaptation. The *News* contains articles about projects, issues, and organizations around the world, reviews of new books, technical abstracts, and a resources bulletin board.

VITA derives its income from government, foundation, and corporate grants; fees for services; contracts; and individual contributions.

For further information write to VITA, 1815 North Lynn Street, Suite 200, Arlington, Virginia 22209 USA.

# Symbols and Abbreviations Used in this Book

| | |
|---|---|
| @ . . . . | at |
| " . . . . | inch |
| ' . . . . | foot |
| C . . . . | degrees Celsius (Centigrade) |
| cc . . . . | cubic centimeter |
| cm. . . . | centimeter |
| cm/sec . . | centimeters per second |
| d or dia . . | diameter |
| F . . . . | degrees Fahrenheit |
| gm. . . . | gram |
| gpm. . . . | gallons per minute |
| HP. . . . | horsepower |
| kg . . . . | kilogram |
| km. . . . | kilometer |
| l . . . . | liter |
| l/pm . . . | liters per minute |
| l/sec . . . | liters per second |
| m . . . . | meter |
| ml . . . . | milliliters |
| mm . . . | millimeters |
| m/m. . . . | meters per minute |
| m/sec. . . | meters per second |
| ppm. . . . | parts per million |
| R . . . . | radius |

# Water Resources

International Resource Center

# Developing Water Sources

There are three main sources of water for small water-supply systems: ground water, surface water, and rainwater. The choice of the source of water depends on local circumstances and the availability of resources to develop the water source.

A study of the local area should be made to determine which source is best for providing water that is (1) safe and wholesome, (2) easily available, and (3) sufficient in quantity. The entries that follow describe the methods for tapping ground water:

- Tubewells
  - Well Casings and Platforms
  - Hand-Operated Drilling Equipment
  - Driven Wells
- Dug Wells
- Spring Development

Once the water is made available, it must be brought from where it is to where it is needed and steps must be taken to be sure that it is pure. These subjects are covered in the major sections that follow:

- Water Lifting and Transport
- Water Storage and Treatment

## GETTING GROUND WATER FROM WELLS & SPRINGS

This section defines ground water, discusses its occurrence, and explains its movement. It describes how to decide on the best site for a well, taking into consideration the nearness to surface water, topography, sediment type, and nearness to pollutants. It also discusses briefly the process of capping and sealing the well and developing the well to assure maximum flow of water.

### Ground Water

Ground water is subsurface water, which fills small openings (pores) of loose sediments (such as sand and gravel) or rocks. For example, if we took a clear glass bowl, filled it with sand, and then poured in some water, we would notice the water "disappear" into the sand (see Figure 1). However, if we looked through the side of the bowl, we would see water in the sand, but below the top of the sand. The sand containing the water is said to be saturated. The top of the saturated sand is called the *water table;* it is the level of the water in the sand.

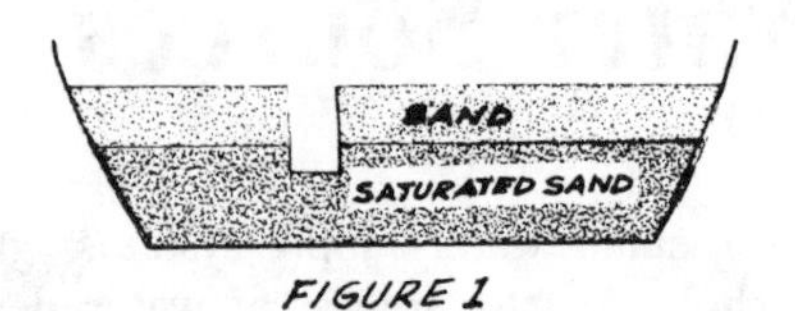


FIGURE 1

The water beneath the water table is true ground water available (by pumping) for human use. There is water in the soil above the water table, but it does not flow into a well and is not available for use by pumping.

If we inserted a straw into the saturated sand in the bowl in Figure 1 and sucked on the straw, we would obtain some water (initially, we would get some sand too). If we sucked long enough, the water table or water level would drop toward the bottom of the bowl. This is exactly what happens when water is pumped from a well drilled below the water table.

The two basic factors in the occurrence of ground water are: (1) the presence of water, and (2) a medium to "house" the water. In nature, water is provided by precipitation (rain and snow) and surface water features (rivers and lakes). The medium is porous rock or loose sediments.

The most abundant ground water reservoir occurs in the loose sands and gravels in river valleys. Here the water table roughly parallels the land surface, that is, the depth to the water table is generally constant. Disregarding any drastic changes in climate, natural ground water conditions are fairly uniform or balanced. In Figure 2, the water poured into the bowl (analogous to precipitation) is balanced by the water discharging out of the bowl at the lower elevation (analogous to discharge into a stream). This movement of ground water is slow, generally just centimeters or inches per day.

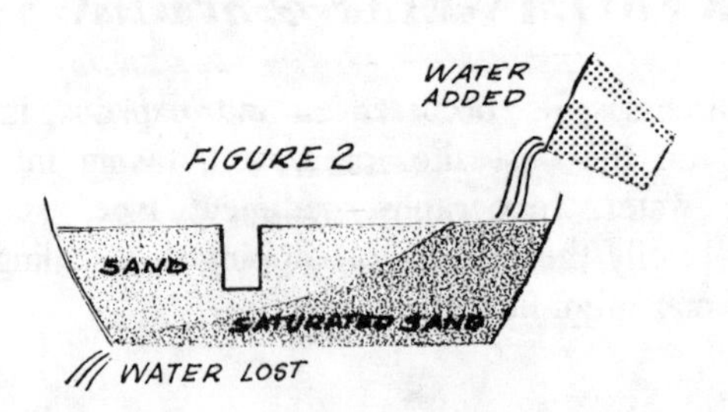


FIGURE 2

When the water table intersects the land surface, springs or swamps are formed (see Figure 3). During a particularly wet season, the water table will come much closer to the land surface than it normally does and many new springs or swampy areas will appear. On the other hand, during a particularly dry season, the water table will be lower than normal and many springs will "dry up." Many shallow wells may also "go dry."

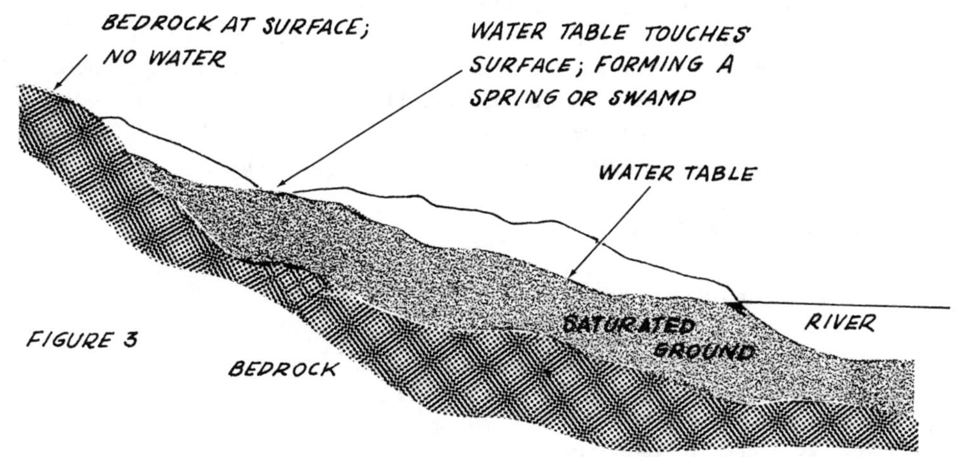

FIGURE 3

## Flow of Water to Wells

A newly dug well fills with water a meter or so (a few feet) deep, but after some hard pumping it becomes dry. Has the well failed? Was it dug in the wrong place? More likely you are witnessing the phenomenon of *drawdown,* an effect every pumped well has on the water table (see Figure 4).

Because water flows through sediments slowly, almost any well can be pumped dry temporarily if it is pumped hard enough. Any pumping will lower the water level to some degree, in the manner shown in Figure 4. A serious problem arises only when the drawdown due to normal use lowers the water table below the level of the well.

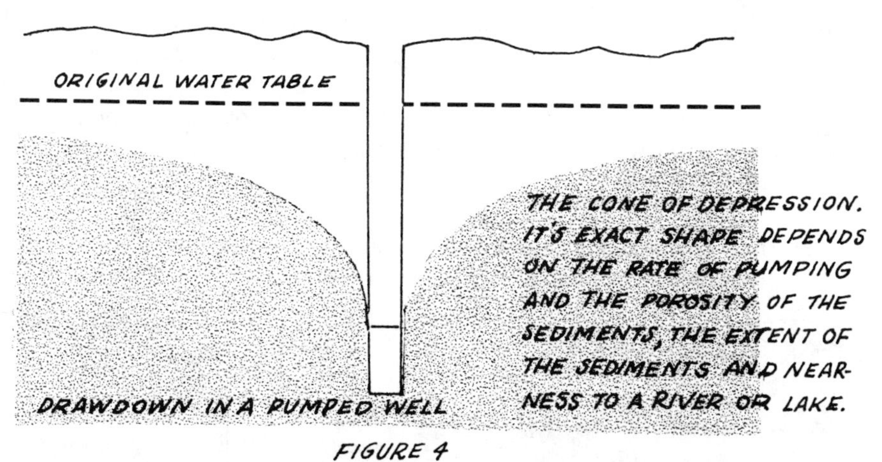

FIGURE 4

After the well has been dug about a meter (several feet) below the water table, it should be pumped at about the rate it will be used to see if the flow into the well is adequate. If it is not sufficient, there may be ways to improve it. Digging the well deeper or wider will not only cut across more of the water-bearing layer to allow more flow into the well, but it will also enable the well to store a greater quantity of the water that may seep in overnight. If the well is still not adequate and can be dug no deeper, it can be widened further, perhaps lengthened in one direction, or more wells can be dug. The goal of all these methods is to intersect more of the water-bearing layers, so that the well will produce more water without lowering the water table to the bottom of the well.

## Where to Dig a Well

Four important factors to consider in choosing a well site are:

- Nearness to Surface Water
- Topography
- Sediment Type
- Nearness to Pollutants

### *Nearness to Surface Water*

If there is surface water nearby, such as a lake or a river, locate the well as near to it as possible. It is likely to act as a source of water and keep the water table from being lowered as much as without it. This does not always work well, however, as lakes and slow-moving bodies of water generally have silt and slime on the bottom, which prevent water from entering the ground quickly.

There may not seem to be much point to digging a well near a river, but the filtering action of the soil will result in water that is cleaner and more free of bacteria. It may also be cooler than surface water. If the river level fluctuates during the year, a well will give cleaner water (than stream water) during the flood season, although ground water often gets dirty during and after a flood. A well will also give more reliable water during the dry season, when the water level may drop below the bed of the river. This method of water supply is used by some cities: a large well is sunk next to a lake or river and horizontal tunnels are dug to increase the flow.

Wells near the ocean, and especially those on islands, may have not only the problem of drawdown, but that of salt water encroachment (see Figure 5). The

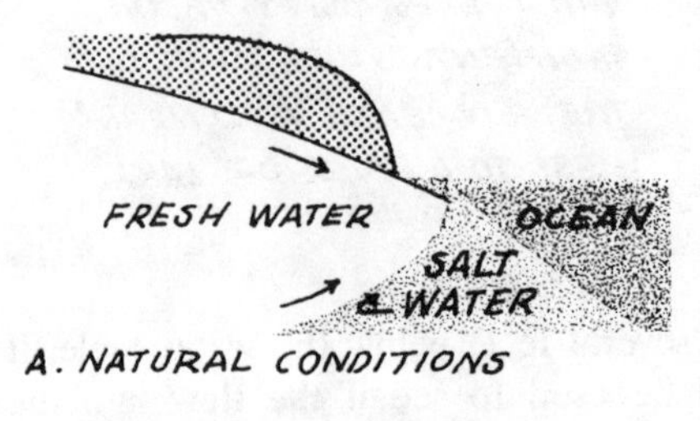


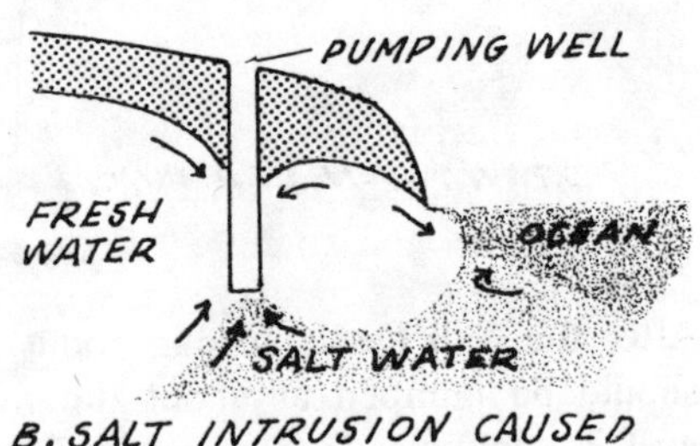


FIGURE 5

underground boundary between fresh and salt water generally slopes inland: Because salt water is heavier than fresh water, it flows in under it. If a well near the shore is used heavily, salt water may come into the well as shown. This should not occur in wells from which only a moderate amount of water is drawn.

## *Topography*

Ground water, being liquid, gathers in low areas. Therefore, the lowest ground is generally the best place to drill or dig. If your area is flat or steadily sloping, and there is no surface water, one place is as good as another to start drilling or digging. If the land is hilly, valley bottoms are the best places to look for water.

You may know of a hilly area with a spring on the side of a hill. Such a spring could be the result of water moving through a layer of porous rock or a fracture zone in otherwise impervious rock. Good water sources can result from such features.

## *Sediment Type*

Ground water occurs in porous or fractured rocks or sediments. Gravel, sand and sandstone are more porous than clay, unfractured shale and granite or "hard rock."

Figure 6 shows in a general way the relationship between the availability of ground water (expressed by typical well discharges) and geologic material (sediments and various rock types). For planning the well discharge necessary for irrigating crops, a good rule of thumb for semi-arid climates–37.5cm (15") of precipitation a year–is a 1500- to 1900-liters (400 to 500 U.S. gallons)-per-minute well that will irrigate about 65 hectares (160 acres) for about six months. From Figure 6, we see that wells in sediments are generally more than adequate. However, enough ground water can be obtained from rock, if necessary, by drilling a number of wells. Deeper water is generally of better quality.

Sand and gravel are normally porous and clay is not, but sand and gravel can contain different amounts of silt and clay, which will reduce their ability to carry water. The only way to find the yield of a sediment is to dig a well and pump it.

In digging a well, be guided by the results of nearby wells and the effects of seasonal fluctuations on nearby wells. And keep an eye on the sediments in your well as it is dug. In many cases you will find that the sediments are in layers, some porous and some not. You may be able to predict where you will hit water by comparing the layering in your well with that of nearby wells.

Figures 7, 8, and 9 illustrate several sediment situations and give guidelines on how deep to dig wells.

**FIGURE 6:** Availability of Ground Water in Water Bearing Sediments or Rock Types

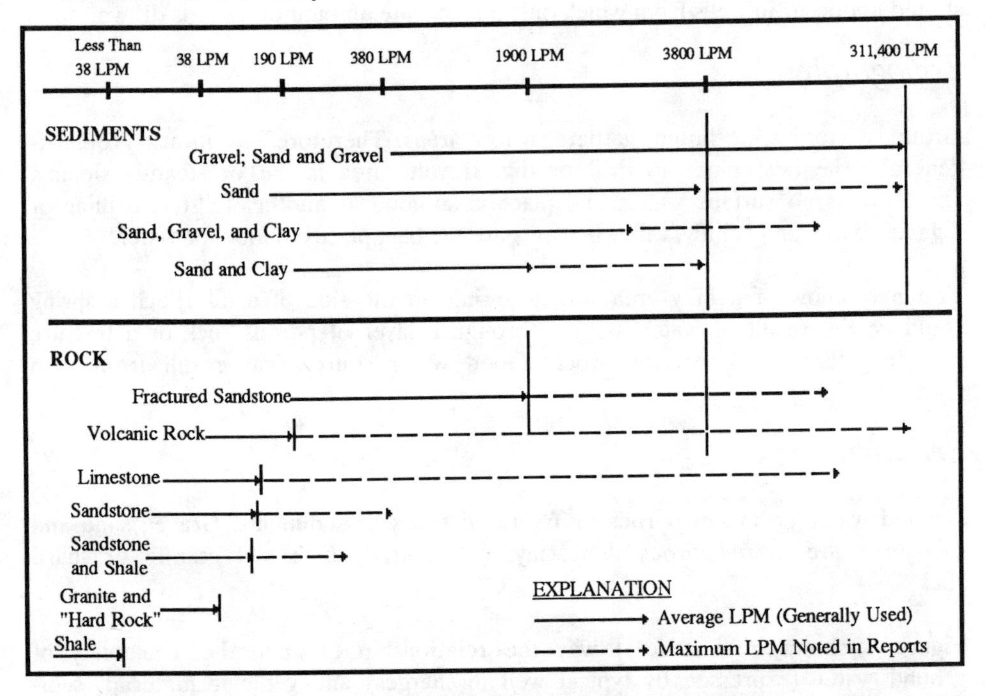


*Aquifers (water bearing sediments) of Sand and Gravel.* Generally yield 11,400 LPM (300 gpm) (but they may yield less depending on pump, well construction, and well development.

*Aquifers of Sand, Gravel, and Clay (Intermixed or Interbedded).* Generally yield between 1900 LPM (500 gpm) and 3800 LPM (1000 gpm), but can yield more —between 3800 LPM (1000 gpm) and 11,400 LPM (3000 gpm)— depending on the percentage of the constituents.

*Aquifers of Sand and Clay.* Generally yield about 1900 LPM (500 gpm) but may yield as much as 3800 LPM (1000 gpm).

*Aquifers of Fractured Sandstone.* Generally yield about 1900 LPM (500 gpm) but may yield more than 3800 LPM (1000 gpm) depending on the thickness of the sandstone and the degree and extent of fracturing (may also yield less than 1900 LPM (500 gpm) if thin and poorly fractured or interbedded with clay or shale).

*Aquifers of Limestone.* Generally yield between 38 LPM (10gpm) but have been known to yield more than 3800 LPM (1000 gpm) due to caverns or nearness of stream, etc.

*Aquifers of Granite and/or "Hard Rock."* Generally yield 38 gpm (10gpm) and may yield less (enough for a small household).

*Aquifers of Shale.* Yield less than 38 LPM (10gpm), not much good for anything except as a last resort.

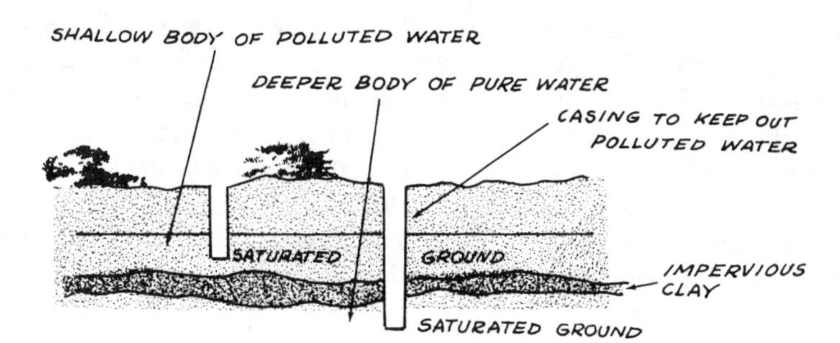


FIGURE 7

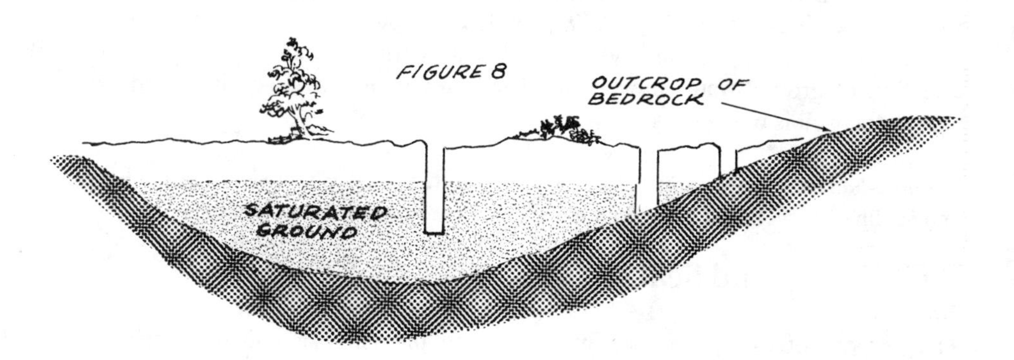


FIGURE 8

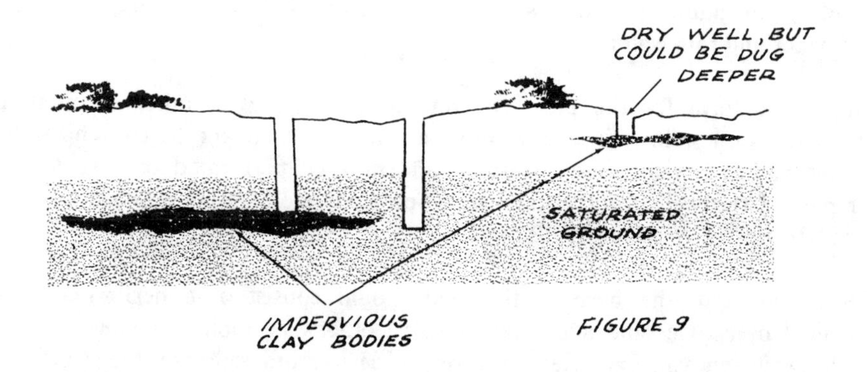


FIGURE 9

## *Nearness to Pollutants*

If pollution is in the ground water, it moves with it. Therefore, a well should always be uphill and 15 to 30 meters (50 to 100 feet) away from a latrine, barnyard, or other source of pollution. If the area is flat, remember that the flow of ground water will be downward, like a river, toward any nearby body of surface water. Locate a well in the upstream direction from pollution sources.

The deeper the water table, the less chance of pollution because the pollutants must travel some distance downward before entering ground water. The water is purified as it flows through the soil.

Extra water added to the pollutants will increase their flow into and through the soil, although it will also help dilute them. Pollution of ground water is more likely during the rainy than the dry season, especially if a source of pollution such as a latrine pit is allowed to fill with water. See also the Overview to the Sanitary Latrines section, p. 149. Similarly, a well that is heavily used will increase the flow of ground water toward it, perhaps even reversing the normal direction of ground-water movement. The amount of drawdown is a guide to how heavily the well is being used.

Polluted surface water must be kept out of the well pit. This is done by casing and sealing the well and providing good drainage around the well cover.

# Well Casing and Seal

The purpose of casing and sealing wells is to prevent contaminated surface water from entering the well or nearby ground water. As water will undoubtedly be spilled from any pump, the top of the well must be sealed with a concrete slab to let the water flow away rather than re-enter the well directly. It is also helpful to build up the pump area with soil to form a slight hill that will help drain away spilled water and rain water.

Casing is the term for the pipe, concrete or grout ring, or other material that supports the well wall. It is usually impermeable in the upper part of the well to keep out polluted water (see Figure 7) and may be perforated or absent in the lower part of the well to let water enter. See also "Well Casing and Platforms," p. 12, and "Reconstructing Dug Wells," p. 57.

In loose sediment, the base of the well should consist of a perforated casing surrounded by coarse sand and small pebbles; otherwise, rapid pumping may bring into the well enough material to form a cavity and collapse the well itself. Packing the area around the well hole in the water-bearing layer with fine gravel will prevent sand from washing in and increase the effective size of the well. The ideal gradation is from sand to 6mm (1/4") gravel next to the well screen. In a drilled well it may be added around the screen after the pump pipe is installed.

## Well Development

Well development refers to the steps taken after a well is drilled to ensure maximum flow and well life by preparing the sediments around the well. The layer of sediments from which the water is drawn often consists of sand and silt. When the well is first pumped, the fine material will be drawn into the well and make the water muddy. You will want to pump out this fine material to keep it from muddying the water later and to make the sediments near the well more porous. However, if the water is pumped too rapidly at first, the fine particles may collect against the perforated casing or the sand grains at the bottom of the well and block the flow of water into it.

A method for removing the fine material successfully is to pump slowly until the water clears, then at successively higher rates until the maximum of the pump or well is reached. Then the water level should be permitted to return to normal and the process repeated until consistently clear water is obtained.

Another method is surging, which is moving a plunger (an attachment on a drill rod) up and down in the well. This causes the water to surge in and out of the sedimentary layer and wash loose the fine particles, as well as any drilling mud stuck on the wall of the well. Coarse sediment washed into the well can be removed by a bailing bucket, or it may be left in the bottom of the well to serve as a filter.

**Sources:**

Anderson, K.E. *Water Well Handbook.* Rolla, Missouri: Missouri Water Wells Drillers Association, 1965.

Baldwin, H.L. and McGuinness, C.L. *A Primer on Ground Water.* Washington, D.C.: U.S. Government Printing Office, 1964.

Davis, S.N. and DeWiest, R.J.M. *Hydrogeology.* New York: Wiley & Sons, 1966.

Todd, D.K. *Ground Water Hydrology.* New York: Wiley & Sons, 1959.

Wagner, E.G. and Lanoix, J.N. *Water Supply for Rural Areas and Small Communities.* Geneva: World Health Organization, 1959.

*Ground Water and Wells.* Saint Paul, Minnesota: Edward E. Johnson, Inc., 1966.

*Small Water Supplies,* Bulletin No. 10. London: The Ross Institute, 1967.

U.S. Army. *Wells.* Technical Manual 5-297. Washington, D.C.: U.S. Government Printing Office, 1957.

# TUBEWELLS

Where soil conditions permit, the tubewells described here will, if they have the necessary casing, provide pure water. They are much easier to install and cost much less than large diameter wells.

Tubewells will probably work well where simple earth borers or earth augers work (i.e., alluvial plains with few rocks in the soil), and where there is a permeable water-bearing layer 15 to 25 meters (50 to 80 feet) below the surface. They are sealed wells, and consequently sanitary, which offer no hazard to small children. The small amounts of materials needed keep the cost down. These wells may not yield enough water for a large group, but they would be big enough for a family of a small group of families.

The storage capacity in small diameter wells is small. Their yield depends largely on the rate at which water flows from the surrounding soil into the well. From a saturated sand layer, the flow is rapid. Water flowing in quickly replaces water drawn from the well. A well that taps such a layer seldom goes dry. But even when water-bearing sand is not reached, a well with even a limited storage capacity may yield enough water for a household.

## Well Casing and Platforms

In home or village wells, casing and platforms serve two purposes: (1) to keep well sides from caving in, and (2) to seal the well and keep any polluted surface water from entering it.

Two low-cost casing techniques are described here:

1. Method A (see Figure 1), from an American Friends Service Committee (AFSC) team in Rasulia, Madhya Pradesh, India.

2. Method B, from an International Voluntary Services (IVS) team in Vietnam.

### *Method A*

#### Tools and Materials

Casing pipe (from pump to water-bearing layer to below minimum water table)–Asbestos cement, tile, concrete, or even galvanized iron pipe will do
Sand
Gravel
Cement
Device for lowering and placing casing (see Figure 2)
Drilling rig - see "Tubewell Boring"
Foot valve, cylinder, pipe, hand pump

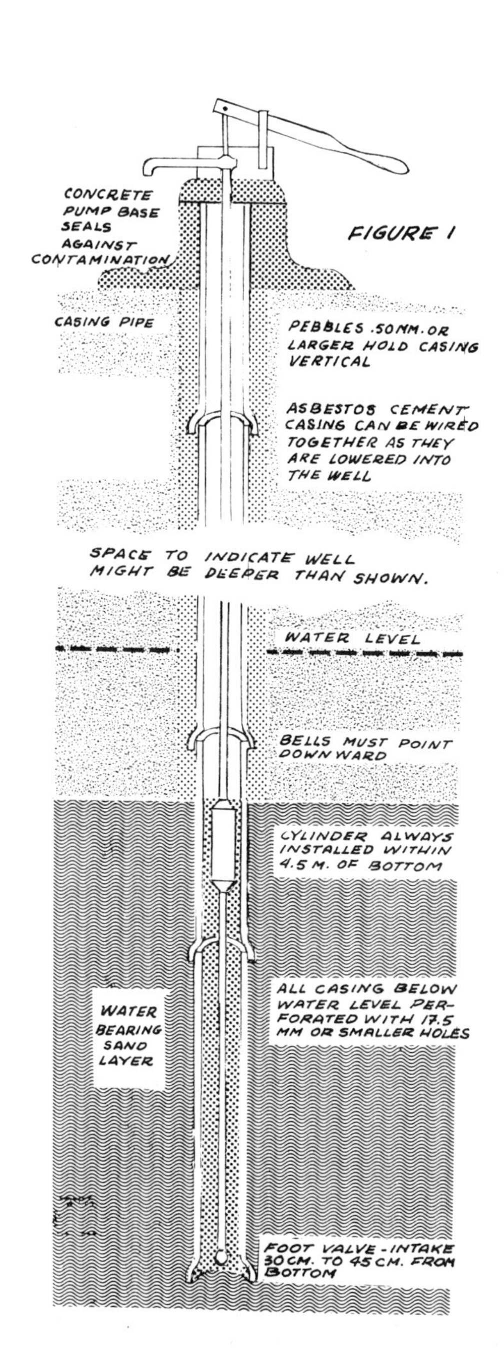


The well hole is dug as deep as possible into the water-bearing strata. The diggings are placed near the hole to make a mound, which later will serve to drain spilled water away from the well. This is important because backwash is one of the few sources of contamination for this type of well. The entire casing pipe below water level should be perforated with many small holes no larger than 5mm (3/16") in diameter. Holes larger than this will allow coarse sand to be washed inside and plug up the well. Fine particles of sand, however, are expected to enter. These should be small enough to be pumped immediately out through the pump. This keeps the well clear. The first water from the new well may bring with it large quantities of fine sand. When this happens, the first strokes should be strong and steady and continued until the water comes clear.

Perforated casing is lowered, bell end downward, into the hole using the device shown in Figure 2. When the casing is properly positioned, the trip cord is pulled and the next section prepared and lowered. Since holes are easily drilled in asbestos cement pipe, they can be wired together at the joint and lowered into the well. Be sure the bells point downward, since this will prevent surface water or backwash from entering the well without the purifying filtration effect of the soil; it will also keep sand and dirt from filling the well. Install the casing vertically and fill the remaining space with pebbles. This will hold the casing plumb. The

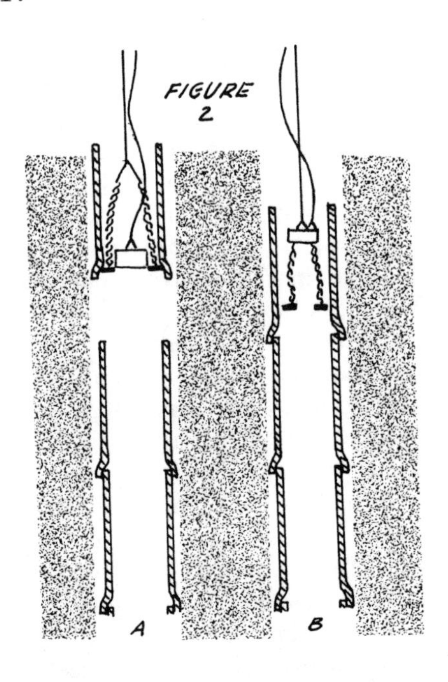


casing should rise 30 to 60cm (1' to 2') above ground level and be surrounded with a concrete pedestal to hold the pump and to drain spilled water away from the hole. Casing joints within 3 meters (10 feet) of the surface should be sealed with concrete or bituminous material.

## *Method B*

Plastic seems to be an ideal casing material, but because it was not readily available, the galvanized iron and concrete casings described here were developed in the Ban Me Thuot area of Vietnam.

**Tools and Materials**

Wooden V-block, 230cm (7 1/2') long (see Figure 3)
Angle iron, 2 sections, 230cm (7 1/2') long
Pipe, 10cm (4") in diameter, 230cm (7 1/2') long
Clamps
Wooden mallet
Soldering equipment
Galvanized sheet metal: 0.4mm x 1m x 2m (0.016" x 39 1/2" x 79")

## Plastic Casing

Black plastic pipe for sewers and drains was almost ideal. Its friction joints could be quickly slipped together and sealed with a chemical solvent. It seemed durable but was light enough to be lowered into the well by hand. It could be easily sawed or drilled to make a screen. Care must be taken to be sure that any plastic used is non-toxic.

## Galvanized Sheet Metal Casing

Galvanized sheet metal was used to make casing similar to downspouting. A thicker gauge than the 0.4mm (0.016") available would have been preferable. Because the sheet metal would not last indefinitely if used by itself, the well hole was made oversize and the ring-shaped space around the casing was filled with a thin concrete mixture which formed a cast concrete casing and seal outside the sheet metal when it hardened.

The 1-meter x 2-meter (39 1/2" x 79") sheets were cut lengthwise into three equal pieces, which yielded three 2-meter (79") lengths of 10cm (4") diameter pipe.

The edges were prepared for making seams by clamping them between the two angle irons, then pounding with a wooden mallet to the shape shown in Figure 3.

FIGURE 3

The seam is made slightly wider at one end than at the other to give the pipe a slight taper, which allows successive lengths to be slipped a short distance inside one another.

The strips are rolled by bridging them over a 2-meter (79") V-shaped wooden block and applying pressure from above with a length of 5cm (2") pipe (see Figure 4). The sheet metal strips are shifted from side to side over the V-block as they are being bent to produce as uniform a surface as possible. When the strip is bent enough, the two edges are hooked together and the 5cm (2") pipe is slipped inside. The ends of the pipe are set up on wooden blocks to form an anvil, and the seam is firmly crimped as shown in Figure 5.

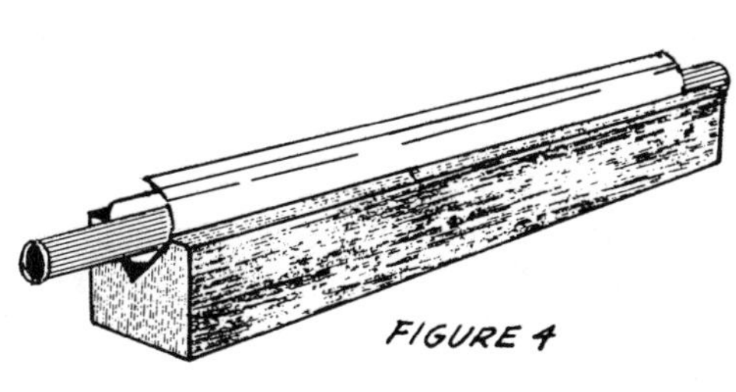
FIGURE 4

After the seam is finished, any irregularities in the pipe are removed by applying pressure by hand or with the wooden mallet and pipe anvil. A local tinsmith and his helper were able to make six to eight lengths (12 to 16 meters) of the pipe per day. Three lengths of pipe were slipped together and soldered as they were made, and the remaining joints had to be soldered as the casing was lowered into the well.

FIGURE 5

The lower end of the pipe was perforated with a hand drill to form a screen. After the casing was lowered to the bottom of the well, fine gravel was packed around the perforated portion of the casing to above the water level.

The cement grouting mortar used around the casings varied from pure cement to a 1:1 1/2 cement : sand ratio mixed with water to a very plastic consistency. The grout was put around the casing by gravity and a strip of bamboo about 10 meters (33 feet) long was used to "rod" the grout into place. A comparison of volume around the casing and volume of grouting used indicated that there may have been some voids left probably below the reach of the bamboo rod. These are not serious however, as long as a good seal is obtained for the first 8 to 10 meters (26 to 33 feet) down from the surface. In general, the greater proportion of cement used and the greater the space around the casing, the better seemed to be the results obtained. However, insufficient experience has been obtained to reach any final conclusions. In addition, economic considerations limit both of these factors.

Care must be taken in pouring the grout. If the sections of casing are not assembled perfectly straight, the casing, as a result, is not centered in the well and the pressure of the grouting is not equal all the way around. The casing may collapse. With reasonable care, pouring the grout in several stages and allowing it to set in-between should eliminate this. The grouting, however, cannot be poured in too many stages because a considerable amount sticks to the sides of the well each time, reducing the space for successive pourings to pass through.

This method can be modified for use in areas where the structure of the material through which the well is drilled is such that there is little or no danger of cave-in. In this situation, the casing serves only one purpose, as a sanitary seal. The well will be cased only about 8 meters (26 feet) down from the ground surface. To do this, the well is drilled to the desired depth with a diameter roughly the same as that of the casing. The well is then reamed out to a diameter 5 to 6cm (2" to 2 1/4") larger than the casing down to the depth the casing will go. A flange fitted at the bottom of the casing with an outside diameter about equal to that of the reamed hole will center the casing in the hole and support the casing on the shoulder where the reaming stopped. Grouting is then poured as in the original method. This modification (1) saves considerable costly material, (2) allows the well to be made a smaller diameter except near the top, (3) lessens grouting difficulties, and (4) still provides adequate protection against pollution.

## Concrete Tile Casing

If the well is enlarged to an adequate diameter, precast concrete tile with suitable joints could be used as casing. This would require a device for lowering the tiles into the well one by one and releasing them at the bottom. Mortar would have to be used to seal the joints above the water level, the mortar being spread on each successive joint before it is lowered. Asbestos cement casing would also be a possibility where it was available with suitable joints.

## No Casing

The last possibility would be to use no casing at all. It is felt that when finances or skills do not permit the well to be cased, there are certain circumstances under which an uncased well would be better than no well at all. This is particularly true in localities where the custom is to boil or make tea out of all water before drinking it, where sanitation is greatly hampered by insufficient water supply, and where small-scale hand irrigation from wells can greatly improve the diet by making gardens possible in the dry season.

The danger of pollution in an uncased well can be minimized by: (1) choosing a favorable site for the well and (2) making a platform with a drain that leads away from the well, eliminating all spilled water.

Such a well should be tested frequently for pollution. If it is found unsafe, a notice to this effect should be posted conspicuously near the well.

## *Well Platform*

In the work in the Ban Me Thuot area, a flat 1.75-meter (5.7') square slab of concrete was used around each well. However, under village conditions, this did not work well. Large quantities of water were spilled, in part due to the enthusiasm of the villagers for having a plentiful water supply, and the areas around wells became quite muddy.

The conclusion was reached that the only really satisfactory platform would be a round, slightly convex one with a small gutter around the outer edge. The gutter should lead to a concreted drain that would take the water a considerable distance from the well. It is worth noting that in Sudan and other very arid areas such spillage from community wells is used to water vegetable gardens or community nurseries.

If the well platform is too big and smooth, there is a great temptation on the part of the villagers to do their laundry and other washing around the well. This should be discouraged. In villages where animals run loose it is necessary to build a small fence around the well to keep out animals, especially poultry and pigs, which are very eager to get water, but tend to mess up the surroundings.

**Sources:**

Koegel, Richard G. Report. Ban Me Thuot, Vietnam: International Voluntary Services, 1959. (Mimeographed.)

Mott, Wendell. *Explanatory Notes on Tubewells.* Philadelphia: American Friends Service Committee, 1956. (Mimeographed.)

# Hand-Operated Drilling Equipment

Two methods of drilling a shallow tubewell with hand-operated equipment are described here: Method A, which was used by an American Friends Service Committee (AFSC) team in India, operates by turning an earth-boring auger. Method B, developed by an International Voluntary Services (IVS) team in Vietnam, uses a ramming action.

## *Earth Boring Auger*

This simple hand-drilling rig can be used to dig wells 15 to 20cm (6" to 8") in diameter up to 15 meters (50') deep.

**Tools and Materials**

Earth auger, with coupling to attach to 2.5cm (1") drill line (see entry on tubewell earth augers)
Standard weight galvanized steel pipe:

**For Drill Line:**

4 pieces: 2.5cm (1") in diameter and 3 meters (10') long (2 pieces have threads on one end only; others need no threads.)
2 pieces: 2.5cm (1") in diameter and 107cm (3 1/2") long

**For Turning Handle:**

2 pieces: 2.5cm (1") in diameter and 61cm (2') long
2.5cm (1") T coupling

**For Joint A:**

4 pieces: 32mm (1 1/4") in diameter and 30cm (1') long

**Sections and Couplings for Joint B:**

23cm (9") Section of 32mm (1 1/4") diameter (threaded at one end only)
35.5cm (14") Section of 38mm (1 1/2") diameter (threaded at one end only)
Reducer coupling: 32mm to 25mm (1 1/4" to 1")
Reducer coupling: 38mm to 25mm (1 1/2" to 1")
8 10mm (3/8") diameter hexagonal head machine steel bolts 45mm (1 3/4") long, with nuts
2 10mm (3/8") diameter hexagonal head machine steel bolts 5cm (2") long, with nuts
9 10mm (3/8") steel hexagonal nuts

**For Toggle Bolt:**

1 3mm (1/8") diameter countersink head iron rivet, 12.5mm (1/2") long
1 1.5mm (1/16") sheet steel, 10mm (3/8") x 25mm (1")

Drills: 3mm (1/8"), 17.5mm (13/16"), 8.75mm (13/32")
Countersink
Thread cutting dies, unless pipe is already threaded
Small Tools: wrenches, hammer, hacksaw, files
For platform: wood, nails, rope, ladder

Basically the method consists of rotating an ordinary earth auger. As the auger penetrates the earth, it fills with soil. When full it is pulled out of the hole and emptied. As the hole gets deeper, more sections of drilling line are added to extend the shaft. Joint A (Figures 1 and 2) is a simple method for attaching new sections.

By building an elevated platform 3 to 3.7 meters (10 to 12 feet) from the ground, a 7.6-meter (25 foot) long section of drill line can be balanced upright. Longer lengths are too difficult to handle. Therefore, when the hole gets deeper than 7.6 meters (25 feet), the drill line must be taken apart each time the auger is removed for emptying. Joint B makes this operation easier. See Figures 1 and 3.

Joint C (see construction details for Tubewell Earth Auger) is proposed to allow rapid emptying of the auger. Some soils respond well to drilling with an auger that has two sides open. These are very easy to empty, and would not require Joint C. Find out what kinds of augers are successfully used in your area, and do a bit of experimenting to find the one best suited to your soil. See the entries on augers.

Joint A has been found to be faster to use and more durable than pipe threaded connectors. The pipe threads become damaged and dirty and are difficult to start. Heavy, expensive pipe wrenches get accidentally dropped into the well and are hard to get out. These troubles can be avoided by using a sleeve pipe fastened with two 10mm (3/8") bolts. Neither a small bicycle wrench nor the inexpensive bolts will obstruct drilling if dropped in. Be sure the 32mm (1 1/4") pipe will fit over your 25mm (1") pipe drill line before purchase. See Figure 2.

Four 3-meter (10') sections and two 107cm (3 1/2') sections of pipe are the most convenient lengths for drilling a 15-meter (50') well. Drill an 8.75mm (13/32") diameter hole through each end of all sections of drill line except those attaching to Joint B and the turning handle, which must be threaded joints. The holes should be 5cm (2") from the end.

When the well is deeper than 7.6 meters (25'), several features facilitate the emptying of the auger, as shown in Figures 3 and 4. First, pull up the full auger

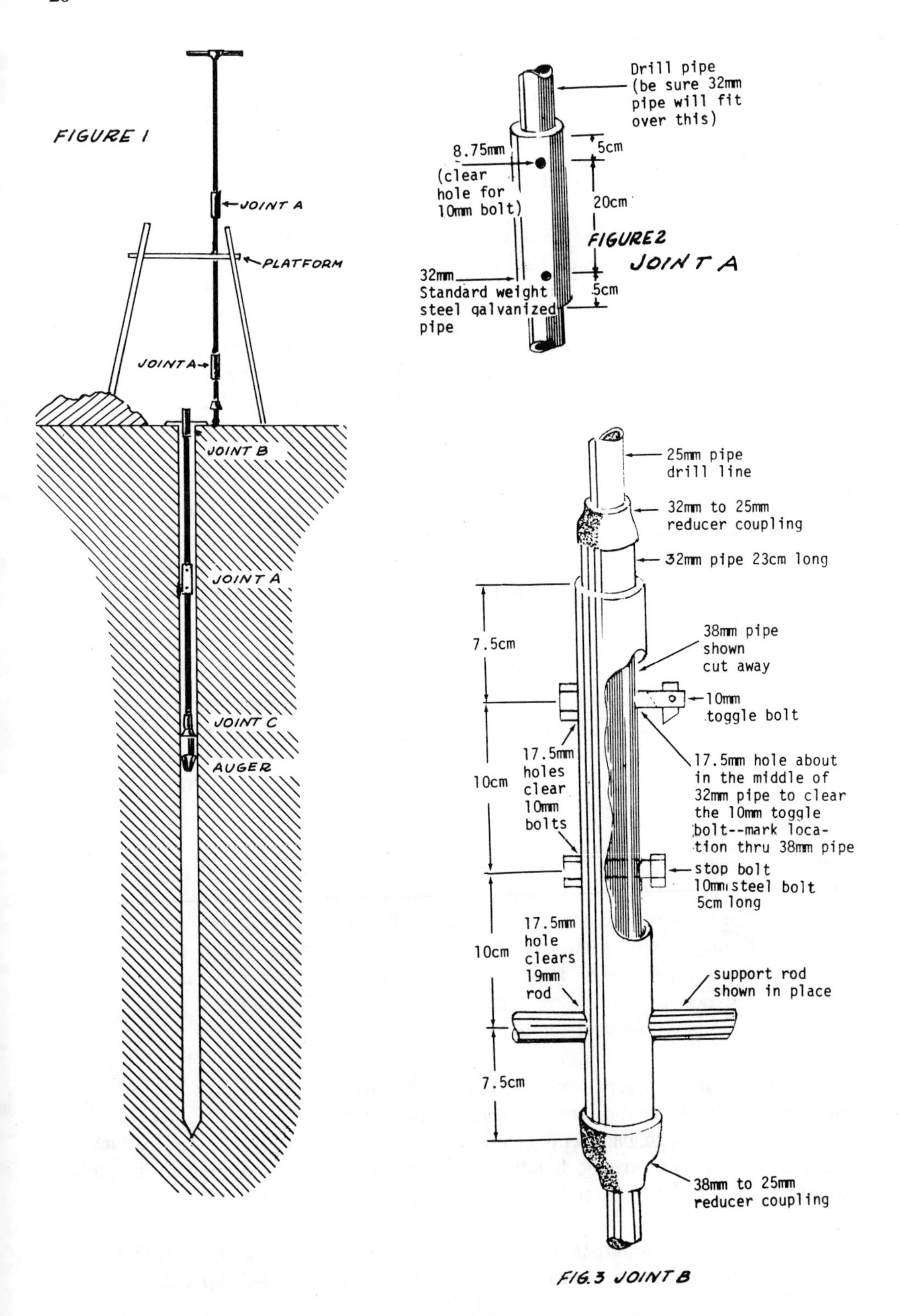
FIGURE 1
JOINT A
PLATFORM
JOINT A
JOINT B
JOINT A
JOINT C
AUGER
Drill pipe
(be sure 32mm
pipe will fit
over this)
8.75mm
(clear
hole for
10mm bolt)
5cm
20cm
FIGURE 2
JOINT A
32mm
Standard weight
steel galvanized
pipe
5cm
25mm pipe
drill line
32mm to 25mm
reducer coupling
32mm pipe 23cm long
7.5cm
38mm pipe
shown
cut away
10mm
toggle bolt
17.5mm
holes
clear
10mm
bolts
10cm
17.5mm hole about
in the middle of
32mm pipe to clear
the 10mm toggle
bolt--mark loca-
tion thru 38mm pipe
stop bolt
10mm steel bolt
5cm long
17.5mm
hole
clears
19mm
rod
10cm
support rod
shown in place
7.5cm
38mm to 25mm
reducer coupling

FIG.3 JOINT B

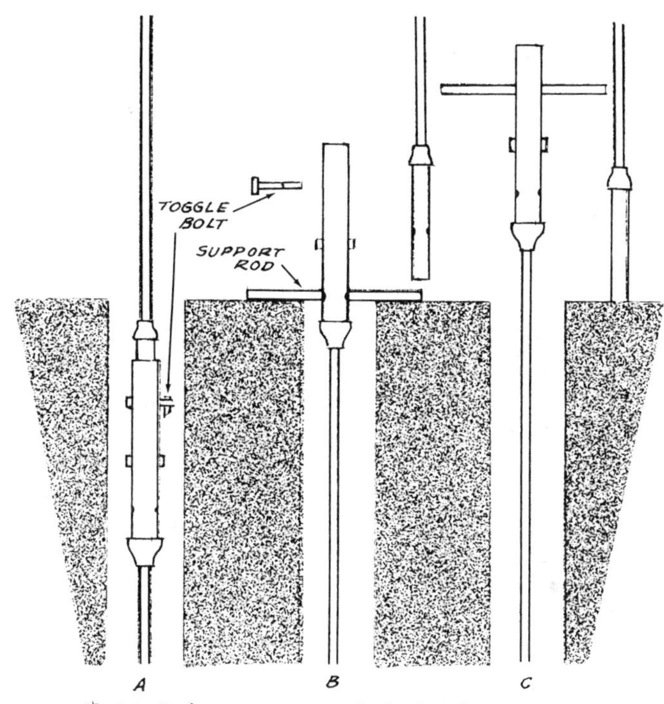


FIGURE 4 JOINT B IN OPERATION

until Joint B appears at the surface. See Figure 4A. Then put a 19mm (3/4") diameter rod through the hole. This allows the whole drill line to rest on it making it impossible for the part still in the well to fall in. Next remove the toggle bolt, lift out the top section of line and balance it beside the hole. See Figure 4B. Pull up the auger, empty it, and replace the section in the hole where it will be held by the 19mm (3/4") rod. See Figure 4C. Next replace the upper section of drill line. The 10mm (3/8") bolt acts as a stop that allows the holes to be easily lined up for reinsertion of the toggle bolt. Finally withdraw the rod and lower the auger for the next drilling. Mark the location for drilling the 8.75mm (13/32") diameter hole in the 32mm (1 1/4") pipe through the toggle bolt hole in the 38mm (1 1/2") pipe. If the hole is located with the 32mm (1 1/4") pipe resting on the stop bolt, the holes are bound to line up.

Sometimes a special tool is needed to penetrate a water-bearing sand layer, because the wet sand caves in as soon as the auger is removed. If this happens a perforated casing is lowered into the well, and drilling is accomplished with an auger that fits inside the casing. A percussion type with a flap, or a rotary type with solid walls and a flap are good possibilities. See the entries describing these devices. The casing will settle deeper into the sand as sand is dug from beneath it. Other sections of casing must be added as drilling proceeds. Try to penetrate the water bearing sand layer as far as possible (at least three feet–one meter). Ten feet (three meters) of perforated casing embedded in such a sandy layer will provide a very good flow of water.

## Tubewell Earth Auger

This earth auger (Figure 5), which is similar to designs used with power drilling equipment, is made from a 15cm (6") steel tube.

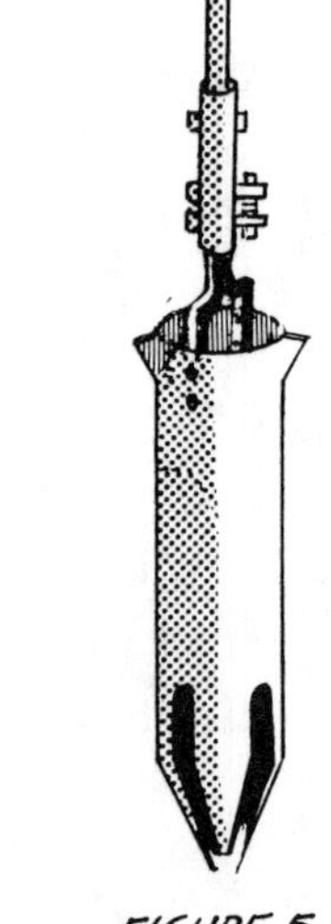

FIGURE 5
TUBEWELL EARTH AUGER

The auger can be made without welding equipment, but some of the bends in the pipe and the bar can be made much more easily when the metal is hot (see Figure 6).

An open earth auger, which is easier to empty than this one, is better suited for some soils. This auger cuts faster than the Tubewell Sand Auger.

**Tools and Materials**

Galvanized pipe: 32mm (1 1/4") in diameter and 21.5cm (8 1/2") long
Hexagonal head steel bolt: 10mm (3/8") in diameter and 5cm (2") long, with nut
2 hexagonal head steel bolts: 10mm (3/8") in diameter and 9.5cm (3 3/4") long
2 Steel bars: 1.25cm x 32mm x 236.5mm (1/2" x 1 1/4" x 9 5/16")
4 Round head machine screws: 10mm (3/8") in diameter and 32mm (1 1/4") long
2 Flat head iron rivets: 3mm (1/8") in diameter and 12.5mm (1/2") long
Steel strip: 10mm x 1.5mm x 2.5cm (3/8" x 1/16" x 1")
Steel tube: 15cm (6") outside diameter, 62.5cm (24 5/8") long
Hand tools

**Source:**

U.S. Army and Air Force. *Wells.* Technical Manual 5-297, AFM 85-23. Washington, D.C.: U.S. Government Printing Office, 1957.

FIGURE 6

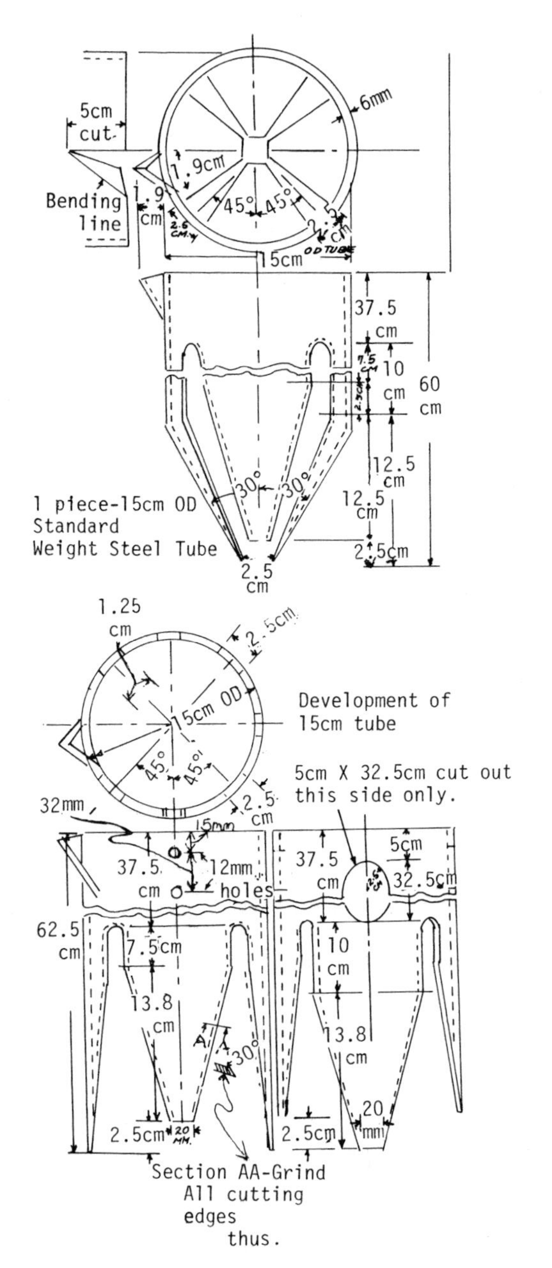
5cm
cut
Bending
line
1.9
cm
1.9cm
6mm
45°
45°
2.5 cm
15cm
OD TUBE
37.5
cm
10
cm
60
cm
12.5
cm
12.5
cm
30°
30°
2.5cm
2.5
cm
1 piece-15cm OD
Standard
Weight Steel Tube
1.25
cm
2.5cm
15cm OD
45°
45°
2.5
cm
Development of
15cm tube
5cm X 32.5cm cut out
this side only.
32mm
15mm
12mm
holes
37.5
cm
62.5
cm
7.5cm
13.8
cm
A
30°
2.5cm
20 MM.
37.5
cm
5cm
32.5cm
10
cm
13.8
cm
20
mm
2.5cm
Section AA-Grind
All cutting
edges
thus.

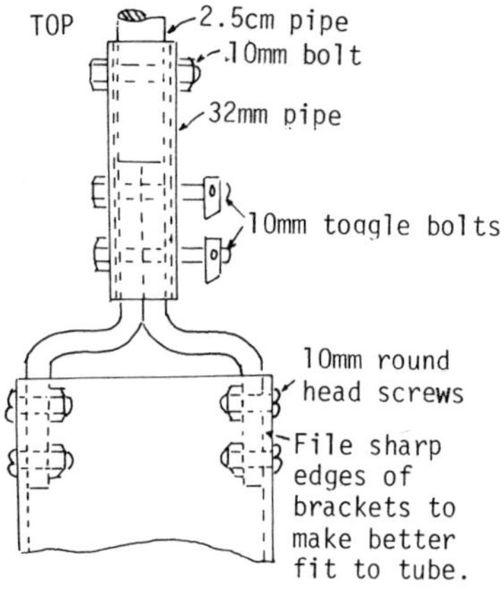
TOP
2.5cm pipe
10mm bolt
32mm pipe
10mm toggle bolts
10mm round
head screws
File sharp
edges of
brackets to
make better
fit to tube.

## Tubewell Sand Auger

This sand auger can be used to drill in loose soil or wet sand, where an earth auger is not effective. The simple cutting head requires less force to turn than the Tubewell Earth Auger, but it is more difficult to empty.

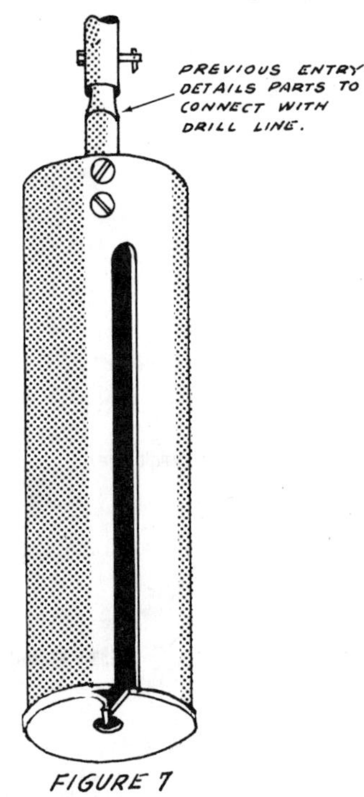


FIGURE 7
TUBEWELL SAND AUGER

A smaller version of the sand auger made to fit inside the casing pipe can be used to remove loose, wet sand.

The tubewell sand auger is illustrated in Figure 7. Construction diagrams are given in Figure 8.

**Tools and Materials**

Steel tube: 15cm (6") outside diameter and 46cm (18") long
Steel plate: 5mm x 16.5cm x 16.5cm (3/16" x 6 1/2" x 6 1/2")
Acetylene welding and cutting equipment
Drill

**Source:**

*Wells,* Technical Manual 5-297, AFM 85-23, U.S. Army and Air Force, 1957.

## Tubewell Sand Bailer

The sand bailer can be used to drill from inside a perforated well casing when a bore goes into loose wet sand and the walls start to cave in. It has been used to make many tubewells in India.

**Tools and Materials**

Steel tube: 12.5cm (5") in diameter and 91.5cm (3') long
Truck innertube or leather: 12.5cm (5") square
Pipe coupling: 15cm to 2.5cm (5" to 1")
Small tools

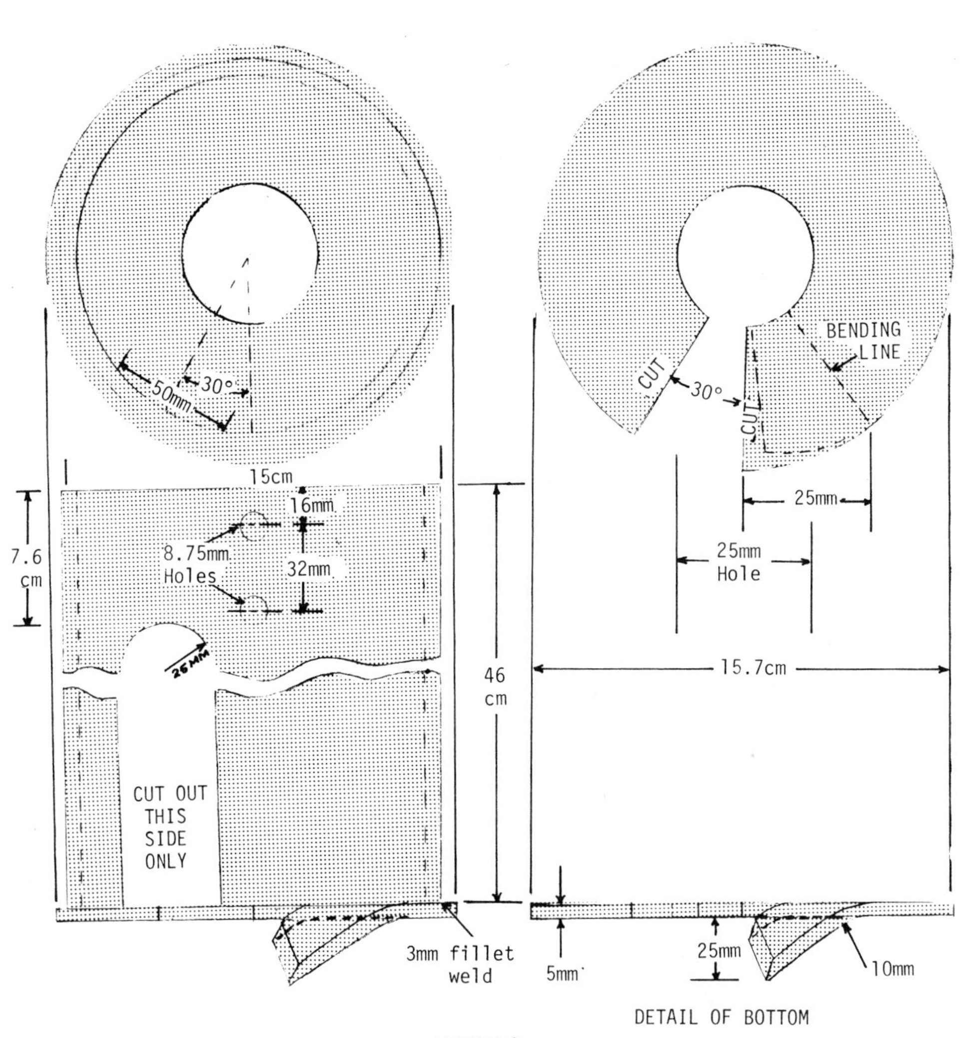

50mm
30°
15cm
16mm
8.75mm
Holes
32mm
7.6
cm
25MM
CUT OUT
THIS
SIDE
ONLY
46
cm
3mm fillet
weld
CUT
30°
CUT
BENDING
LINE
25mm
25mm
Hole
15.7cm
5mm
25mm
10mm
DETAIL OF BOTTOM


FIGURE 8

Repeatedly jamming this "bucket" into the well will remove sand from below the perforated casing, allowing the bucket to settle deeper into the sand layer. The casing prevents the walls from caving in. The bell is removed from the first section of casing; at least one other section rests on top of it to help force it down as digging proceeds. Try to penetrate the water bearing sand layer as far as possible: 3 meters (10') of perforated casing embedded in such a sandy layer will usually provide a very good flow of water.

Be sure to try your sand "bucket" in wet sand before attempting to use it at the bottom of your well.

**Source:**

*Explanatory Notes on Tubewells,* Wendell Mott, American Friends Service Committee, Philadelphia, Pennsylvania, 1956 (Mimeographed).

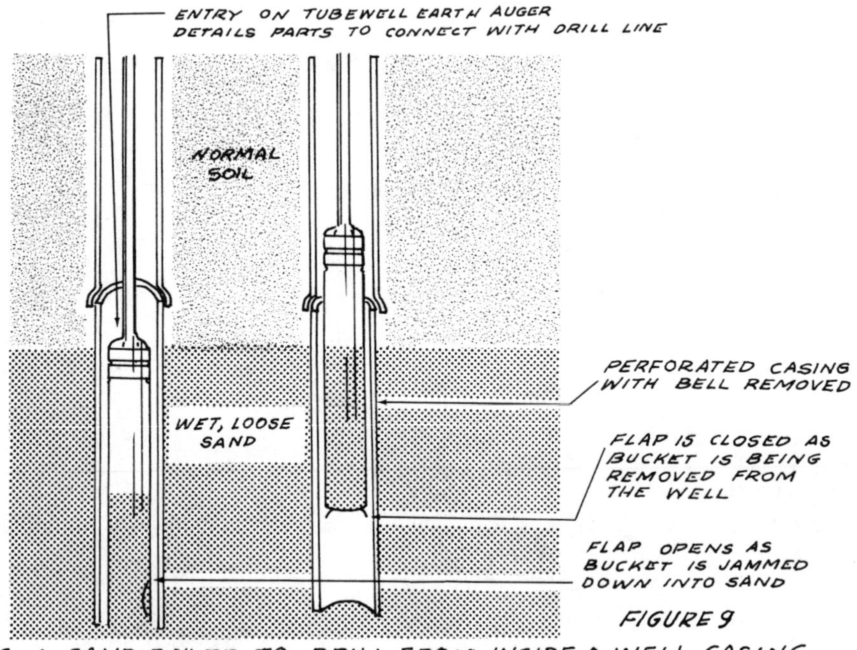


FIGURE 9

USING A SAND BAILER TO DRILL FROM INSIDE A WELL CASING

# *Ram Auger*

The equipment described here has been used successfully in the Ban Me Thuot area of Vietnam. One of the best performances was turned in by a crew of three inexperienced mountain tribesmen who drilled 20 meters (65') in a day and a half. The deepest well drilled was a little more than 25 meters (80'); it was completed, including the installation of the pump, in six days. One well was drilled through about 11 meters (35') of sedimentary stone.

**Tools and Materials**

**For tool tray:**

Wood: 3cm x 3cm x 150cm (1 1/4" x 1 1/4" x 59")
Wood: 3cm x 30cm x 45cm (1 1/4" x 12" x 17 3/4")

**For safety rod:**

Steel rod: 1cm (3/8") in diameter, 30cm (12") long
Drill
Hammer
Anvil
Cotter pin

**For auger support:**

Wood: 4cm x 45cm x 30cm (1 1/3" x 17 3/4" x 12")
Steel: 10cm x 10cm x 4mm (4" x 4" x 5/32")

## Location of the Well

Two considerations are especially important for the location of village wells: (1) the average walking distance for the village population should be as short as possible; (2) it should be easy to drain spilled water away from the site to avoid creating a mudhole.

In the Ban Me Thuot area, the final choice of location was in all cases left up to the villagers. Water was found in varying quantities at all the sites chosen. (See "Getting Ground Water from Wells and Springs.")

## Starting to Drill

A tripod is set up over the approximate location for the well (see Figure 1). Its legs are set into shallow holes with dirt packed around them to keep them from moving. To make sure the well is started exactly vertically, a plumb bob (a string with a stone tied to it is good enough) is hung from the auger guide on the

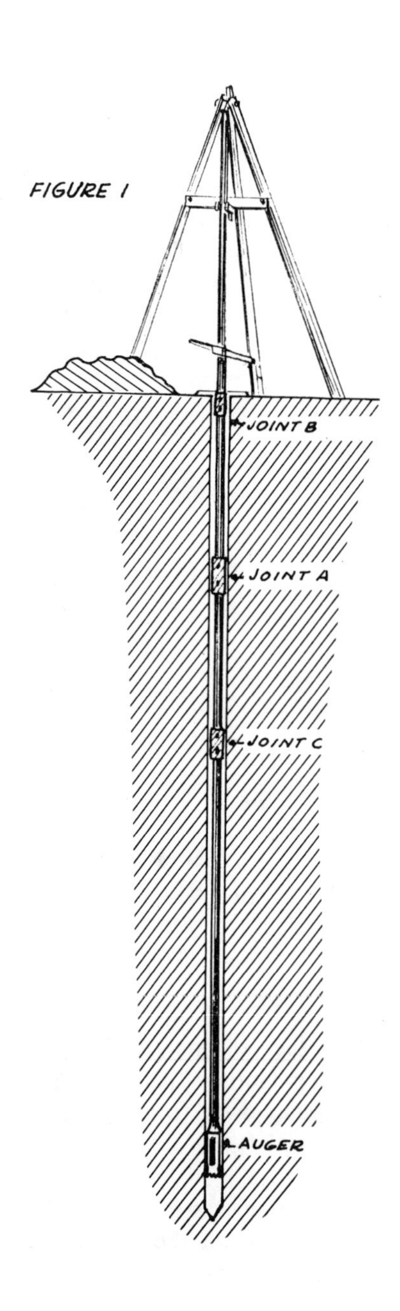


tripod's crossbar to locate the exact starting point. It is helpful to dig a small starting hole before setting up the auger.

## Drilling

Drilling is accomplished by ramming the auger down to penetrate the earth and then rotating it by its wooden handle to free it in the hole before lifting it to repeat the process. This is a little awkward until the auger is down 30cm to 60cm (1' to 2') and should be done carefully until the auger starts to be guided by the hole itself. Usually two or three people work together with the auger. One system that worked out quite well was to use three people, two working while the third rested, and then alternate.

As the auger goes deeper it will be necessary from time to time to adjust the handle to the most convenient height. Any wrenches or other small tools used should be tied by means of a long piece of cord to the tripod so that if they are accidentally dropped in the well, they can easily be removed. Since the soil of the Ban Me Thuot area would stick to the auger, it was necessary to keep a small amount of water in the hole at all times for lubrication.

## Emptying the Auger

Each time the auger is rammed down and rotated, it should be noted how much penetration has been obtained. Starting with an empty auger the penetration is greatest on the first stroke and becomes successively less on each following one as the earth packs more and more tightly inside the auger. When progress becomes too slow it is time to raise the auger to the surface and empty it. Depending on the material being penetrated, the auger may be completely full or have 30cm (1') or less of material in it when it is emptied. A little experience will give one a "feel" for the most efficient time to bring up the auger for emptying. Since the material in the auger is hardest packed at the bottom, it is usually easiest to empty the auger by inserting the **auger cleaner** through the slot in the side of the auger part way down and pushing the material out through the top of the auger in several passes. When the auger is brought out of the hole for emptying, it is usually leaned up against the tripod, since this is faster and easier than trying to lay it down.

## Coupling and Uncoupling Extensions

The **extensions** are coupled by merely slipping the small end of one into the large end of the other and pinning them together with a 10mm (3/8") bolt. It has been found sufficient and time-saving to just tighten the nut finger-tight instead of using a wrench.

Each time the auger is brought up for emptying, the extensions must be taken apart. For this reason the extensions have been made as long as possible to minimize the number of joints. Thus at a depth of 18.3 meters (60'), there are only two joints to be uncoupled in bringing up the auger.

For the sake of both safety and speed, use the following procedure in coupling and uncoupling. When bringing up the auger, raise it until a joint is just above the ground and slip the auger support (see Figures 2 and 3) into place, straddling the extension so that the bottom of the coupling can rest on the small metal plate. The next step is to put the safety rod (see Figure 4) through the lower side in the coupling and secure it with either a cotter pin or a piece of wire. The purpose of the safety rod is to keep the auger from falling into the well if it should be knocked off the auger support or dropped while being raised.

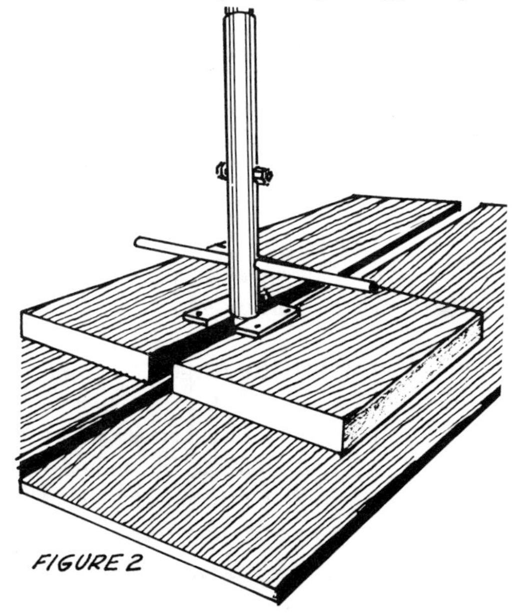
FIGURE 2

Once the safety rod is in place, remove the coupling bolt and slip the upper extension out of the lower. Lean the upper end of the extension against the tripod be-

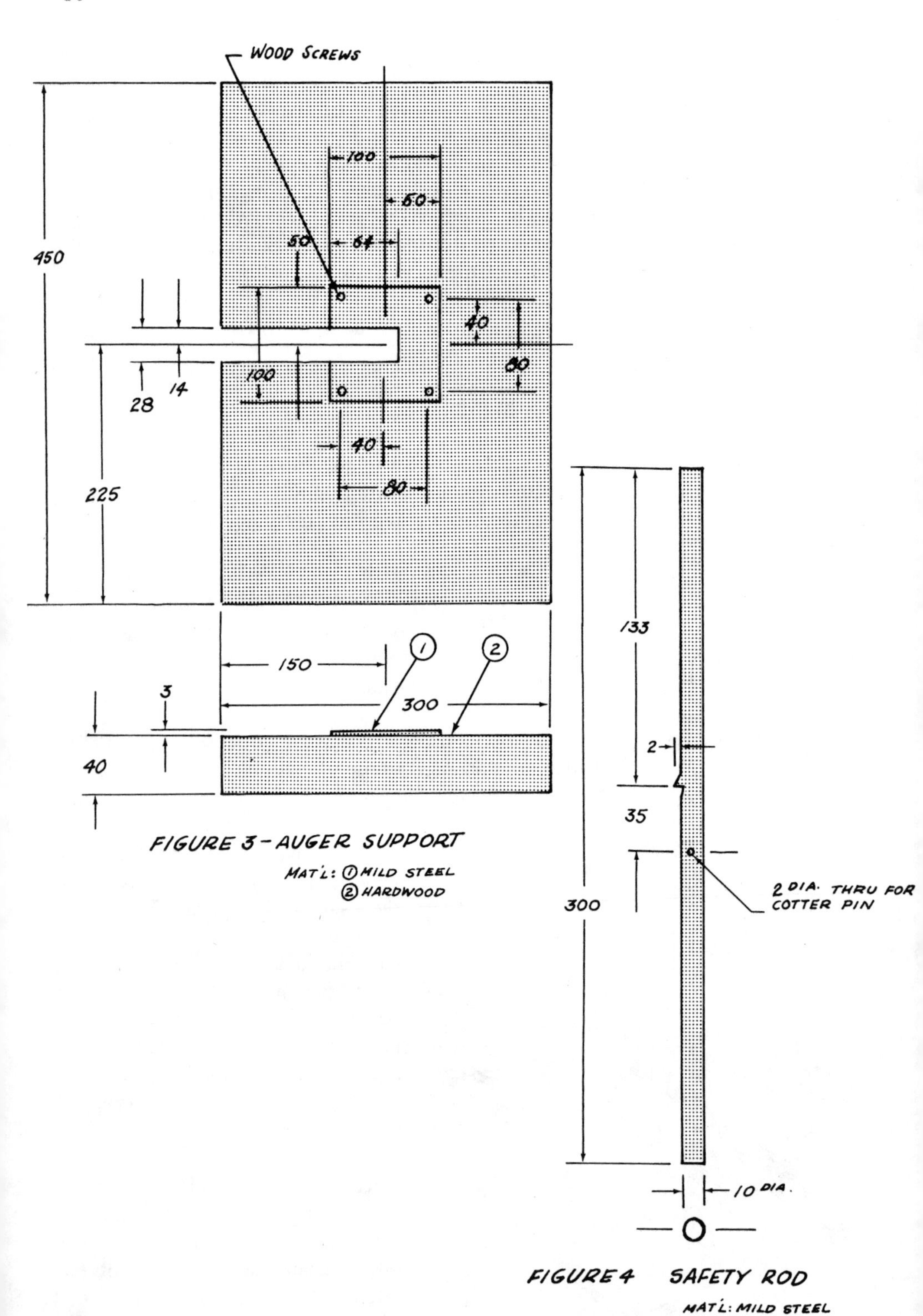


FIGURE 3 - AUGER SUPPORT

MAT'L: ① MILD STEEL
② HARDWOOD

FIGURE 4 SAFETY ROD

MAT'L: MILD STEEL

tween the two wooden pegs in the front legs, and rest the lower end on the tool tray (see Figures 5 and 6). The reason for putting the extensions on the tool tray is to keep dirt from sticking to the lower ends and making it difficult to put the extensions together and take them apart.

To couple the extensions after emptying the auger, the procedure is the exact reverse of uncoupling.

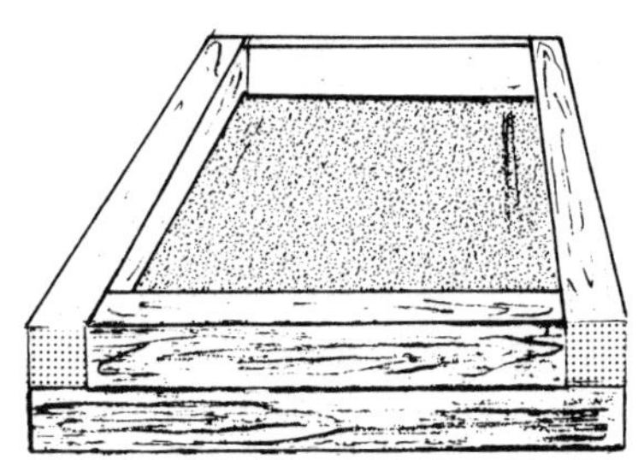

FIGURE 5 TOOL TRAY

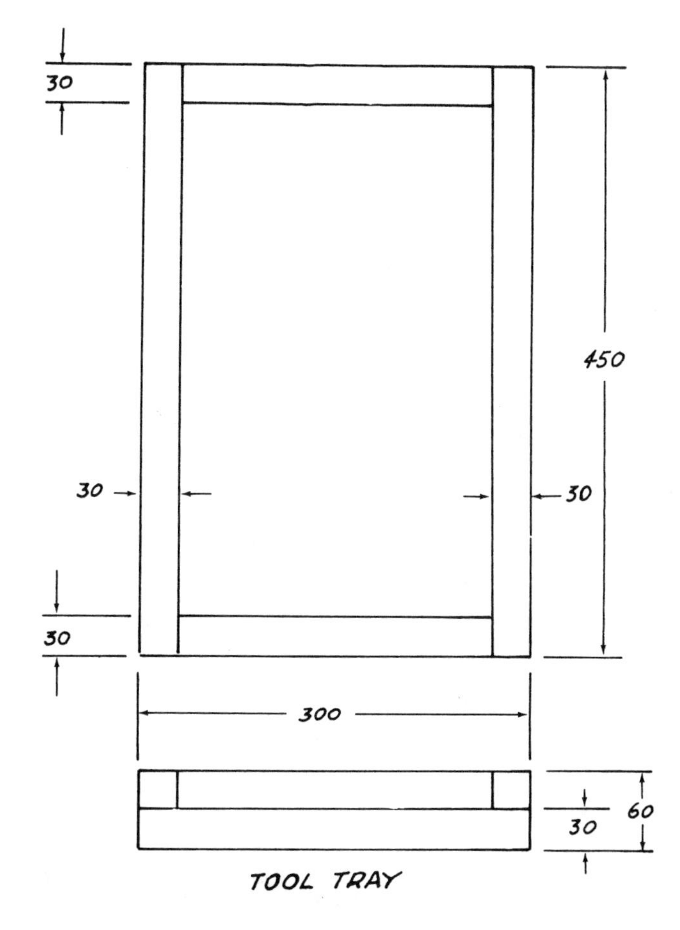


TOOL TRAY

FIGURE 6

## Drilling Rock

When stone or other substances the auger cannot penetrate are met, a heavy drilling bit must be used.

## Depth of Well

The rate at which water can be taken from a well is roughly proportional to the depth of the well below the water table as long as the well keeps going into water-bearing ground. However, in village wells where water can only be raised slowly by handpump or bucket, this is not usually of major importance. The important point is that in areas where the water table varies from one time of year to another the well must be deep enough to give sufficient water at all times.

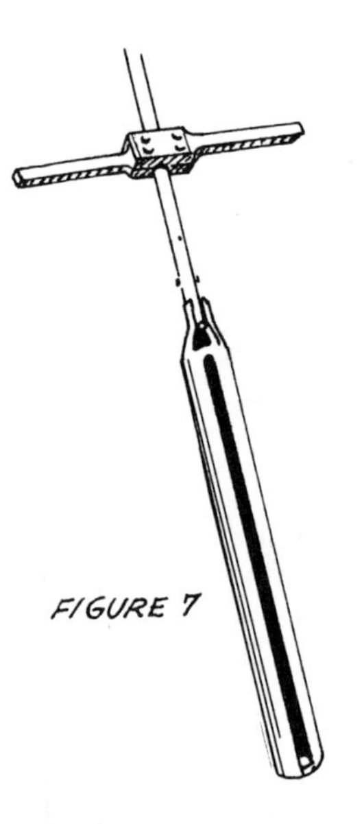

FIGURE 7

Information on the water table variation may be obtained from already existing wells, or it may be necessary to drill a well before any information can be obtained. In the latter case the well must be deep enough to allow for a drop in the water table.

**Source:**

Report by Richard G. Koegel, International Voluntary Services, Ban Me Thuot, Vietnam, 1959 (Mimeographed).

## Equipment

The following section gives construction details for the well-drilling equipment used with the ram auger:

- Auger, Extensions, and Handle
- Auger Cleaner
- Demountable Reamer
- Tripod and Pulley
- Bailing Bucket
- Bit for Drilling rock

### Auger, Extensions, and Handle

The auger is hacksawed out of standard-weight steel pipe about 10cm (4") in diameter (see Figure 8). Lightweight tubing is not strong enough. The extensions (see Figure 9) and handle (see Figure 10) make it possible to bore deep holes.

**Tools and Materials**

Pipe: 10cm (4") in diameter, 120cm (47 1/4") long, for auger
Pipe: 34mm outside diameter (1" inside diameter); 3 or 4 pieces 30cm (12") long, for auger and extension socket
Pipe: 26mm outside diameter (3/4" inside diameter); 3 or 4 pieces 6.1 or 6.4 meters (20' or 21') long, for drill extensions
Pipe: 10mm outside diameter (1/2" inside diameter); 3 or 4 pieces 6cm (2 3/8") long
Hardwood: 4cm x 8cm x 50cm (1 1/2" x 3 1/8" x 19 3/4"), for handle
Mild steel: 3mm x 8cm x 15cm (1/8" x 3 1/8" x 6")
4 Bolts: 1cm (3/8") in diameter and 10cm (4") long
4 Nuts

Hand tools and welding equipment

In making the auger, a flared-tooth cutting edge is cut in one end of the 10cm pipe. The other end is cut, bent, and welded to a section of 34mm outside-diameter (1" inside-diameter) pipe, which forms a socket for the drill line extensions. A slot that runs nearly the length of the auger is used for removing soil from the auger. Bends are made stronger and more easily and accurately when the steel is hot. At first, an auger with two cutting lips similar to a post-hole auger was used; but it became plugged up and did not cut cleanly. In some soils, however, this type of auger may be more effective.

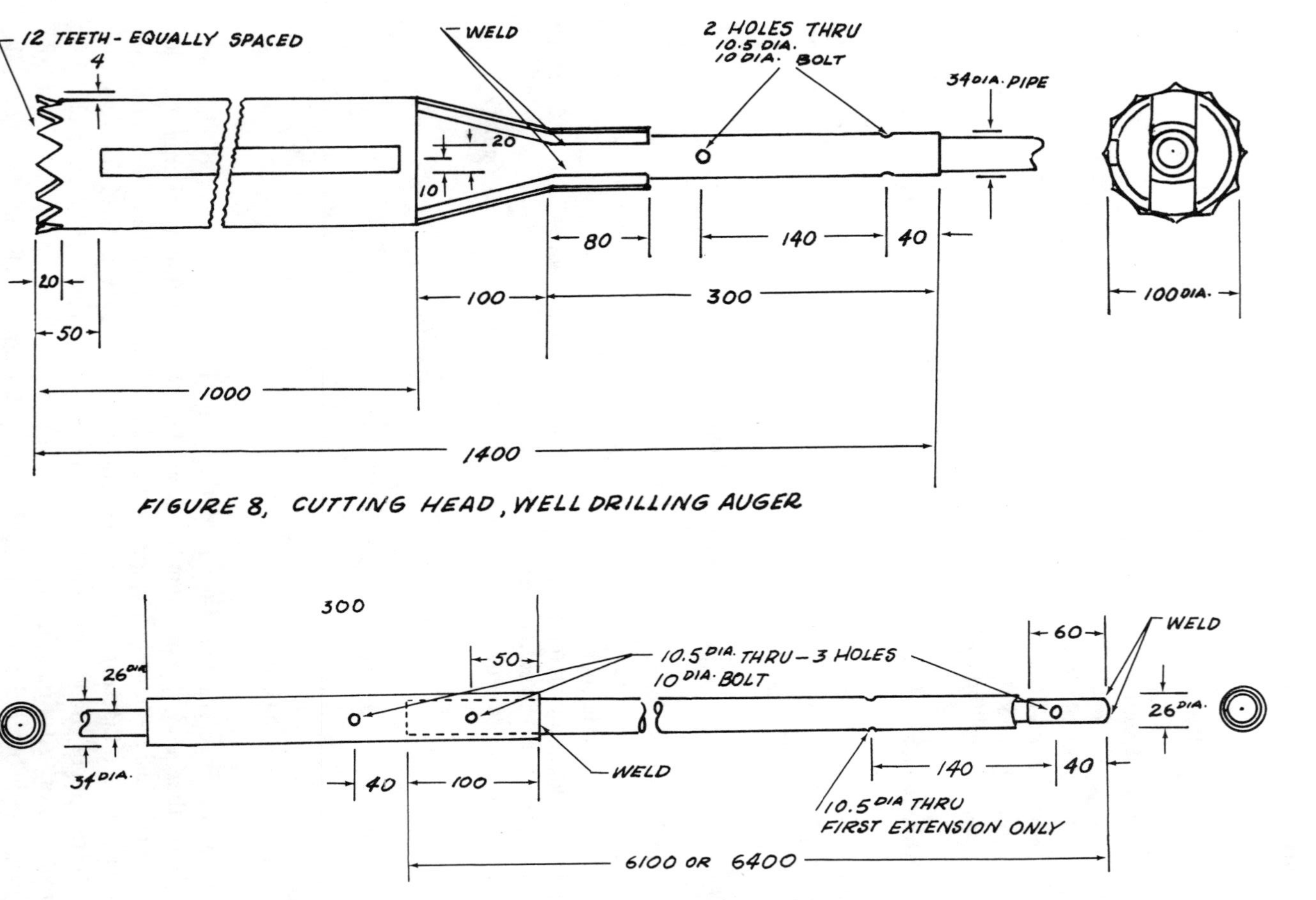


FIGURE 8, CUTTING HEAD, WELL DRILLING AUGER

FIGURE 9 EXTENSION, WELL DRILLING AUGER

NOTE: 34 DIA. COUPLING MAY BE OMITTED ON LAST EXTENSION

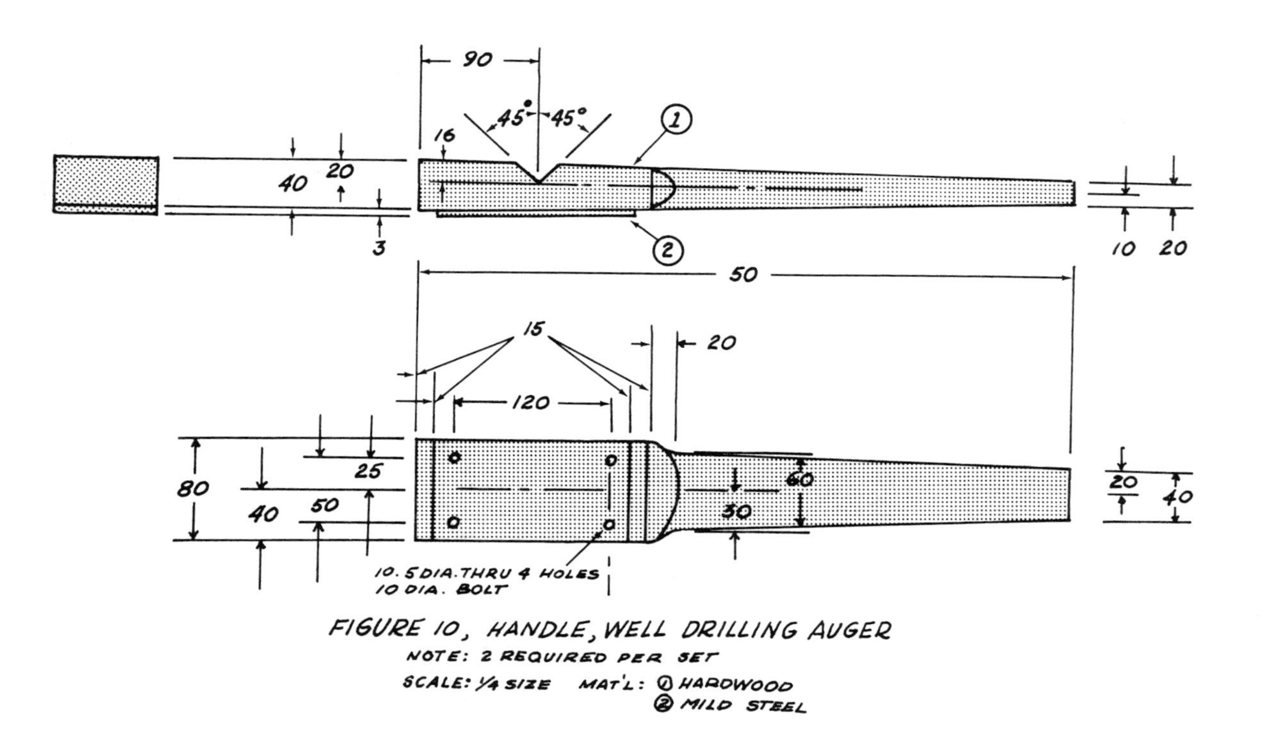


FIGURE 10, HANDLE, WELL DRILLING AUGER

NOTE: 2 REQUIRED PER SET

SCALE: 1/4 SIZE MAT'L: ① HARDWOOD ② MILD STEEL

**Auger Cleaner**

Soil can be removed rapidly from the auger with this auger cleaner (see Figure 11). Figure 12 gives construction details.

**Tools and Materials**

Mild steel: 10cm (4") square and 3mm (1/8") thick
Steel rod: 1cm (3/8") in diameter and 52cm (20 1/2") long
Welding equipment
Hacksaw
File

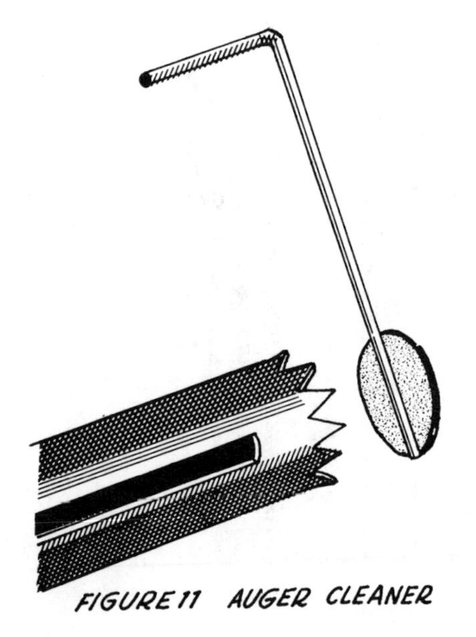

FIGURE 11 AUGER CLEANER

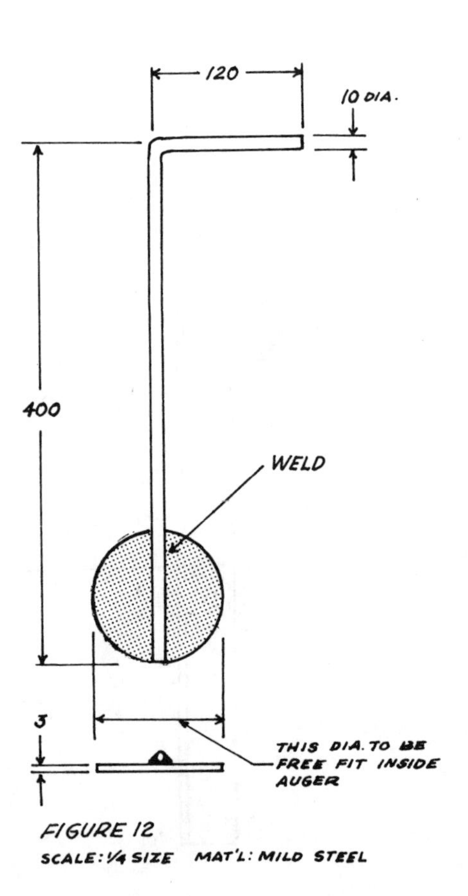


FIGURE 12
SCALE: 1/4 SIZE MAT'L: MILD STEEL

**Demountable Reamer**

If the diameter of a drilled hole has to be made bigger, the demountable reamer described here can be attached to the auger.

**Tools and Materials**

Mild steel: 20cm x 5cm x 6mm (6" x 2" x 1/4"), to ream a well diameter of 19cm (7 1/2")
2 Bolts: 8mm (5/16") in diameter and 10cm (4") long
Hacksaw
Drill
File
Hammer
Vise

The reamer is mounted to the top of the auger with two hook bolts (see Figure 13). It is made from a piece of steel 1cm (1/2") larger than the desired well diameter (see Figure 14).

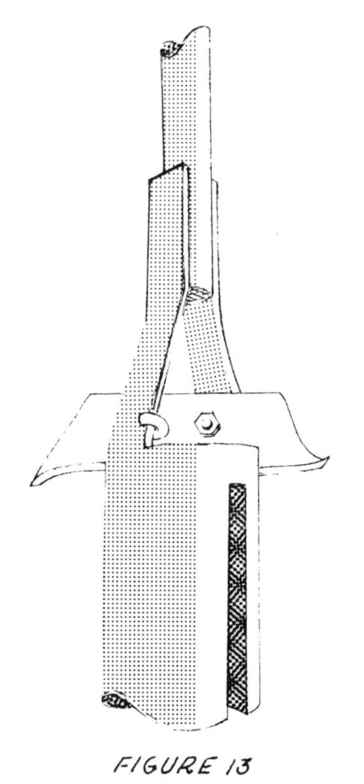

FIGURE 13

After the reamer is attached to the top of the auger, the bottom of the auger is plugged with some mud or a piece of wood to hold the cuttings inside the auger.

In reaming, the auger is rotated with only slight downward pressure. It should be emptied before it is too full so that not too many cuttings will fall to the bottom of the well when the auger is pulled up.

Because the depth of a well is more important than the diameter in determining the flow and because doubling the diameter means removing four times the amount of earth, larger diameters should be considered only under special circumstances. (See "Well Casing and Platforms," page 12.)

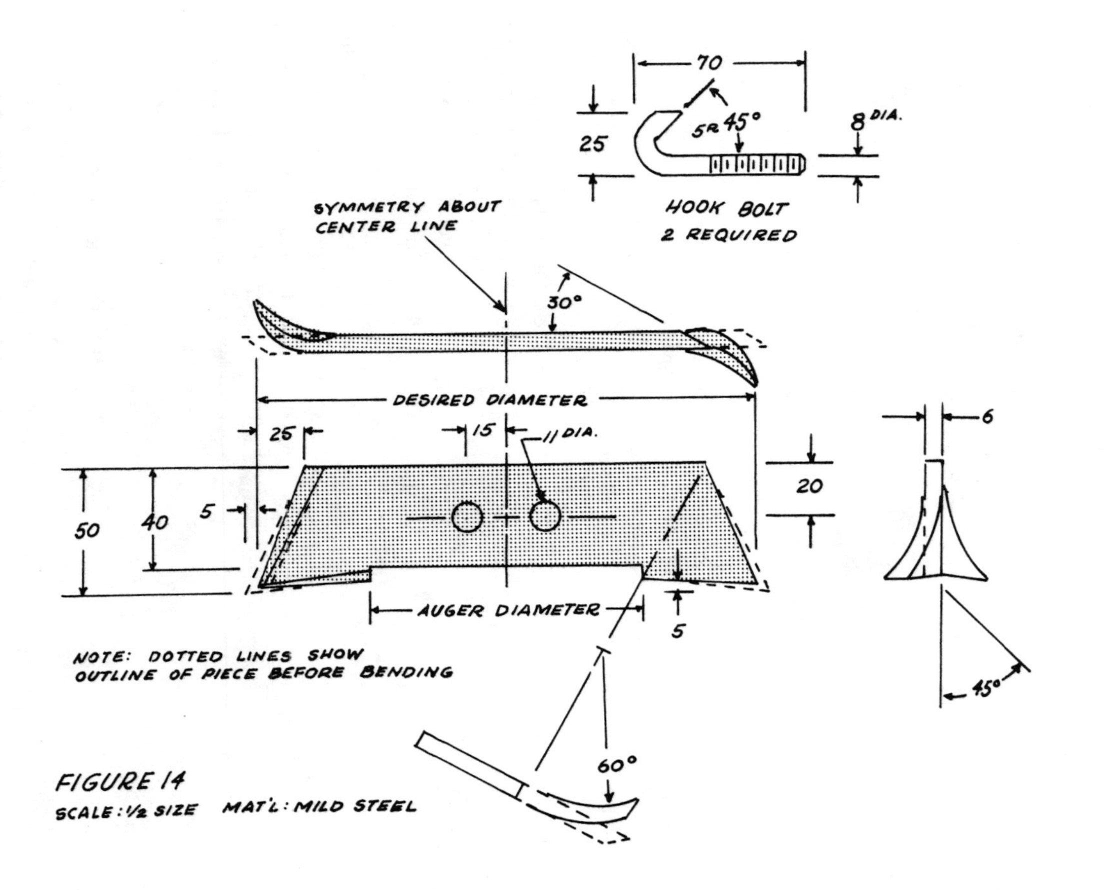


FIGURE 14
SCALE: 1/2 SIZE MAT'L: MILD STEEL

**Tripod and Pulley**

The tripod (see Figures 15 and 16), which is made of poles and assembled with 16mm (5/8") bolts, serves three purposes: (1) to steady the extension of the auger when it extends far above ground; (2) to provide a mounting for the pulley (see Figures 17 and 19) used with the drill bit and bailing bucket; and (3) to provide a place for leaning long pieces of casing, pipe for pumps, or auger extensions while they are being put into or taken out of the well.

When a pin or bolt is put through the holes in the two ends of the "L"-shaped pulley bracket (see Figures 15 and 18) that extend horizontally beyond the front of the tripod crossbar, a loose guide for the upper part of the auger extensions is formed.

To keep the extensions from falling when they are leaned against the tripod, two 30cm (12") long wooden pegs are driven into drilled holes near the top of the tripod's two front legs (see Figure 19).

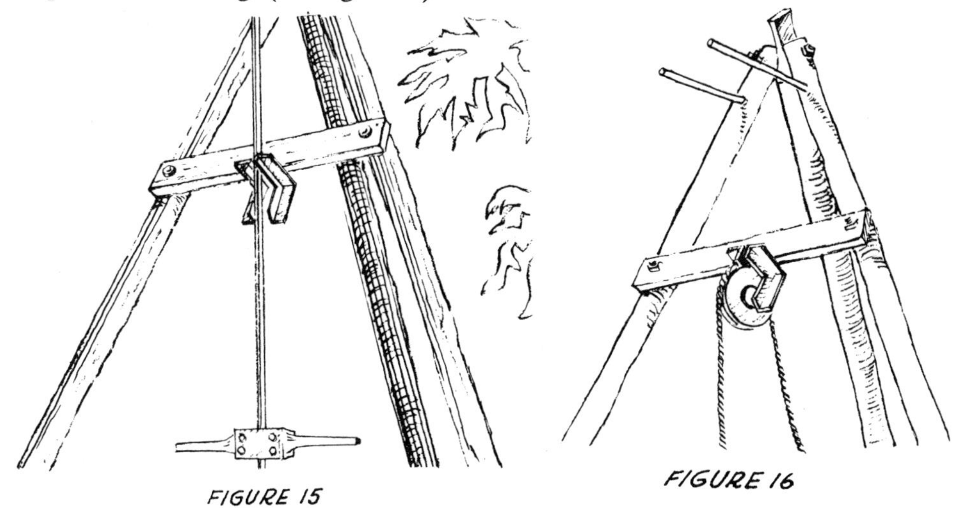

FIGURE 15

FIGURE 16

**Tools and Materials**

3 Poles: 15cm (3") in diameter and 4.25 meters (14') long
Wood for cross bar: 1.1 meter (43 1/2") x 12cm (4 3/4") square
For pulley wheel:
Wood: 25cm (10") in diameter and 5cm (2") thick
Pipe: 1.25cm (1/2") inside diameter, 5cm (2") long
Axle bolt: to fit close inside 1.25cm (1/2") pipe
Angle iron: 80cm (31 1/2") long, 50cm (19 3/4") webs, 5mm (3/16") thick
4 Bolts: 12mm (1/2") in diameter, 14cm (5 1/2") long; nuts and washers
Bolt: 16mm (5/8") in diameter and 40cm (15 3/4") long; nuts and washer
2 Bolts: 16mm (5/8") in diameter and 25cm (9 7/8") long; nuts and washers
Bore 5 places through center of poles for assembly with 16mm bolts

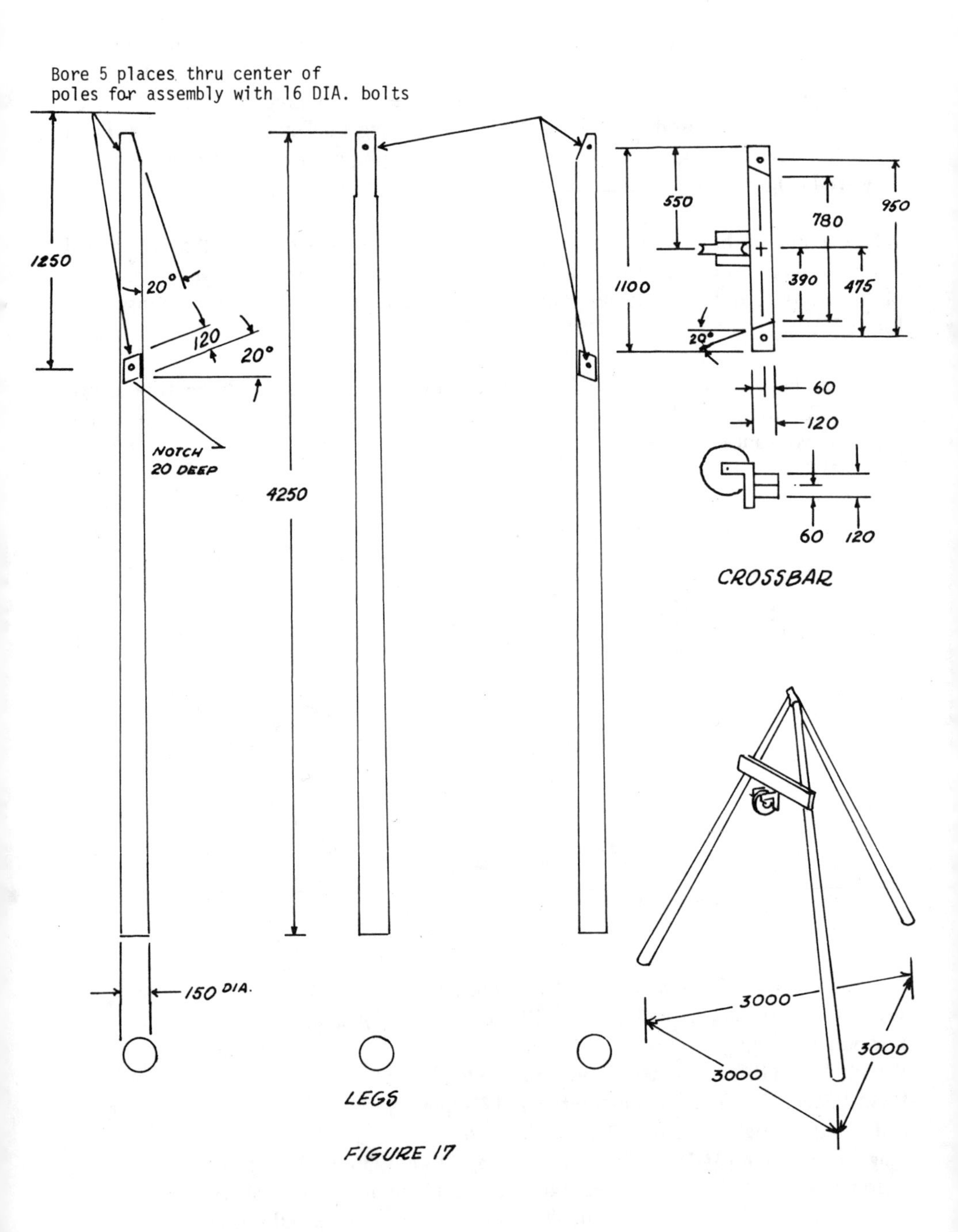
Bore 5 places thru center of poles for assembly with 16 DIA. bolts
1250
20°
120
20°
NOTCH 20 DEEP
4250
150 DIA.
550
780
950
1100
390
475
20°
60
120
60
120
CROSSBAR
3000
3000
3000
LEGS

FIGURE 17

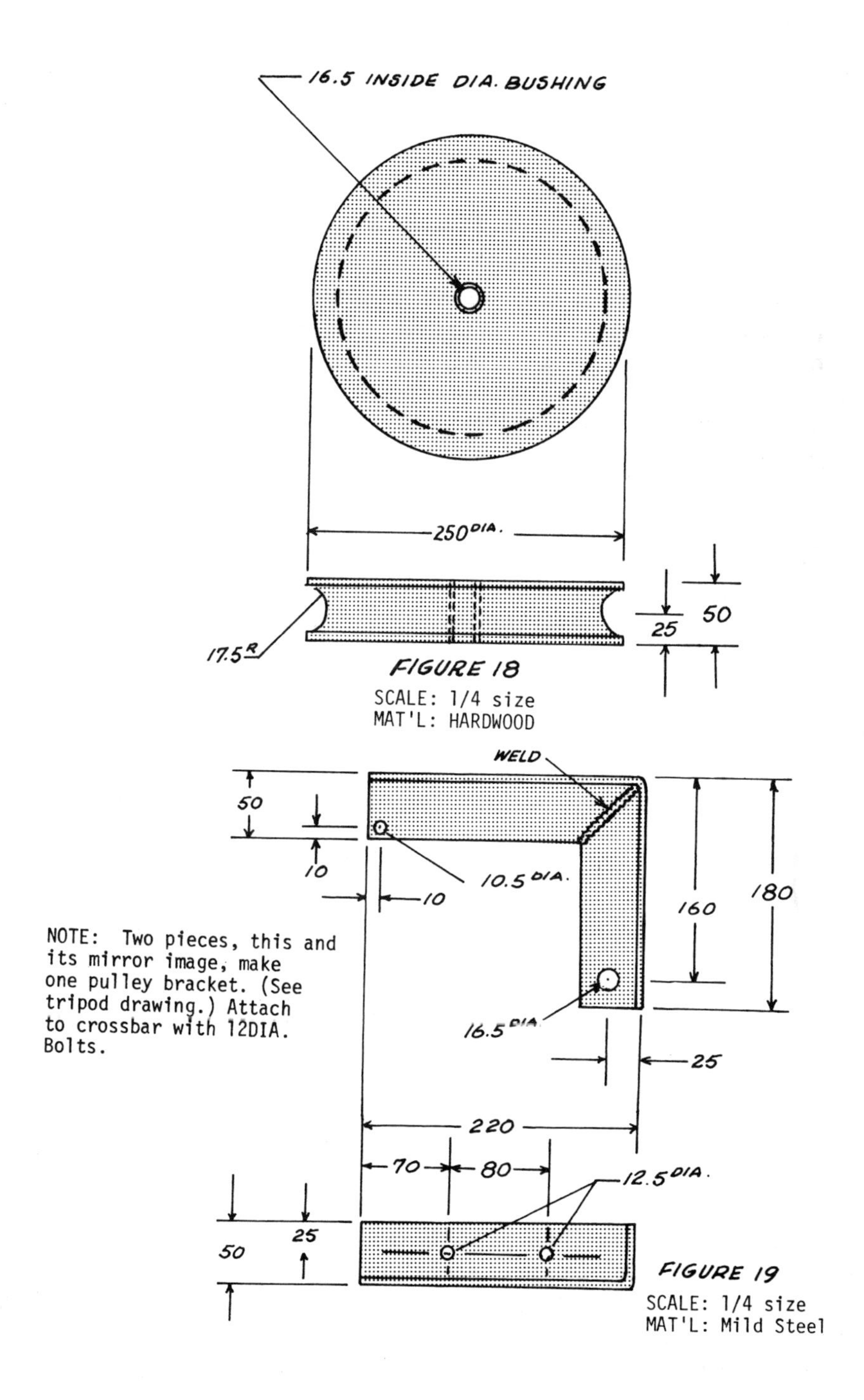


FIGURE 18
SCALE: 1/4 size
MAT'L: HARDWOOD

NOTE: Two pieces, this and its mirror image, make one pulley bracket. (See tripod drawing.) Attach to crossbar with 12DIA. Bolts.

FIGURE 19
SCALE: 1/4 size
MAT'L: Mild Steel

**Bailing Bucket**

The bailing bucket can be used to remove soil from the well shaft when cuttings are too loose to be removed with the auger.

**Tools and Materials**

Pipe: about 8.5cm ( 3 3/8") in diameter, 1 to 2cm (1/2" to 3/4") smaller in diameter than the auger, 180cm (71") long
Steel rod: 10mm (3/8") in diameter and 25cm (10") long; for bail (handle)
Steel plate: 10cm (4") square, 4mm (5/32") thick
Steel bar: 10cm x 1cm x 5mm (4" x 3/8" x 3/16")
Machine screw: 3mm (1/8") in diameter by 16mm (5/8") long; nut and washer
Truck innertube: 4mm (5/32") thick, 10mm (3/8") square
Welding equipment
Drill
Hacksaw
Hammer
Vise
File
Rope

Both standard weight pipe and thin-walled tubing were tried for the bailing bucket. The former, being heavier, was harder to use, but did a better job and stood up better under use. Both the steel bottom of the bucket and the rubber valve should be heavy because they receive hard usage. The metal bottom is reinforced with a crosspiece welded in place (see Figures 20 and 21).

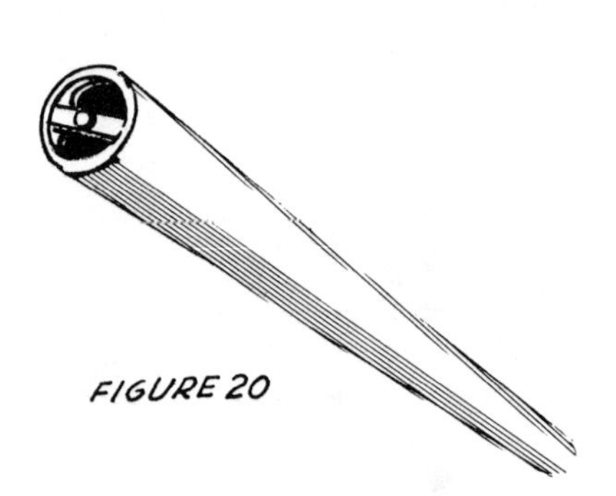

FIGURE 20

When water is reached and the cuttings are no longer firm enough to be brought up in the auger, the bailing bucket must be used to clean out the well as work progresses.

For using the bailing bucket the pulley is mounted in the pulley bracket with a 16mm (5/8") bolt as axle. A rope attached to the bailing bucket is then run over the pulley and the bucket is lowered into the well. The pulley bracket is so designed that the rope coming off the pulley lines up vertically with the well, so that there is no need to shift the tripod.

The bucket is lowered into the well, preferably by two people and allowed to drop

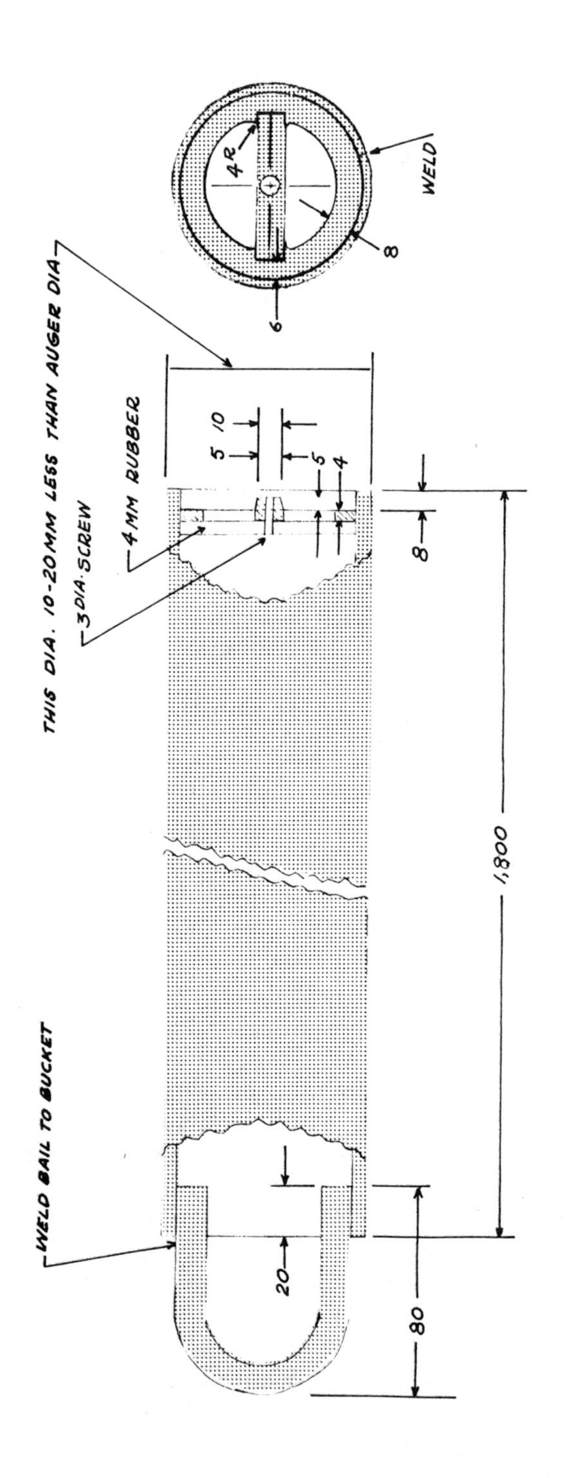


FIGURE 21
MAT'L: MILD STEEL

the last meter or meter and one-half (3 to 5 feet) so that it will hit the bottom with some speed. The impact will force some of the loose soil at the bottom of the well up into the bucket. The bucket is then repeatedly raised and dropped 1 to 2 meters (3 to 6 feet) to pick up more soil. Experience will show how long this should be continued to pick up as much soil as possible before raising and emptying the bucket. Two or more people can raise the bucket, which should be dumped far enough from the well to avoid messing up the working area.

If the cuttings are too thin to be brought up with the auger but too thick to enter the bucket, pour a little water down the well to dilute them.

**Bit for Drilling Rock**

The bit described here has been used to drill through layers of sedimentary stone up to 11 meters (36') thick.

**Tools and Materials**

Mild steel bar: about 7cm (2 3/4") in diameter and about 1.5 meters (5') long, weighing about 80kg (175 pounds)
Stellite (a very hard type of tool steel) insert for cutting edge
Anvil and hammers, for shaping
Steel rod: 2.5cm x 2cm x 50cm (1" x 3/4" x 19 3/4") for bail
Welding equipment

The drill bit for cutting through stone and hard formations is made from the 80kg (175-pound) steel bar (see Figures 22 and 23). The 90-degree cutting edge is hard-surfaced with stellite. A bail (or handle) for attaching a rope or cable is welded to the top. The bail should be large enough to make "fishing" easy if the rope breaks. A 2.5cm (1") rope was used at first, but this was subject to much wear when working in mud and water. A 1cm (3/8") steel cable was substituted for the rope, but it was not used enough to be able to show whether the cable or the rope is better. One advantage of rope is that it gives a snap at the end of the fall which rotates the bit and keeps it from sticking. A swivel can be mounted between the bit and the rope or cable to let the bit rotate.

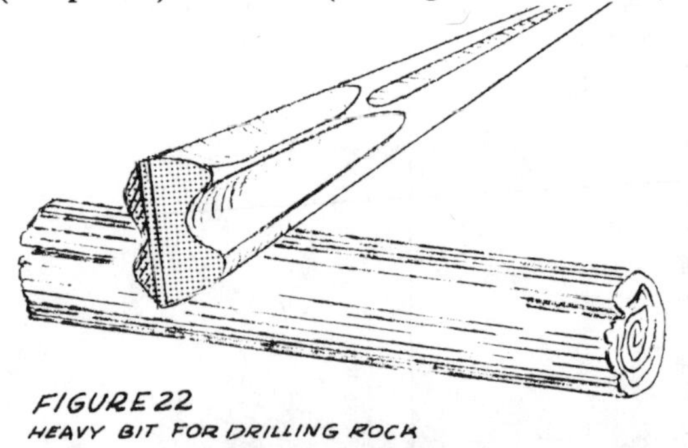

FIGURE 22
HEAVY BIT FOR DRILLING ROCK

If a bar this size is difficult to find or too expensive, it may be possible, depending on the circumstances, to make one by welding a short steel cutting end onto a piece of pipe, which is made heavy enough by being filled with concrete.

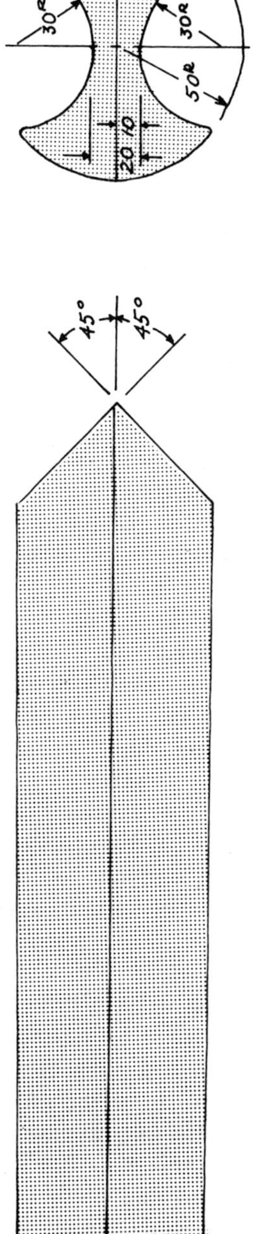

NOTE: TAPER ALL SURFACES TO BLEND INTO SURFACE OF REMAINDER OF BAR 400-500 MM BEHIND CUTTING HEAD



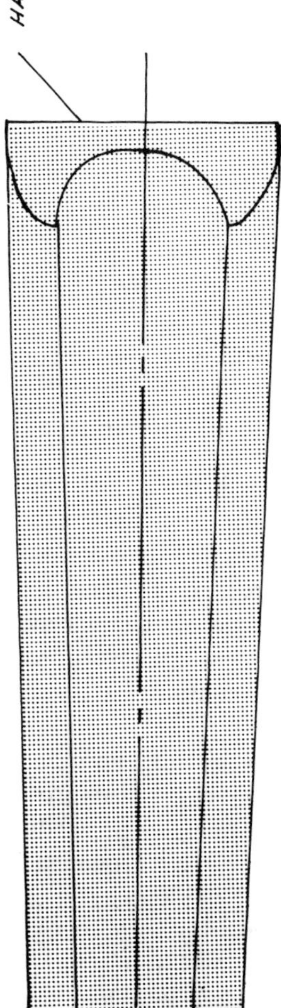

FIGURE 23

In using the drilling bit, put the pulley in place as with the bailing bucket, attach the bit to its rope or cable, and lower it into the well. Since the bit is heavy, wrap the rope once or twice around the back leg of the tripod so that the bit cannot "get away" from the workers with the chance of someone being hurt or the equipment getting damaged. The easiest way to raise and drop the bit is to run the rope through the pulley and then straight back to a tree or post where it can be attached at shoulder height or slightly lower. Workers line up along the rope and raise the bit by pressing down on the rope; they drop it by allowing the rope to return quickly to its original position (see Figure 24). This requires five to seven workers, occasionally more. Frequent rests are necessary, usually after every 50 to 100 strokes. Because the work is harder near the ends of the rope than in the middle, the positions of the workers should be rotated to distribute the work evenly.

FIGURE 24

A small amount of water should be kept in the hole for lubrication and to mix with the pulverized stone to form a paste that can be removed with a bailing bucket. Too much water will slow down the drilling.

The speed of drilling, of course, depends on the type of stone encountered. In the soft water-bearing stone of the Ban Me Thuot area it was possible to drill several meters (about 10 feet) per day. However, when hard stone such as basalt is encountered, progress is measured in centimeters (inches). The decision must then be made whether to continue trying to penetrate the rock or to start over in a new location. Experience in the past has indicated that one should not be too hasty in abandoning a location, since on several occasions what were apparently thin layers of hard rock were penetrated and drilling then continued at a good rate.

Occasionally the bit may become stuck in the well and it will be necessary to use a lever arrangement consisting of a long pole attached to the rope to free it (see Figure 25). Alternatively, a windlass may be used, consisting of a horizontal pole used to wrap the rope around a vertical pole pivoted on the ground and held in place by several workers (see Figure 26). If these fail, it may be necessary to rent or borrow a chain hoist. A worn rope or cable may break when trying to retrieve a stuck bit. If this happens, fit a hook to one of the auger extensions, attach enough extensions together to reach the desired depth, and after hooking the bit, pull with the chain hoist. A rope or cable may also be used for this purpose, but are considerably more difficult to hook onto the bit.

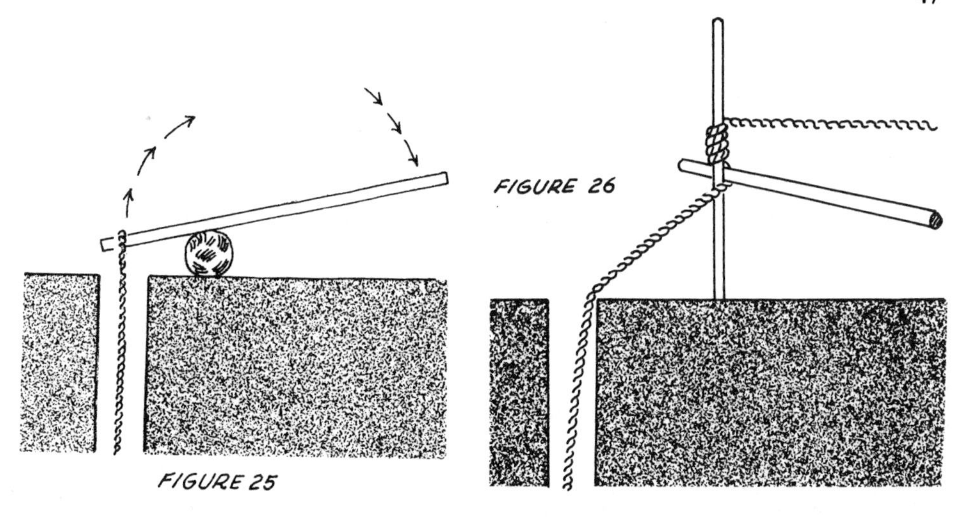


## Drilling Mechanically

The following method can be used for raising and dropping the bit mechanically:

- Jack up the rear wheel of a car and replace the wheel with a small drum (or use the rim as a pulley).
- Take the rope that is attached to the bit, come from the tripod on the pulley, and wrap the rope loosely around the drum.
- Pull the unattached end of the rope taut and set the drum in motion. The rope will move with the drum and raise the bit.
- Let the end of the rope go slack quickly to drop the bit.

It will probably be necessary to polish and/or grease the drum.

# Dry Bucket Well Drilling

The dry bucket method is a simple and quick method of drilling wells in dry soil that is free of rocks. It can be used for 5cm to 7.5cm (2" to 3") diameter wells in which steel pipe is to be installed. For wells that are wider in diameter, it is a quick method of removing dry soil before completing the bore with a wet bucket, tubewell sand bailer, or tubewell sand auger.

A 19.5-meter (64') hole can be dug in less than three hours with this method, which works best in sandy soil, according to the author of this entry, who has drilled 30 wells with it.

**Tools and Materials**

Dry bucket
Rope: 16mm (5/8") or 19mm (3/4") in diameter and 6 to 9 meters (20' to 30') longer than the deepest well to be drilled
3 Poles: 20cm (4") in diameter at large end and 3.6 to 4.5 meters (12' to 15') long
Chain, short piece
Pulley
Bolt: 12.5mm (1/2") in diameter and 30 to 35cm (12" to 14") long (long enough to reach through the upper ends of the three poles)

A dry bucket is simply a length of pipe with a bail or handle welded to one end and a slit cut in the other.

The dry bucket is held about 10cm (several inches) above the ground, centered above the hole location and then dropped (see Figure 1). This drives a small amount of soil up into the bucket. After this is repeated two or three times, the bucket is removed, held to one side and tapped with a hammer or a piece of iron to dislodge the soil. The process is repeated until damp soil is reached and the bucket will no longer remove soil.

To make the dry bucket, you will need the following tools and materials:

Hacksaw
File
Iron rod: 10mm (3/8") or 12.5mm (1/2") in diameter and 30cm (1') long
Iron pipe: slightly larger in diameter than the largest part of casing to be put in the well (usually the coupling) and 152cm (5') long

Bend the iron rod into a U-shape small enough to slide inside the pipe. Weld it in place as in Figure 2.

File a gentle taper on the inside of the opposite end to make a cutting edge (see Figure 3).

Cut a slit in one side of the sharpened end of the pipe (see Figure 2).

**Source:**

John Brelsford, VITA Volunteer, New Holland, Pennsylvania

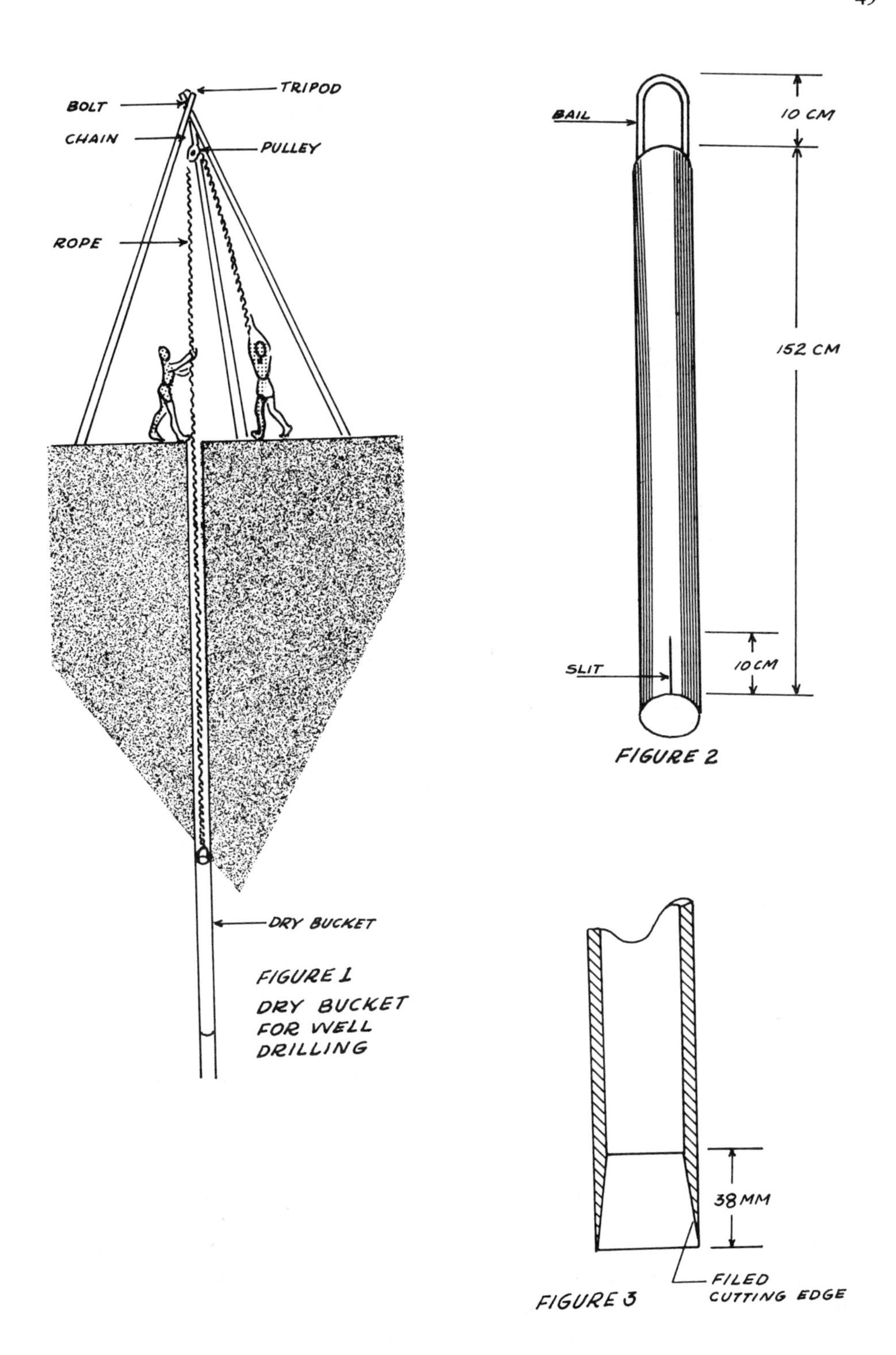


FIGURE 1
DRY BUCKET FOR WELL DRILLING

FIGURE 2

FIGURE 3

# Driven Wells

A pointed strainer called a well point, properly used, can quickly and cheaply drive a sanitary well, usually less than 7.6 meters (25') deep. In soils where the driven well is suitable, it is often the cheapest and fastest way to drill a sanitary well. In heavy soils, particularly clay, drilling with an earth auger is faster than driving with a well point.

**Tools and Materials**

Well point and driving cap (see Figure 1): usually obtainable through mail order houses from the United States and elsewhere
Pipe: 3cm (1") in diameter
Heavy hammer and wrenches
Pipe compound
Special pipe couplings and driving arrangements are desirable but not necessary

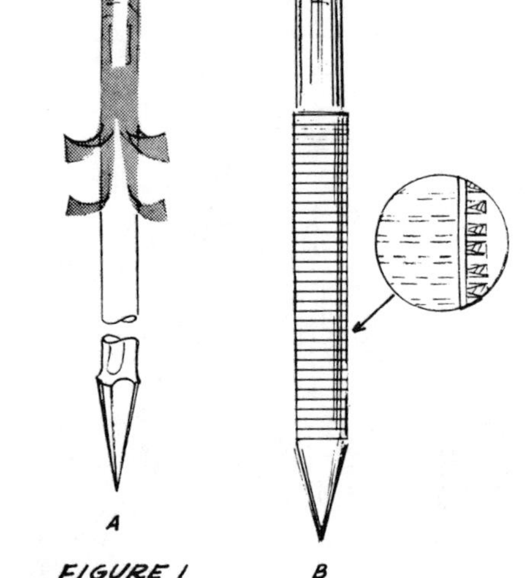


FIGURE 1

Driven wells are highly successful in coarse sand where there are not too many rocks and the water table is within 7 meters (23') of the surface. They are usually used as shallow wells where the pump cylinder is at ground level. If conditions for driving are very good, 10cm (4") diameter points and casings that can accept the cylinder of a deep well can be driven to depths of 10 - 15 meters (33' to 49'). (Note that suction pumps generally cannot raise water beyond 10 meters.)

The most common types of well points are:

- a pipe with holes covered by a screen and a brass jacket with holes. For general use, a #10 slot or 60 mesh is recommended. Fine sand requires a finer screen, perhaps a #6 slot or 90 mesh;

- a slotted steel pipe with no covering screen, which allows more water to enter but is less rugged.

Before starting to drive the point, make a hole at the site with hand tools. The hole should be plumb and slightly larger in diameter than the well point.

The joints of the drive pipe must be carefully made to prevent thread breakage and assure airtight operation. Clean and oil the threads carefully and use joint compound and special drive couplings when available. To ensure that joints stay tight, give the pipe a fraction of a turn **after** each blow, until the top joint is permanently set. Do not twist the whole string and do not twist and pound at the same time. The latter may help get past stones, but soon will break the threads and make leaky joints.

Be sure the drive cap is tight and butted against the end of the pipe (see Figure 2). check with a plumb bob to see that the pipe is vertical. Test it occasionally and keep it straight by pushing on the pipe while driving. Hit the drive cap squarely each time or you may damage the equipment.

Several techniques can help avoid damage to the pipe. The best way is to drive with a steel bar that is dropped inside the pipe and strikes against the inside of the steel well point. It is retrieved with a cable of rope. Once water enters the well, this method does not work.

Another way is to use a driver pipe, which makes sure that the drive cap is hit squarely. A guide rod can be mounted on top of the pipe and weight dropped over it, or the pipe itself can be used to guide a falling weight that strikes a special drive clamp.

FIGURE 2

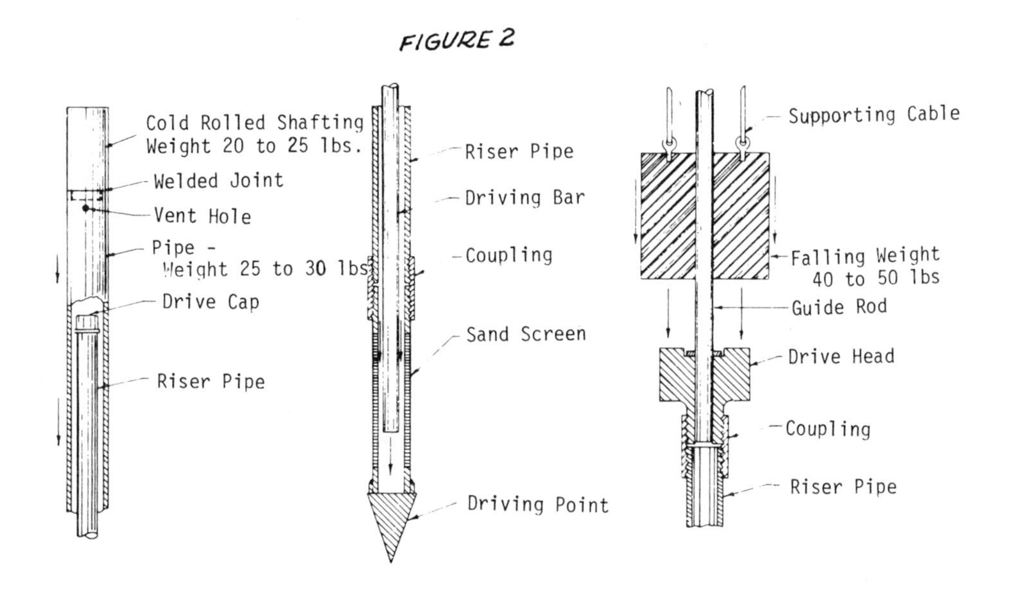


The table in Figure 3 will help identify the formations being penetrated. Experience is needed, but this may help you to understand what is happening. When you think that the water-bearing layer has been reached, stop driving and attach a handpump to try the well.

| Type of Formation | Driving Conditions | Rate of Descent | Sound of Blow | Rebound | Resistance to Rotation |
|---|---|---|---|---|---|
| Soft moist clay | Easy driving | Rapid | Dull | None | Slight but continuous |
| Tough hardened clay | Difficult driving | Slow but steady | None | Frequent rebounding | Considerable |
| Fine sand | Difficult driving | Varied | None | Frequent rebounding | Slight |
| Coarse sand | Easy driving (especially when saturated with water). | Unsteady irregular penetration for successive blows. | Dull | None | Rotation is easy and accompanied by a gritty sound |
| Gravel | Easy driving | Unsteady irregular penetration for successive blows. | Dull | None | Rotation is irregular and accompanied by a gritty sound |
| Boulder and rock | Almost impossible | Little or none | Loud | Sometimes of both hammer and pipe | Dependent on type of formation pre viously passed through by pipe |

*From: Wells, TM5-297/AFM 85-23, 1957 Army Technical Manual, p.24.*

**FIGURE 3**

Usually, easier driving shows that the water-bearing level has been reached, especially in coarse sand. If the amount of water pumped is not enough, try driving a meter or so (a few feet) more. If the flow decreases, pull the point back until the point of greatest flow is found. The point can be raised by using a lever arrangement like a fence-post jack, or, if a drive-monkey is used, by pounding the pipe back up.

Sometimes sand and silt plug up the point and the well must be "developed" to clear this out and improve the flow. First try hard, continuous pumping at a rate faster than normal. Mud and fine sand will come up with the water, but this should clear in about an hour. It may help to allow the water in the pipe to drop back down, reversing the flow periodically. With most pitcher pumps this is easily accomplished by lifting the handle very high; this opens the check valve, allowing air to enter, and the water rushes back down the well.

If this does not clear up the flow, there may be silt inside the point. This can be removed by putting a 19mm (3/4") pipe into the well and pumping on it. Either use the pitcher pump or quickly and repeatedly raise and lower the 19mm (3/4") pipe. By holding your thumb over the top of the pipe on the upstroke, a jet of muddy water will result on each downstroke. After getting most of the material out, return to direct pumping. Clean the sand from the valve and cylinder of the pump after developing the well. If you have chosen too fine a screen, it may not be possible to develop the well successfully. A properly chosen screen allows the fine material to be pumped out, leaving a bed of coarse gravel and sand that provides a highly porous and permeable water-gathering area.

The final step is to fill in the starting borehole with puddle clay or, if clay is not available, with well-tamped earth. Make a solid, water-proof pump platform (concrete is best) and provide a place for spilled water to drain away.

**Source:**

Wagner, E.G. and Lanoix, J.N. *Water Supply for Rural Areas and Small Communities.* Geneva: World Health Organization, 1959.

# DUG WELLS

A village well must often act as a reservoir, because at certain hours of the day the demand for water is heavy, whereas during the night and the heat of the day there is no call on the supply. What is suggested here is to make the well large enough to allow the water slowly percolating in to accumulate when the well is not in use in order to have an adequate supply when demand is heavy. For this reason wells are usually made 183 to 213cm (6' to 7') in diameter.

Wells cannot store rainy season water for the dry season, and there is seldom any reason for making a well larger in diameter than 213cm (7').

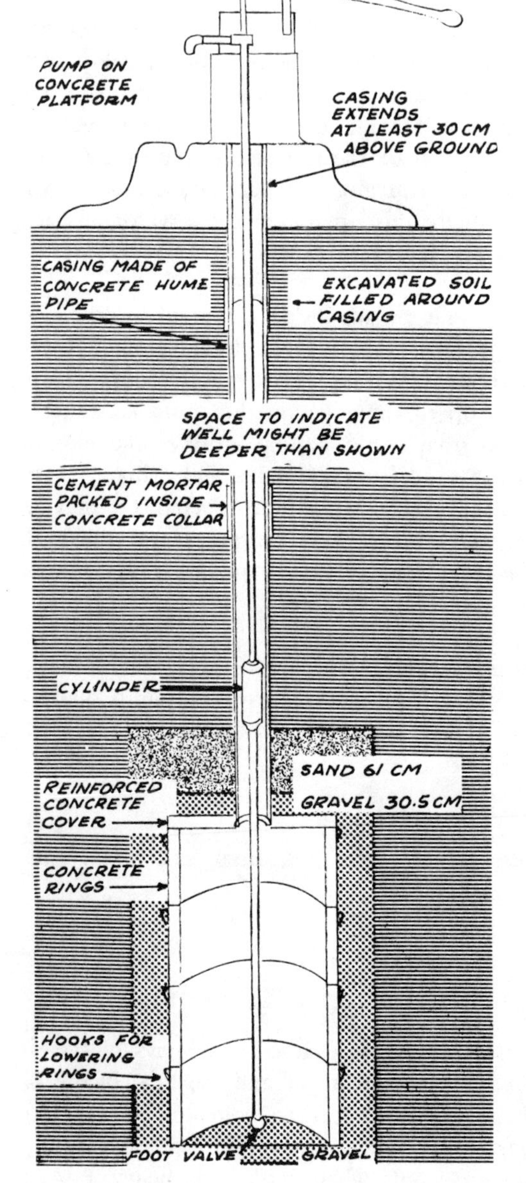


The depth of a well is much more important than the diameter in determining the amount of water that can be drawn when the water level is low. A deep, narrow well will often provide more water than a wide shallow one.

Remember that tubewells are much easier to construct than dug wells, and should be used if your region allows their construction and an adequate amount of water can be drawn from them during the busy hours (see section on Tubewells).

Deep dug wells have several disadvantages. The masonry lining needed is very expensive. Construction is potentially very dangerous; workers should not dig deeper than one and a half meters without shoring up the hole. An open well is very easily contaminated by organic matter that falls in from the surface and by the buckets used to lift the water. There is an added problem of disposing of the great quantity of soil removed from a deep dug well.

## Sealed Dug Well

The well described here has an underground concrete tank that is connected to the surface with a casing pipe, rather than a large-diameter lining as described in the preceding entry. The advantages are that it is relatively easy to build, easy to seal, takes up only a small surface area, and is low in cost.

Many of these wells were installed in India by an American Friends Service Committee team there; they perform well unless they are not deep enough or sealed and capped properly.

**Tools and Materials**

4 reinforced concrete rings with iron hooks for lowering, 91.5cm (3') in diameter
1 reinforced concrete cover with a seating hole for casing pipe
Washed gravel to surround tank: 1.98 cubic meters (70 cubic feet)
Sand for top of well: 0.68 cubic meters (24 cubic feet)
Concrete pipe: 15cm (6") in diameter, to run from the top of the tank cover to at least 30.5cm (1') above ground
Concrete collars: for joints in the concrete pipe
Cement: 4.5kg (10 pounds) for mortar for pipe joints
Deep-well pump and pipe
Concrete base for pump
Tripod, pulleys, rope for lowering rings
Special tool for positioning casing when refilling, see "Positioning Casing Pipe," below
Digging tools, ladders, rope

A villager in Barpali, India, working with an American Friends Service Committee unit there, suggested that they make a masonry tank at the bottom of the well, roof it over, and draw the water from it with a pump. The resulting sealed well has many advantages:

- It provides pure water, safe for drinking.
- It presents no hazard of children falling in.
- Drawing water is easy, even for small children.
- The well occupies little space, a small courtyard can accommodate it.
- The cost of installation is greatly reduced.
- The labor involved is much reduced.
- There is no problem of getting rid of excavated soil, since most of it is replaced.
- The casing enables the pump and pipe to be easily removed for servicing.
- The gravel and sand surrounding the tank provide an efficient filter to prevent silting, allow a large surface area for percolating water to fill the tank, and increase the effective stored volume in the tank.

On the other hand, compared to a well where people draw their own buckets or other containers of water, there are three minor disadvantages: only one person can pump at a time, the pump requires regular maintenance, and a certain amount of technical skill is required to make the parts used in the well and to install them properly.

A well is dug 122cm (4') in diameter and about 9 meters (30') deep. The digging should be done in the dry season, after the water table has dropped to its lowest level. There should be a full 3 meter (10') reaccumulation of water within 24 hours after the well has been bailed or pumped dry. Greater depth is, of course, desirable.

Spread 15cm (6") of clean, washed gravel or small rock over the bottom of the well. Lower the four concrete rings and cover into the well and position them there to form the tank. A tripod of strong poles with block and tackle is needed to lower the rings, because they weigh about 180kg (400 pounds) each. The tank formed by the rings and cover is 183cm (6') high and 91.5cm (3') in diameter. The cover has a round opening which forms a seat for the casing pipe and allows the suction pipe to penetrate to about 15cm (6") from the gravel bottom.

The first section of concrete pipe is positioned in the seat and grouted (mortared) in place. It is braced vertically by a wooden plug with four hinged arms to brace against the sides of the wall. Gravel is packed around the concrete rings and over the top of the cover till the gravel layer above the tank is at least 15cm (6") deep. This is then covered with 61cm (2') of sand. Soil removed from the well is then shoveled back until the shaft is filled within 15cm (6") of the top of the first section of casing. The next section of casing is then grouted in place, using a concrete collar made for this purpose. The well is filled and more sections of casing added until the casing extends at least 30cm (1') above the surrounding soil level.

The soil that will not pack back into the well can be used to make a shallow hill around the casing to encourage spilled water to drain away from the pump. A concrete cover is placed on the casing and a pump installed.

If concrete or other casing pipe cannot be obtained, a chimney made of burned bricks and sand-cement mortar will suffice. The pipe is somewhat more expensive, but much easier to install.

**Source:**

*A Safe Economical Well.* Philadelphia: American Friends Service Committee, 1956 (Mimeographed).

# Deep Dug Well

Untrained workers can safely dig a deep sanitary well with simple, light equipment, if they are well supervised. The basic method is outlined here.

**Tools and Materials**

Shovels, mattocks
Buckets
Rope–deep wells require wire rope
Forms–steel, welded and bolted together
Tower with winch and pulley
Cement
Reinforcing rod
Sand
Aggregate
Oil

The hand dug well is the most widespread of any kind of well. Unfortunately, in many places these wells are dug by people unfamiliar with good sanitation methods and become infected by parasitic and bacterial disease. By using modern methods and materials, dug wells can safely be made 60 meters (196.8') deep and will give a permanent source of pure water.

Experience has shown that for one person, the average width of a round well for best digging speed is 1 meter (3 1/4'). However, 1.3 meters (4 1/4') is best for two workers digging together and they dig more than twice as fast as one person. Thus, two workers in the larger hole is usually best.

Dug wells always need a permanent lining (except in solid rock, where the best method is usually to drill a tubewell).

The lining prevents collapse of the hole, supports the pump platform, stops entrance of contaminated surface water, and supports the well intake, which is the part of the well through which water enters. It is usually best to build the lining while digging, since this avoids temporary supports and reduces danger of cave-ins.

Dug wells are lined in two ways: (1) where the hole is dug and the lining is built in its permanent place and (2) where sections of lining are added to the top and the whole lining moves down as earth is removed from beneath it. The second method is called caissoning; often a combination of both is best (Figure 2.)

If possible, use concrete for the lining because it is strong, permanent, and made mostly of local materials. It can also be handled by unskilled workers with good speed and results. (See section on Concrete Construction).

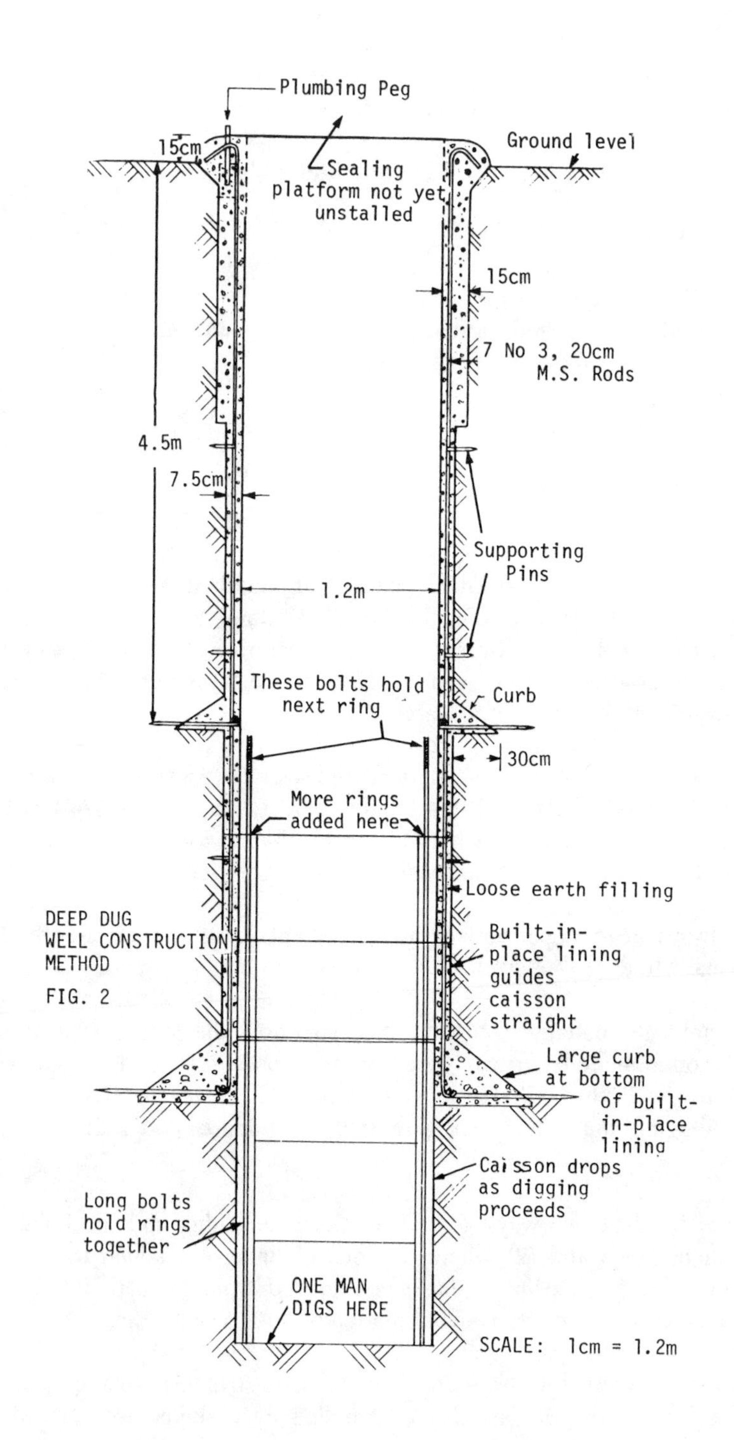


DEEP DUG WELL CONSTRUCTION METHOD

FIG. 2

Masonry and brickwork are widely used in many countries and can be very satisfactory if conditions are right. In bad ground, however, unequal pressures can make them bulge or collapse. Building with these materials is slow and a thicker wall is required than with concrete. There is also always the danger of movement during construction in loose sands or swelling shale before the mortar has set firmly between the bricks or stones.

Wood and steel are not good for lining wells. Wood requires bracing, tends to rot and hold insects, and sometimes makes the water taste bad. Worst of all, it will not make the well watertight against contamination. Steel is seldom used because it is expensive, rusts quickly, and if it is not heavy enough is subject to bulging and bending.

The general steps in finishing the first 4.6 meters (15') are:

- set up a tripod winch over cleared, level ground and mark reference points for plumbing and measuring the depth of the well.

- have two workers dig the well while another raises and unloads the dirt until the well is exactly 4.6 meters (15') deep.

- trim the hole to size using a special jig mounted on the reference points.

- place the forms carefully and fill one by one with tamped concrete.

After this is done, dig to 9.1 meters (30'), trim and line this part also with concrete. A 12.5cm (5") gap between the first and second of these sections is filled with pre-cut concrete that is grouted (mortared) in place. Each lining is self-supporting as it has a curb. The top of the first section of lining is thicker than the second section and extends above the ground to make a good foundation for the pump housing and to make a safe seal against ground water.

This method is used until the water-bearing layer is reached; there an extra-deep curb is constructed. From this point on, caissoning is used.

Caissons are concrete cylinders fitted with bolts to attach them together. They are cast and cured on the surface in special molds, prior to use. Several caissons are lowered into the well and assembled together. As workers dig, the caissons drop lower as earth is removed from beneath them. The concrete lining guides the caissons.

If the water table is high when the well is dug, extra caissons are bolted in place so that the well can be finished by a small amount of digging, and without concrete work, during the dry season.

Details on plans and equipment for this process are found in *Water Supply for Rural Areas and Small Communities,* by E. G. Wagner and J. N. Lanoix, World Health Organization, 1959.

## Reconstructing Dug Wells

Open dug wells are not very sanitary, but they can often be rebuilt by relining the top 3 meters (10') with a watertight lining, digging and cleaning the well and covering it. This method involves installation of a buried concrete slab; see Figure 3 for construction details.

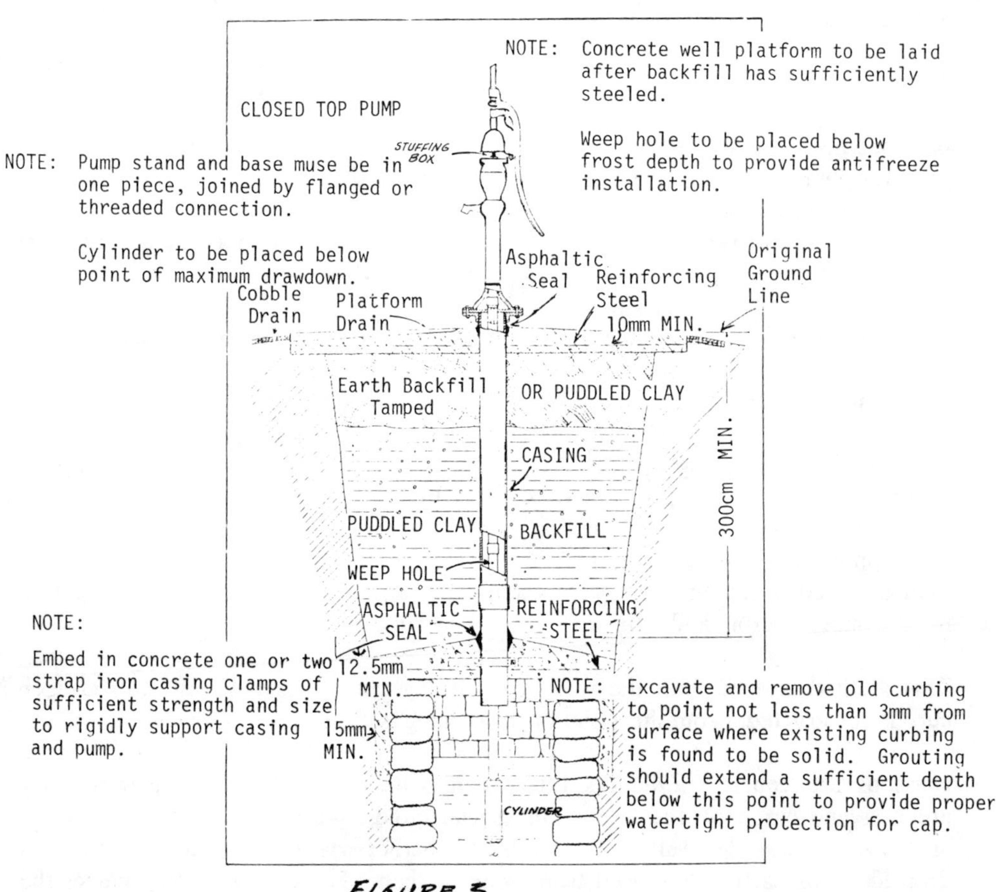


FIGURE 3

**Tools and Materials**

Tools and materials for reinforced concrete
A method for entering the well
Pump and drop pipe

**Before starting,** check the following:

- o Is the well dangerously close to a privy or other source of contamination? Is it close to a water source? Is it desirable to dig a new well elsewhere instead of cleaning this one? Could a privy be moved, instead?

- o Has the well ever gone dry? Should you deepen it as well as clean it?

- o Surface drainage should generally slope away from the well and there should be effective disposal of spilled water.

- o What method will you use to remove the water and what will it cost?

- o Before entering the well to inspect the old lining, check for a lack of oxygen by lowering a lantern or candle. If the flame remains lit, it is reasonably safe to enter the well. If the flame goes out, the well is dangerous to enter. Tie a rope around the person entering the well and have two strong workers on hand to pull him out in case of accident.

## *Relining the Wall*

The first job is to prepare the upper 3 meters (10') of the lining for concrete by removing loose rock and chipping away old mortar with a chisel, as deep as possible (see Figure 4). The next task is to clean out and deepen the well, if that is necessary. All organic matter and silt should be bailed out. The well may be dug deeper, particularly during the dry season, with the methods outlined in "Deep Dug Wells." One way to increase the water yield is to drive a well point deeper into the water-bearing soil. This normally will not raise the level of water in the well, but may make the water flow into the well faster. The well point can be piped directly to the pump, but this will not make use of the reservoir capacity of the dug well.

The material removed from the well can be used to help form a mound around the well so water will drain away from the opening. Additional soil will usually be needed for this mound. A drain lined with rock should be provided to take spilled water away from the concrete apron that covers the well.

Reline the well with concrete troweled in place over wire mesh reinforcement. The largest aggregate should be pea-sized gravel and the mix should be fairly rich with concrete, using no more than 20-23 liters (5 1/2 to 6 gallons) of water to a

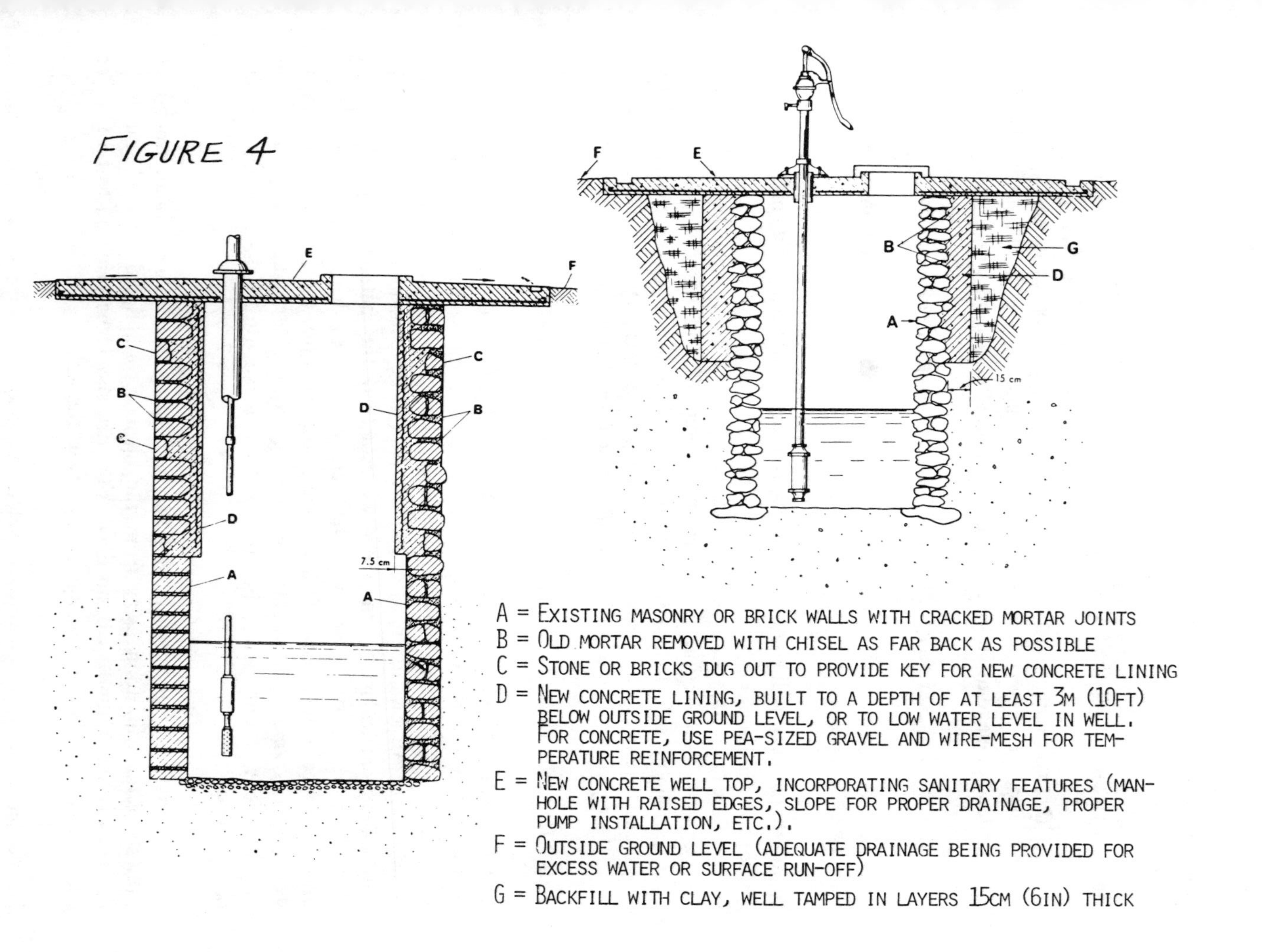
E
F
C
B
D
A
7.5 cm
G
15 cm
A = EXISTING MASONRY OR BRICK WALLS WITH CRACKED MORTAR JOINTS
B = OLD MORTAR REMOVED WITH CHISEL AS FAR BACK AS POSSIBLE
C = STONE OR BRICKS DUG OUT TO PROVIDE KEY FOR NEW CONCRETE LINING
D = NEW CONCRETE LINING, BUILT TO A DEPTH OF AT LEAST 3M (10FT) BELOW OUTSIDE GROUND LEVEL, OR TO LOW WATER LEVEL IN WELL. FOR CONCRETE, USE PEA-SIZED GRAVEL AND WIRE-MESH FOR TEMPERATURE REINFORCEMENT.
E = NEW CONCRETE WELL TOP, INCORPORATING SANITARY FEATURES (MANHOLE WITH RAISED EDGES, SLOPE FOR PROPER DRAINAGE, PROPER PUMP INSTALLATION, ETC.).
F = OUTSIDE GROUND LEVEL (ADEQUATE DRAINAGE BEING PROVIDED FOR EXCESS WATER OR SURFACE RUN-OFF)
G = BACKFILL WITH CLAY, WELL TAMPED IN LAYERS 15CM (6IN) THICK

FIGURE 4

43kg (94 pound) sack of cement. Extend the lining 70cm (27 1/2") above the original ground surface.

## *Installing the Cover and Pump*

Cast the well cover so that it makes a watertight seal with the lining to keep surface impurities out. The cover will also support the pump. Extend the slab out over the mound about a meter (a few feet) to help drain water away from the site. Make a manhole and space for the drop pipe of the pump. Mount the pump off center so there is room for the manhole. The pump is mounted on bolts cast into the cover. The manhole must be 10cm (4") higher than the surface of the slab. The manhole cover must overlap by 5cm (2") and should be fitted with a lock to prevent accidents and contamination. Be sure that the pump is sealed to the slab.

## *Disinfecting the Well*

Disinfect the well by using a stiff brush to wash the walls with a very strong solution of chlorine. Then add enough chlorine in the well to make it about half the strength of the solution used on the walls. Sprinkle this last solution all over the surface of the well to distribute it evenly. Cover the well and pump up the water until the water smells strongly of chlorine. Let the chlorine remain in the pump and well for one day and then pump it until the chlorine is gone.

Have the well water tested several days after disinfection to be sure that it is pure. If it is not, repeat the disinfection and testing. If it is still not pure, get expert advice.

**Sources:**

Wagner, E.G. and Lanoix, J.N. *Water Supply for Rural Areas and Small Communities.* Geneva: World Health Organization, 1959.

*Manual of Individual Water Supply Systems,* Public Health Service Publication No. 24. Washington, D.C.: Department of Health and Human Services.

# SPRING DEVELOPMENT

Springs, particularly in sandy soil, often make excellent water sources, but they should be dug deeper, sealed, protected by a fence, and piped to the home. Proper development of a spring will increase the flow of ground water and lower the chances of contamination from surface water. If fissured rock or limestone are present, get expert advice before attempting to develop the spring.

Springs occur where water, moving through porous and saturated underground layers of soil (aquifer), emerges at the ground surface. They can be either:

- o **Gravity seepage,** where the water bearing soil reaches the surface over an impermeable layer, or

- o **Pressure** or **artesian,** where the water, under pressure and trapped by a hard layer of soil, finds an opening and rises to the surface. (In some parts of the world, all springs are called **artesian.**)

The following steps should be considered in developing springs:

1) Observe the seasonal flow variations over a period of a year if possible.

2) Determine the type of spring–seepage or artesian–by digging a small hole. An earth auger with extensions is the most suitable tool for that job. It may not be possible to reach the underlying impermeable layer.

3) Have chemical and biological tests made on samples of the water.

Dig a small hole near the spring to learn the depth of the hard layer of soil and to find out whether the spring is gravity seepage or pressure. Check uphill and nearby for sources of contamination. Test the water to see if it must be purified before being used for drinking. A final point: Find out if the spring runs during long dry spells.

For gravity-fed springs, the soil is usually dug to the hard, underlying layers and a tank is made with watertight concrete walls on all but the uphill side (see Figures 1 and 2). The opening on the uphill side should be lined with porous concrete or stone without mortar, so that it will admit the gravity seepage water. It can be backfilled with gravel and sand, which helps to keep fine materials in the water-bearing soil from entering the spring. If the hard soil cannot be reached easily, a concrete cistern is built that can be fed by a perforated pipe placed in the water-bearing layer of earth. With a pressure spring, all sides of the tank are made of watertight reinforced concrete, but the bottom is left open. The water enters through the bottom.

Read the section in this handbook on cisterns before developing your spring. No matter how the water enters your tank, you must make sure the water is pure by:

- o building a complete cover to stop surface pollution and keep out sunlight, which causes algae to grow.

- o installing a locked manhole with at least a 5cm (2") overlap to prevent entrance of polluted ground water.

Fig. 1. PROPERLY PROTECTED SPRING (I)

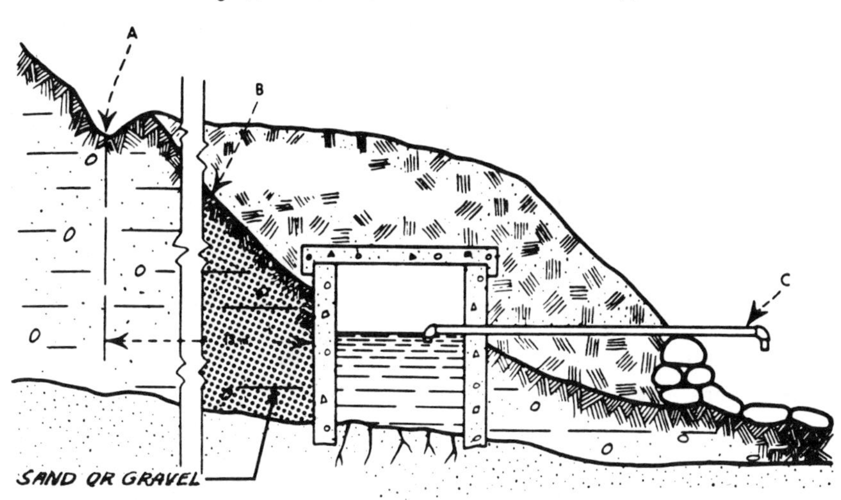


**A** = Protective drainage ditch to keep drainage water a safe distance from spring
**B** = Original slope and ground line
**C** = Screened outlet pipe : can discharge freely or be piped to village or residence

Springs can offer an economical and safe source of water. A thorough search should be made for signs of ground-water outcropping. Springs that can be piped to the user by gravity offer an excellent solution. Rainfall variation may influence the yield, so dry-weather flow should be checked.

Fig. 2. PROPERLY PROTECTED SPRING (II)

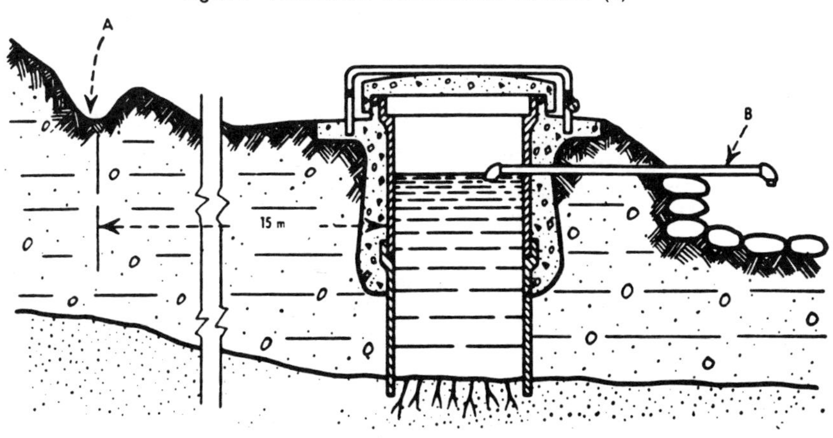


**A** = Protective drainage ditch to keep drainage water a safe distance from spring
**B** = Screened outlet pipe : to discharge freely or be piped to village or residence

- installing a screened overflow that discharges at least 15cm (6") above the ground. The water must land on a cement pad or rock surface to keep thé water from making a hole in the ground and to ensure proper drainage away from the spring.

- arranging the spring so that surface water must filter through at least 3 meters (10') of soil before reaching the ground water. Do this by making a diversion ditch for surface water about 15 meters (50') or more from the spring. Also, if necessary, cover the surface of the ground near the spring with a heavy layer of soil or clay to increase the distances that rainwater must travel, thus ensuring that it has to filter through 3 meters (10') of soil.

- making a fence to keep people and animals away from the spring's immediate surroundings. The suggested radius is 7.6 meters (25').

- installing a pipeline from the overflow to the place where the water is to be used.

Before using the spring, disinfect it thoroughly by adding chlorine or chlorine compounds. Shut off the overflow to hold the chlorine solution in the well for 24 hours. If the spring overflows even though the water is shut off, arrange to add chlorine so that it remains strong for at least 30 minutes, although 12 hours would be much safer. After the chlorine is flushed from the system have the water tested. (See section on "Superchlorination.")

**Sources:**

Wagner, E.G. and Lanoix, J.N. *Water Supply for Rural Areas and Small Communities.* Geneva: World Health Organization, 1959.

*Manual of Individual Water Supply Systems,* Public Health Service Publication No. 24. Washington, D.C.: U.S. Department of Health and Human Services.

**Acknowledgements**

John M. Jenkins III, VITA Volunteer, Marrero, Louisiana
Ramesh Patel, VITA Volunteer, Albany, New York
William P. White, VITA Volunteer, Brooklyn, Connecticut

# Water Lifting and Transport

## OVERVIEW

Once a source of water has been found and developed, four basic questions must be answered:

1. What is the rate of flow of the water in your situation?
2. Between what points must the water be transported?
3. What kind and size of piping is needed to transport the required flow?
4. What kind of pump, if any, is necessary to produce the required flow?

The information in this section will help you to answer the third and fourth questions, once you have determined the answers to the first two.

## Moving Water

The first three entries in this section discuss the flow of water in small streams, partially filled pipes, and when the height of the reservoir and size of pipe are known. They include equations and alignment charts (also called nomographs) that give simple methods of estimating the flow of water under the force of **gravity**, that is, without pumping. The fourth tells how to measure flow by observing the spout from a horizontal pipe.

Four entries follow on piping, including a discussion of pipes made of bamboo.

You will note that in the alignment charts here and elsewhere, the term "nominal diameter, inches, U.S. Schedule 40" is used along with the alternate term, "inside diameter in centimeters," in referring to pipe size.

Pipes and fittings are usually manufactured to a standard schedule of sizes. U.S. Schedule 40, the most common in the United States, is also widely used in other countries. When one specifies "2-inch Schedule 40," one automatically specifies the pressure rating of the pipe and its inside and outside diameters (neither of which, incidentally, is actually 2"). If the schedule is not known, measure the **inside diameter** and use this for flow calculations.

## Lifting Water

Next, seveal entries follow the steps required to design a water-pumping system with piping. The first entry in this group, "Pump Specifications: Choosing or

Evaluating a Pump," presents all the factors that must be considered in selecting a pump. Fill out the form included there and make a piping sketch, whether you plan to send it to a consultant for help or do the design and selection yourself.

The first pieces of information needed for selecting pump type and size are: (1) the flow rate of water needed and (2) the **head** or pressure to be overcome by the pump. The head is composed of two parts: the height to which the liquid must be raised, and the resistance to flow created by the pipe walls (friction-loss).

The **friction-loss head** is the most difficult factor to measure. The entry "Determining Pump Capacity and Horsepower Requirements" describes how to select the economic pipe size(s) for the flow desired. With the pipe(s) selected one must then calculate the friction-loss head. The entry "Estimating Flow Resistance of Pipe Fittings" makes it possible to estimate extra friction caused by constrictions of pipe fittings. With this information and the length of pipe, it is possible to estimate the pump power requirement using the entry, "Determining Pump Capacity and Horsepower Requirements."

These entries have another very important use. You may already have a pump and wonder "Will it do this job?" or "What size motor should I buy to do this job with the pump I have?" The entry "Pump Specifications: Choosing or Evaluating a Pump" can be used to collect all the information on the pump and on the job you want it to do. With this information, you can ask a consultant or VITA if the pump can be used or not.

There are many varieties of pumps for lifting water from where it is to where it is to be delivered. But for any particular job, there are probably one or two kinds of pumps that will serve better than others. We will discuss here only two broad classes of pumps: lift pumps and force pumps.

A **lift** or suction pump is located at the top of a well and raises water by suction. Even the most efficient suction pump can create a negative pressure of only 1 atmosphere: theoretically, it could raise a column of water 10.3m (34') at sea level. But because of friction losses and the effects of temperature, a suction pump at sea level can actually lift water only 6.7m to 7.6m (22' to 25'). The entry "Determining Lift Pump Capability" explains how to find out the height a lift pump will raise water at different altitudes with different water temperatures.

When a lift pump is not adequate, a **force** pump must be used. With a force pump, the pumping mechanism is placed at or near the water level and pushes the water up. Because it does not depend on atmospheric pressure, it is not limited to a 7.6m (25') head.

Construction details are given for two irrigation pumps that can be made at the village level. An easy-to-maintain pump handle mechanism is described. Use of the hydraulic ram, a self-powered pump, is described.

Finally, there are entries on Reciprocating Wire Power Transmission for Water Pumps, and on Wind Energy for Water Pumping. Further details on pumps can be found in the publications listed below and in the Reference section at the back of the book.

Margaret Crouch, ed. *Six Simple Pumps.* Arlington, Virginia: Volunteers in Technical Assistance, 1982.

Molenaar, Aldert. *Water Lifting Devices for Irrigation.* Rome: Food and Agriculture Organization, 1956.

*Small Water Supplies.* London: The Ross Institute, The London School of Hygiene and Tropical Medicine, 1967.

# WATER TRANSPORT

## Estimating Small Stream Water Flow

A rough but very rapid method of estimating water flow in small streams is given here. In looking for water sources for drinking, irrigation, or power generation, one should survey all the streams available. If sources are needed for use over a long period, it is necessary to collect information throughout the year to determine flow changes–especially high and low flows. The number of streams that must be used and the flow variations are important factors in determining the necessary facilities for utilizing the water.

**Tools and Materials**

Timing device, preferably watch with second hand
Measuring tape
Float (see below)
Stick for measuring depth

FIGURE 1

The following equation will help you to measure flow quickly:

$$Q = K \times A \times V,$$

where:

Q (Quantity) = flow in liters per minute

A (Area) = cross-section of stream, perpendicular to flow, in square meters

V (Velocity) = stream velocity, meters per minute

K (Constant) = a corrected conversion factor. This is used because surface flow is normally faster than average flow. For normal stages use K = 850; for flood states use K = 900 to 950.

## *To Find Area of a Cross-Section*

The stream will probably have different depths along its length so select a place where the depth of the stream is average.

- Take a measuring stick and place it upright in the water about one-half meter (1 1/2') from the bank.
- Note the depth of water.
- Move the stick 1 meter (3') from the bank in a line directly across the stream. Note the depth.
- Move the stick 1.5 meters (4 1/2') from the bank, note the depth, and continue moving it at half-meter (1 1/2') intervals until you cross the stream.

Note the depth each time you place the stick upright in the stream. Draw a grid, like the one in Figure 2, and mark the varying depths on it so that a cross-section of the stream is shown. A scale of 1cm to 10cm is often used for such grids. By counting the grid squares and fractions of squares, the area of the water can be estimated. For example, the grid shown here has a little less than 4 square meters of water.

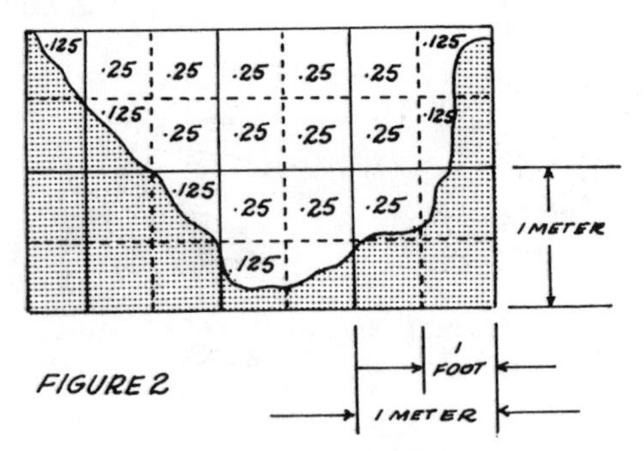


FIGURE 2

## To Find Velocity

Put a float in the stream and measure the distance of travel in one minute (or fraction of a minute, if necessary.) The width of the stream where the velocity is being measured should be as constant as possible and free of rapids.

A light surface float, such as a chip, will often change course because of wind or surface currents. A weighted float, which sits upright in the water, will not change course so easily. A lightweight tube or tin can, partly filled with water or gravel so that it floats upright with only a small part showing above water, makes a good float for measuring.

## Measuring Wide Streams

For a wide, irregular stream, it is better to divide the stream into 2- or 3-meter sections and measure the area and velocity of each. Q is then calculated for each section and the Qs added together to give a total flow.

Example (see Figure 2):

Cross section is 4 square meters

Velocity of float = 6 meters traveled in 1/2 minute

Stream flow is normal

$$Q = 850 \times 4 \times \frac{6 \text{ meters}}{.5 \text{ minute}}$$

Q = 40,800 liters per minute or 680 liters per second

**Using English Units**

If English units of measurement are used, the equation for measuring stream flow is: Q = K x A x V, where:

Q = flow in U.S. gallons per minute

A = cross-section of stream, perpendicular to flow, in square feet

V = stream velocity in feet per minute

K = a corrected conversion factor: 6.4 for normal stages; 6.7 to 7.1 for flood stages

The grid used would be like the one in Figure 3; a common scale is 1" to 12".

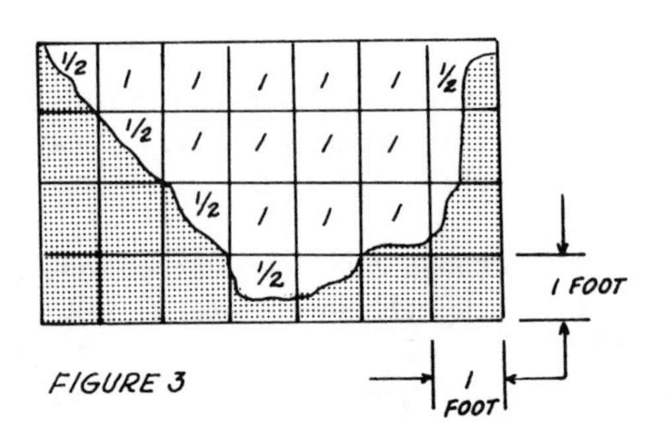


FIGURE 3

Example:

Cross-section is 15 square feet

Float velocity = 20' in 1/2 minute

Stream flow is normal

$$Q = 6.4 \times 15 \times \frac{20 \text{ feet}}{.5 \text{ minute}}$$

Q = 3,800 gallons per minute

**Source:**

Clay, C.H. *Design of Fishways and Other Fish Facilities.* Ottawa: P.E. Department of Fisheries of Canada, 1961.

# Measuring Water Flow in Partially-Filled Pipes

The flow of water in partially-filled horizontal pipes (Figure 1) or circular channels can be determined–if you know the inside diameter of the pipe and the depth of the water flowing–by using the alignment chart (nomograph) in Figure 2.

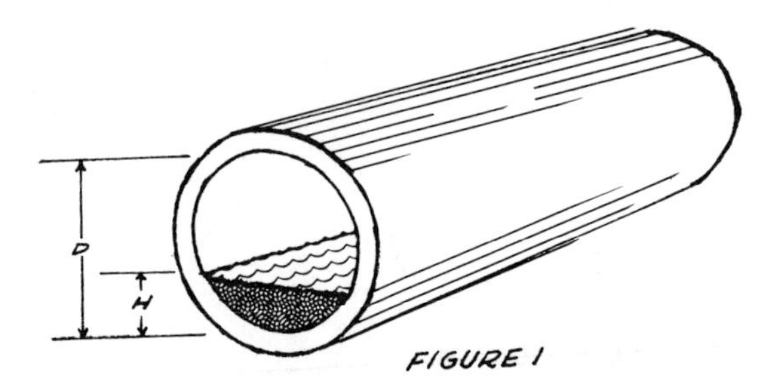


FIGURE 1

This method can be checked for low flow rates and small pipes by measuring the time required to fill a bucket or drum with a weighed quantity of water. A liter of water weighs lkg (1 U.S. gallon of water weighs 8.33 pounds).

**Tools and Materials**

Ruler to measure water depth (if ruler units are inches, multiply by 2.54 to convert to centimeters)
Straight edge, to use with alignment chart

The alignment chart applies to pipes with 2.5cm to 15cm inside diameters, 20 to 60% full of water, and having a reasonably smooth surface (iron, steel, or concrete sewer pipe). The pipe or channel must be reasonably horizontal if the result is to be accurate. The eye, aided by a plumb line to give a vertical reference, is a sufficiently good judge. If the pipe is not horizontal another method will have to be used. To use the alignment chart, simply connect the proper point on the "K" scale with the proper point on the "d" scale with the straight edge. The flow rate can then be read from the "q" scale.

q = rate of flow of water, liters per minute 8.33 pounds = 1 gallon.

d = internal diameter of pipe in centimeters.

K = decimal fraction of vertical diameter under water. Calculate K by measuring the depth of water (h) in the pipe and dividing it by the pipe diameter (d), or $K = \frac{h}{d}$ (see Figure 1).

Example:

What is the rate of flow of water in a pipe with an internal diameter of 5cm, running 0.3 full? A straight line connecting 5 on the d-scale with 0.3 on the K-scale intersects the q-scale at flow of 18 liters per minute.

**Source:**

Greve Bulletin 32, Volume 12, No. 5, Purdue University, 1928.

FIGURE 2

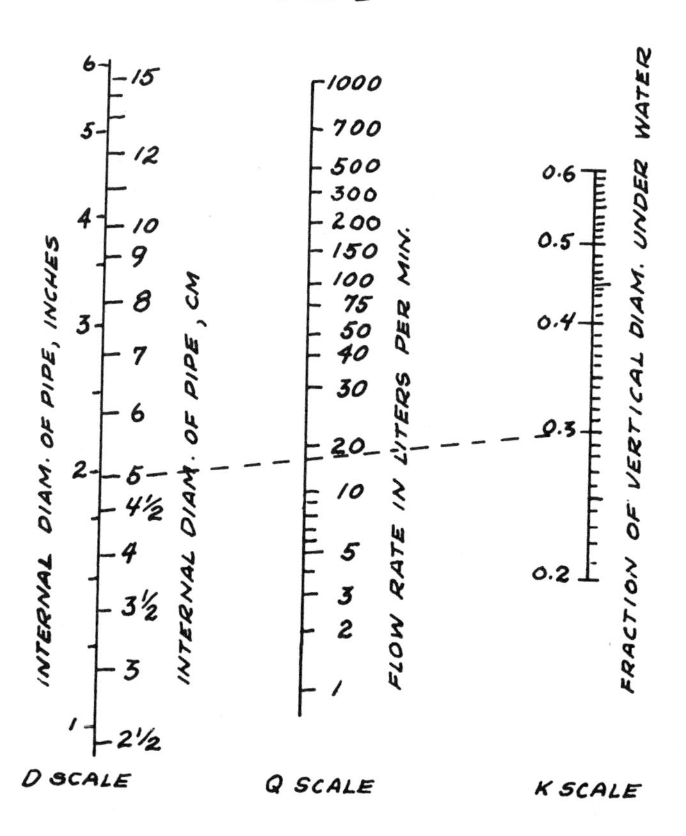

# Determining Probable Water Flow with Known Reservoir Height and Size and Length of Pipe

The alignment chart in Figure 1 gives a reasonably accurate determination of water flow when pipe size, pipe length, and height of the supply reservoir are known. The example given here is for the analysis of an existing system. To design a new system, assume a pipe diameter and solve for flow rate, repeating the procedure with new assumed diameters until one of them provides a suitable flow rate.

## Tools and Materials

Straightedge, for use with alignment chart
Surveying instruments, if available

The alignment chart was prepared for clean, new steel pipe. Pipes with rougher surfaces or steel or cast iron pipe that has been in service for a long time may give flows as low as 50 percent of those predicted by this chart.

The **available head** (h) is in meters and is taken as the difference in elevation between the supply reservoir and the point of demand. This may be crudely estimated by eye, but for accurate results some sort of surveying instruments are necessary.

For best results, the **length of pipe** (L) used should include the equivalent lengths of fittings as described in the section, "Estimating Flow Resistance of Pipe Fittings," p. 76. This length (L) divided by the **pipe internal diameter** (D) gives the necessary "L/D" ratio. In calculating L/D, note that the units of measuring both "L" and "D" must be the same, e.g., feet divided by feet; meters divided by meters; centimeters by centimeters.

Example:

Given available head (h) of 10 meters, pipe internal diameter (D) of 3cm, and equivalent pipe length (L) of 30 meters (3,000cm).

$$\text{Calculate } L/D = \frac{3{,}000\text{cm}}{3\text{cm}} = 1{,}000$$

The alignment chart solution is in two steps:

1. Connect internal diameter 3cm to available head (10 meters), and make a mark on the Index Scale. (In this step, disregard "Q" scale)

2. Connect mark on Index Scale with L/D (1,000), and read flow rate (Q) of approximately 140 liters per minute.

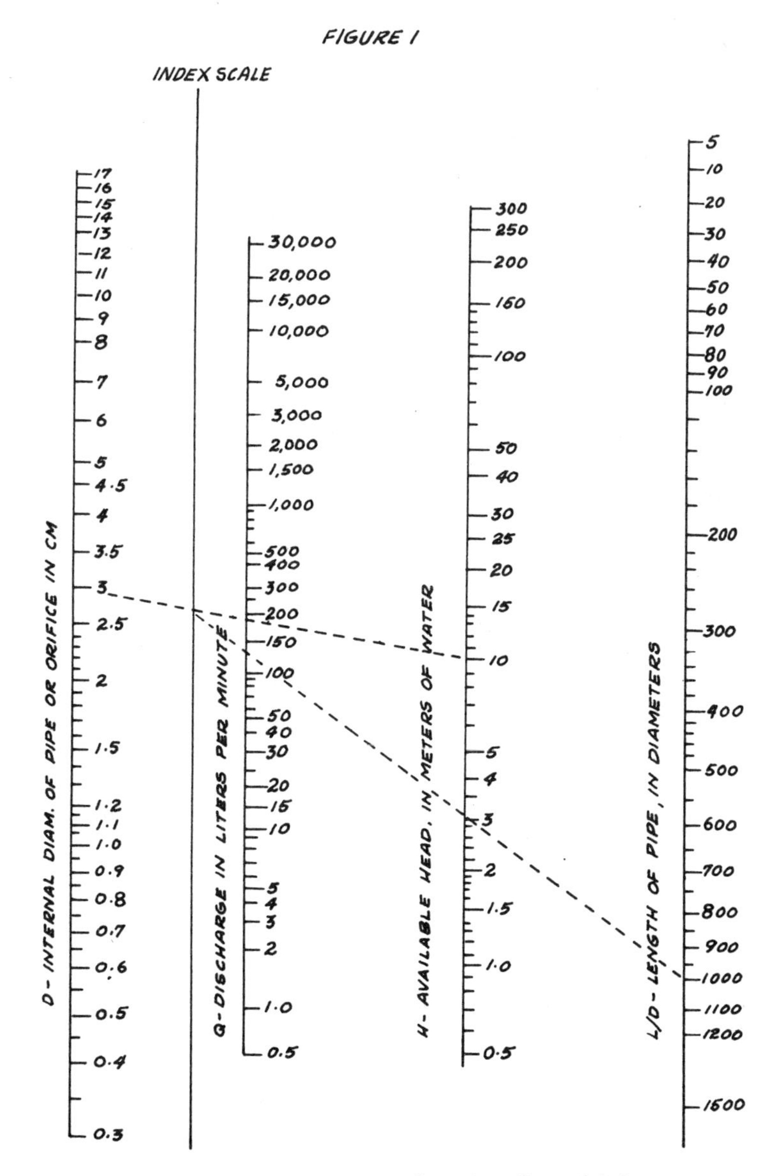


Alignment chart for determining probable water flow with known reservoir height and size and length of pipe.

**Source:**

Crane Company Technical Paper #407, pages 54-55.

# Estimating Water Flow from Horizontal Pipes

If a horizontal pipe is discharging a full stream of water, you can estimate the rate of flow from the alignment chart in Figure 2. This is a standard engineering technique for estimating flows; its results are usually accurate to within 10 percent of the actual flow rate.

### Tools and Materials

Straightedge and pencil, to use alignment chart
Tape measure
Level
Plumb bob

The water flowing from the pipe must completely fill the pipe opening (see Figure 1). The results from the chart will be most accurate when there is no constricting or enlarging fitting at the end of the pipe.

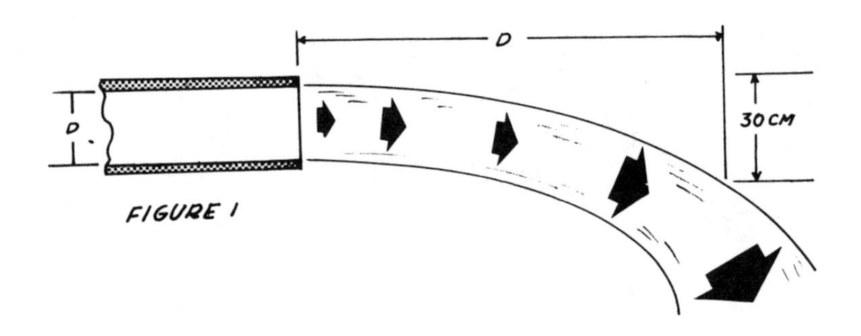

FIGURE 1

Example:

> Water is flowing out of a pipe with an inside diameter (d) of 3cm (see Figure 1). The stream drops 30cm at a point 60cm from the end of the pipe.
>
> Connect the 3cm inside diameter point on the "d" scale in Figure 2 with the 60cm point on the "D" scale. This line intersects the "q" scale at about 100 liters per minute, the rate at which water is flowing out of the pipe.

**Source:**

Duckworth, Clifford C. "Flow of Water from Horizontal Open-end Pipes." *Chemical Processing,* June 1959, p. 73.

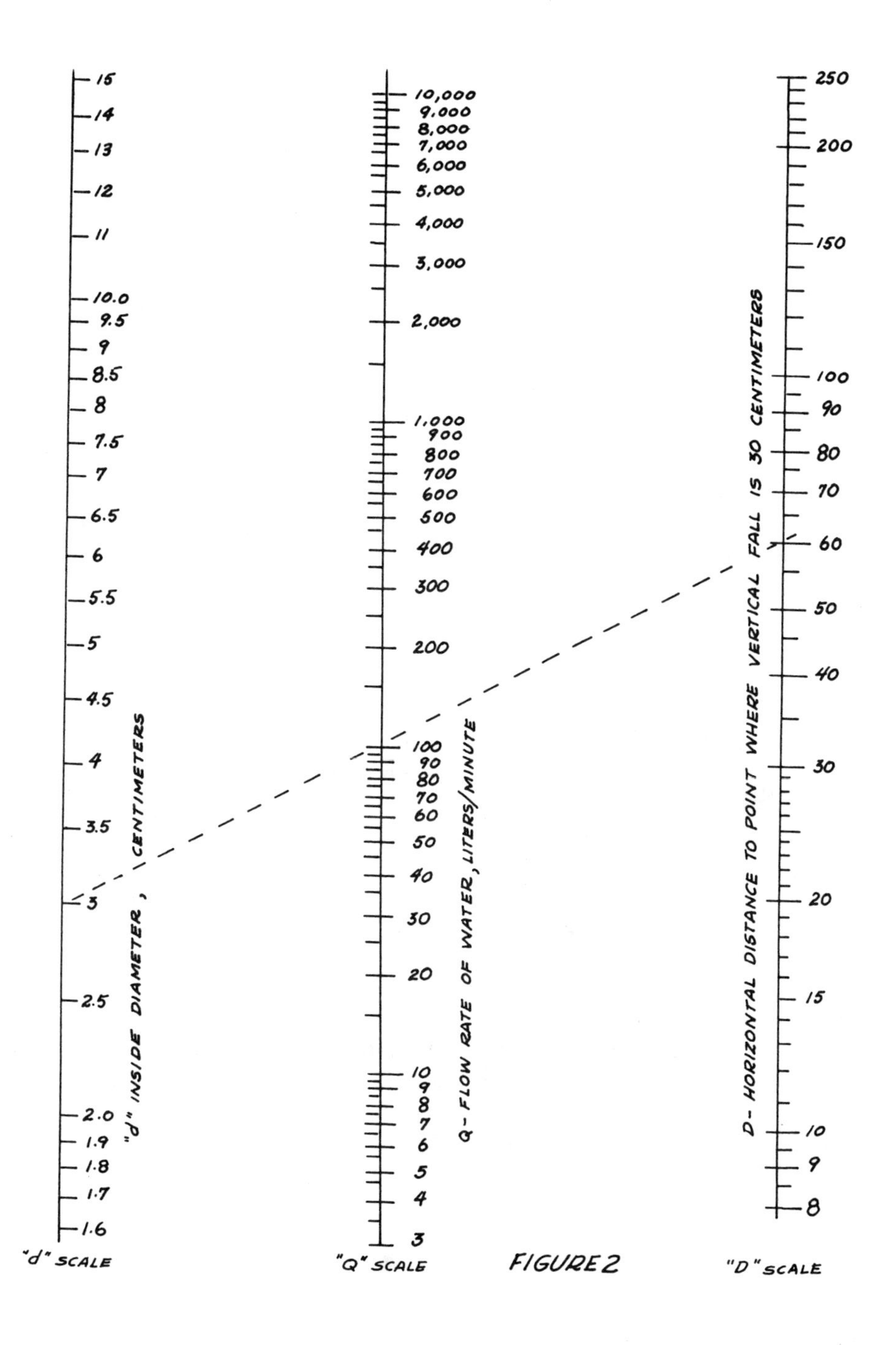

"d" INSIDE DIAMETER, CENTIMETERS
Q- FLOW RATE OF WATER, LITERS/MINUTE
D- HORIZONTAL DISTANCE TO POINT WHERE VERTICAL FALL IS 30 CENTIMETERS
"d" SCALE
"Q" SCALE
"D" SCALE


FIGURE 2

# Determining Pipe Size or Velocity of Water in Pipes

The choice of pipe size is one of the first steps in designing a simple water system.

The alignment chart in Figure 1 can be used to compute the pipe size needed for a water system when the water velocity is known. The chart can also be used to find out what water velocity is needed with a given pipe size to yield the required rate of flow.

**Tools and Materials**

Straightedge
Pencil

Practical water systems use water velocities from 1.2 to 1.8 meters (3.9 to 5.9 feet) per second. Very fast velocity requires high pressure pumps, which in turn require large motors and use excessive power. Velocities that are too low are expensive because larger pipe diameters must be used.

It may be advisable to calculate the cost of two or more systems based on different pipe sizes. Remember, it is usually wise to choose a little larger pipe if higher flows are expected in the next 5 to 10 years. In addition, water pipes often build up rust and scale, reducing the diameter and thereby increasing the velocity and pump pressure required to maintain flow at the original rate. If extra capacity is designed into the piping system, more water can be delivered by adding to the pump capacity without changing all the piping.

To use the chart, locate the flow (liters per minute) you need on the Q-scale. Draw a line from that point, through 1.8m/sec velocity on the V-scale, to the d-scale. Choose the nearest standard size pipe.

For example, suppose you need a flow of 50 liters per minute at the time of peak demand. Draw a line from 50 liters per minute on the Q-scale through 1.8m/sec on the V-scale. Notice that this intersects the d-scale at about 2.25. The correct pipe size to choose would be the next largest standard pipe size, e.g., 1" nominal diameter, U.S. Schedule 40. If pumping costs (electricity or fuel) are high, it would be well to limit velocity to 1.2m/sec and install a slightly larger pipe size.

**Source:**

Crane Company Technical Paper #409, pages 46-47.

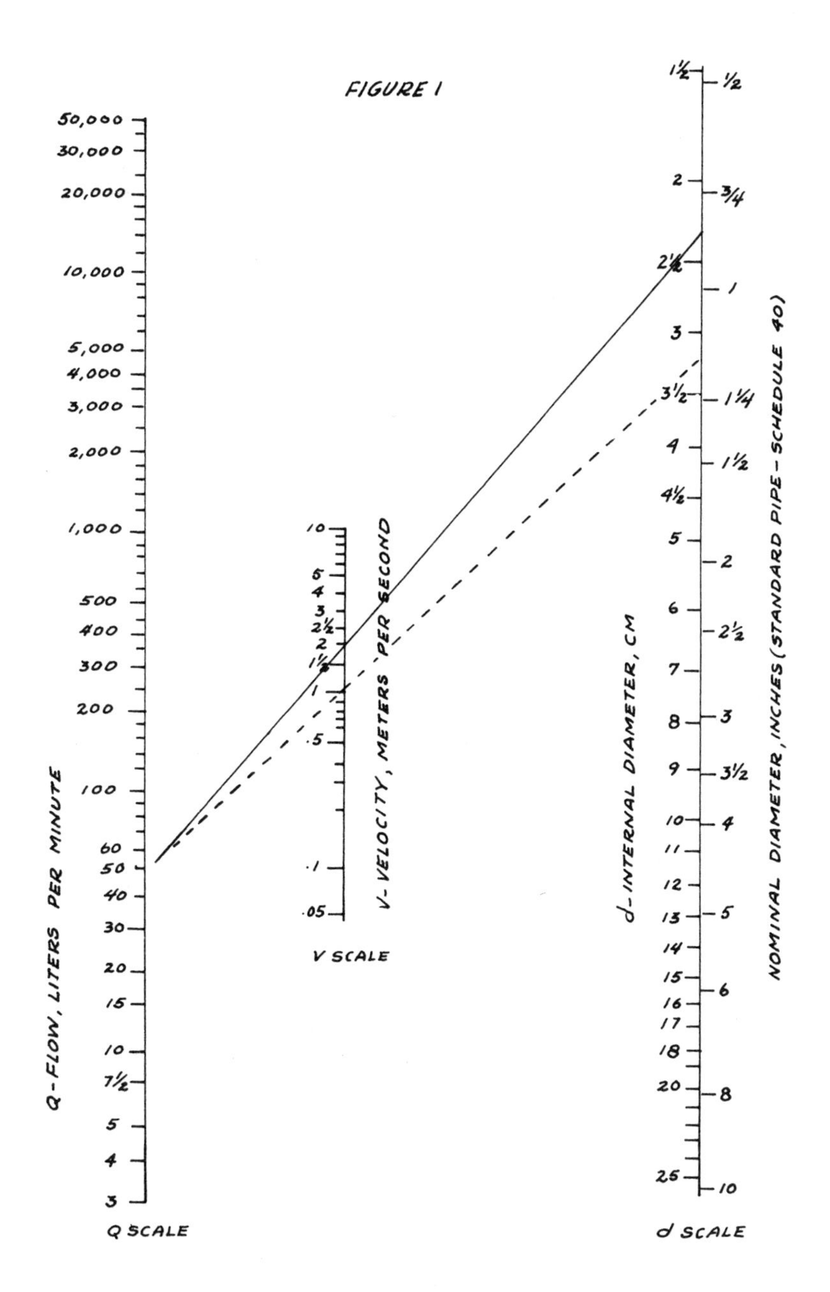
FIGURE I
Q-FLOW, LITERS PER MINUTE
Q SCALE
V-VELOCITY, METERS PER SECOND
V SCALE
d-INTERNAL DIAMETER, CM
NOMINAL DIAMETER, INCHES (STANDARD PIPE-SCHEDULE 40)
d SCALE

# Estimating Flow Resistance of Pipe Fittings

One of the forces a pump must overcome to deliver water is the friction/resistance of pipe fittings and valves to the flow of water. Any bends, valves, constrictions, or enlargements (such as passing through a tank) add to friction.

The alignment chart in Figure 1 gives a simple but reliable way to estimate this resistance: it gives the equivalent length of straight pipe that would have the same resistance. The sum of these equivalent lengths is then added to the actual length of pipe. This gives the **total equivalent pipe length,** which is used in the entry, "Determining Pump Capacity and Horsepower Requirements," to determine total friction loss.

Rather than calculate the pressure drop for each valve or fitting separately, Figure 1 gives the equivalent length of straight pipe.

## *Valves*

Note the difference in equivalent length depending on how far the valve is open.

1. Gate Valve: full opening valve; can see through it when open; used for complete shut off of flow.

2. Globe Valve: cannot see through it when open; used for regulating flow.

3. Angle Valve: like the globe, used for regulating flow.

4. Swing Check Valve: a flapper opens to allow flow in one direction but closes when water tries to flow in the opposite direction.

**Example 1:**

**Pipe with 5cm inside diameter**

| | Equivalent Length in Meters |
|---|---|
| a. Gate Valve (fully open) | .4 |
| b. Flow into line - ordinary entrance | 1.0 |
| c. Sudden enlargement into 10cm pipe (d/D = 1/2) | 1.0 |
| d. Pipe length | 10.0 |
| Total Equivalent Pipe Length | 12.4 |

# Resistance of Valves and Fittings to Flow of Fluids

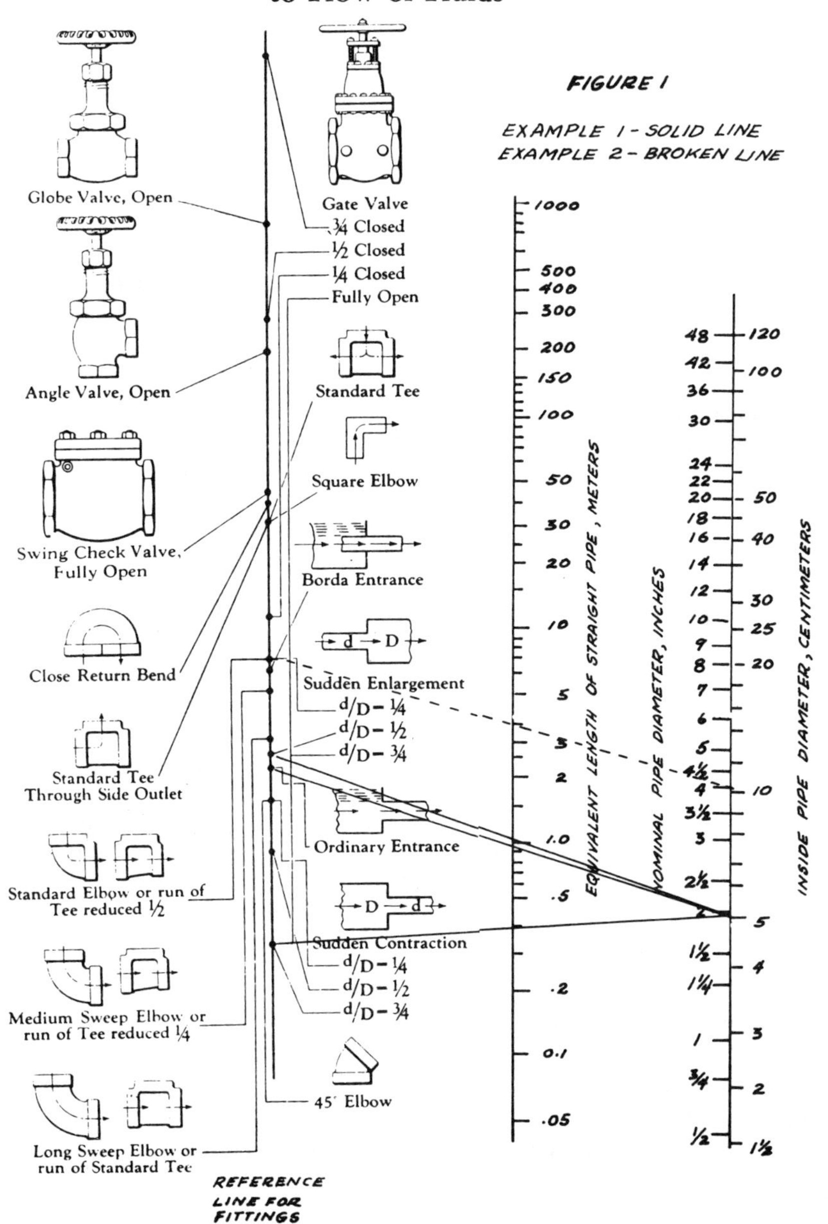

**Example 2:**

**Pipe with 10cm inside diameter**

| | Equivalent Length in Meters |
|---|---|
| a. Elbow (standard) | 4.0 |
| b. Pipe length | 10.0 |
| Total Equivalent Pipe Length | 14.0 |

## *Fittings*

Study the variety of tees and elbows: note carefully the direction of flow through the tee. To determine the equivalent length of a fitting, (a) pick proper dot on "fitting" line, (b) connect with inside diameter of pipe, then using a straight edge read equivalent length of straight pipe in meters, and (c) add the fitting equivalent length to the actual length of pipe being used.

**Source:**

Crane Company Technical Paper #409, pages 20-21.

# Bamboo Piping

Where bamboo is readily available, it seems to be a good substitute for metal pipe. Bamboo pipe is easy to make with unskilled labor and local materials. The important features of the design and construction of a bamboo piping system are given here.

Bamboo pipe is extensively used in Indonesia to transport water to villages. In many rural areas of Taiwan, bamboo is commonly used in place of galvanized iron for deep wells up to a maximum depth of 150 meters (492'). Bamboos of 50mm (2") diameter are straightened by means of heat, and the inside nodes knocked out. The screen is made by punching holes in the bamboo and wrapping that section with a fibrous mat-like material from a palm tree, *Chamaerops humilis.* In fact, such fibrous screens are also used in many galvanized iron tube wells.

Bamboo piping can hold pressure up to two atmospheres (about 2.1kg per square centimeter or 30 pounds per square inch). It cannot, therefore, be used as pressure piping. It is most suitable in areas where the source of supply is higher than the area to be served and the flow is under gravity.

Figure 1 is a sketch of a bamboo pipe water supply system for a number of villages. Figure 2 shows a public water fountain.

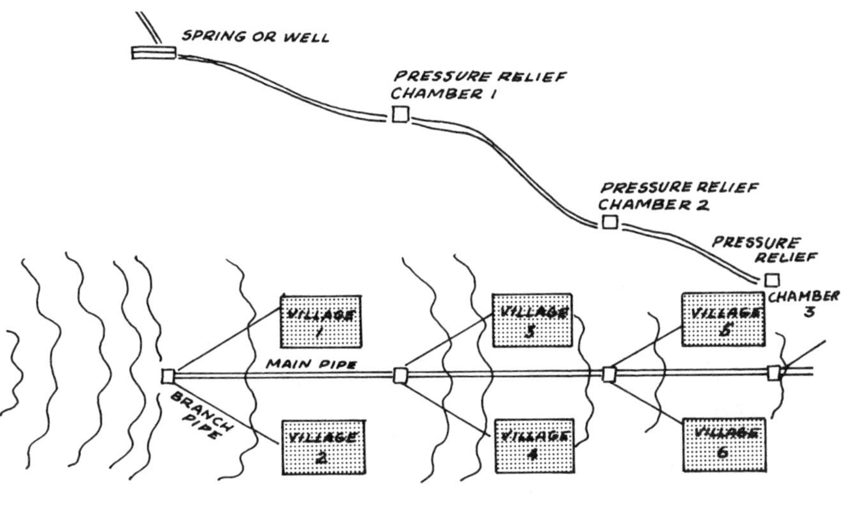


FIGURE 1

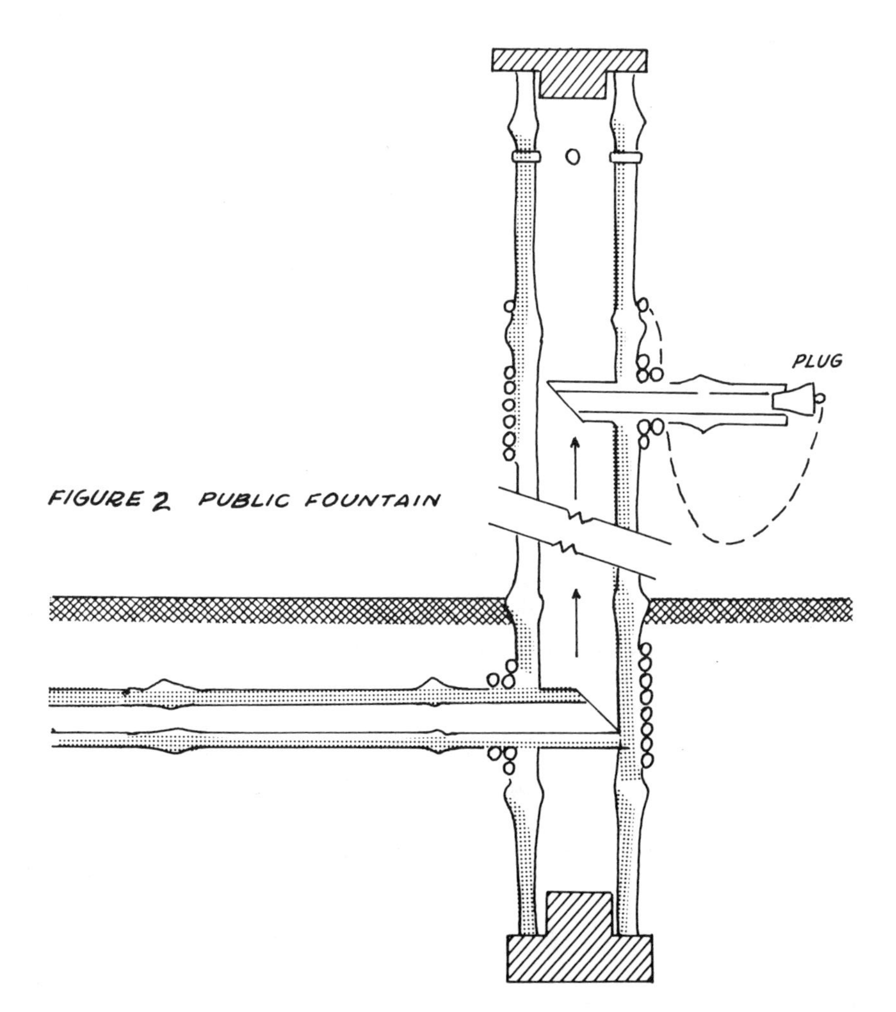


FIGURE 2 PUBLIC FOUNTAIN

## Health Aspects

If bamboo piping is to carry water for drinking purposes, the only preservative treatment recommended is boric acid: borax in a 1:1 ratio by weight. The recommended treatment is to immerse green bamboo completely in a solution of 95 percent water and 5 percent boric acid.

After a bamboo pipe is put into operation it gives an undesirable odor to the water. This, however, disappears after about three weeks. If chlorination is done before discharge to the pipe, a reservoir giving sufficient contact time for effective disinfection is required since bamboo pipe removes chlorine compounds and no residual chlorine will be maintained in the pipe. To avoid possible contamination by ground water, an ever present danger, it is desirable to maintain the pressure within the pipe at a higher level than any water pressure outside the pipe. Any leakage will then be from the pipe, and contaminated water will not enter the pipe.

## Design and Construction

### Tools and Materials

Chisels (see text and Figure 3)
Nail, cotter pin, or linchpin
Caulking materials
Tar
Rope

Bamboo pipe is made of lengths of bamboo of the desired diameter by boring out the dividing membrane at the joints. A circular chisel for this purpose is shown in Figure 3. One end of a short length of steel pipe is belled out to increase the diameter and the edge sharpened. A length of bamboo pipe of sufficiently small diameter to slide into the pipe is used as a boring bar and secured to the pipe by drilling a small hole through the assembly and driving a nail through the hole. (A cotter pin or linchpin could be used instead of the nail.) Three or more chisels ranging from smallest to the maximum desired diameter are required. At each joint the membrane is removed by first boring a hole with the smallest diameter chisel, then progressively enlarging the hole with the larger diameter chisels.

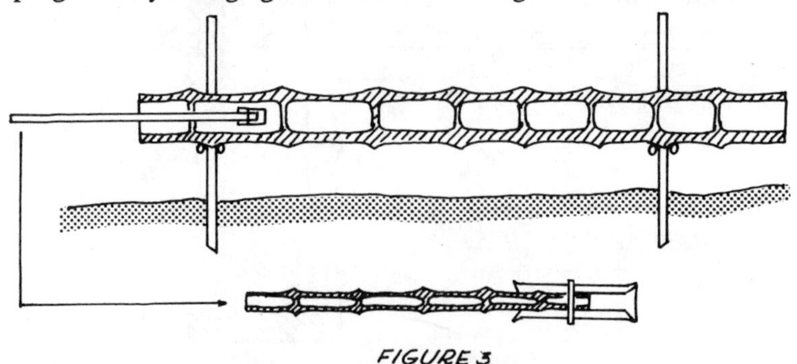

FIGURE 3

Bamboo pipe lengths are joined in a number of ways, as shown in Figure 4. Joints are made watertight by caulking with cotton wool mixed with tar, then tightly binding with rope soaked in hot tar.

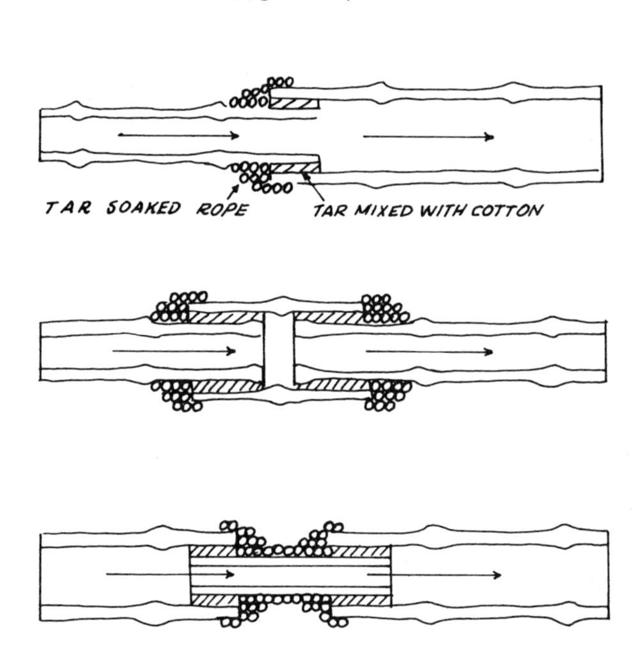


Bamboo pipe is preserved by laying the pipe below ground level and ensuring a continuous flow in the pipe. Where the pipe is laid above ground level, it is protected by wrapping it with layers of palm fiber with soil between the layers. This treatment will give a life expectancy of about 3 to 4 years to the pipe; some bamboo will last up to 5-6 years. Deterioration and failure usually occur at the natural joints, which are the weakest parts.

Where the depth of the pipe below the water source is such that the maximum pressure will be exceeded, pressure relief chambers must be installed. A typical chamber is shown in Figure 5. These chambers are also installed as reservoirs for branch supply lines to villages en route.

Size requirements for bamboo pipe may be determined by using the pipe capacity alignment chart in Figure 6.

**Source:**

*Water Supply Using Bamboo Pipe.* AID-UNC/IPSED Series Item No. 3, International Program in Sanitary Engineering Design, University of North Carolina, 1966.

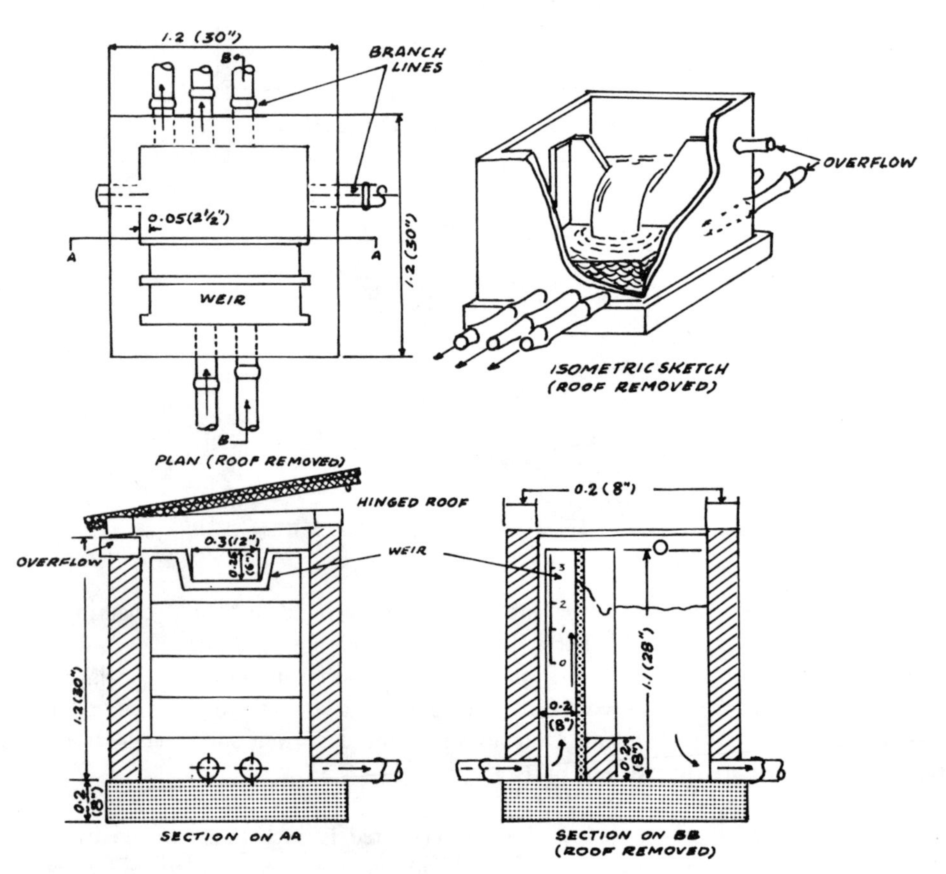
1.2 (30")
BRANCH LINES
0.05(2½")
A
A
1.2 (30")
WEIR
B
B
PLAN (ROOF REMOVED)
OVERFLOW
ISOMETRIC SKETCH
(ROOF REMOVED)
HINGED ROOF
OVERFLOW
0.3(12")
WEIR
1.2(30")
0.2 (8")
SECTION ON AA
0.2 (8")
1.1 (28")
0.2 (8")
SECTION ON BB
(ROOF REMOVED)

FIGURE 5

FIGURE 6

NOMOGRAPH FOR FLOW IN BAMBOO PIPE

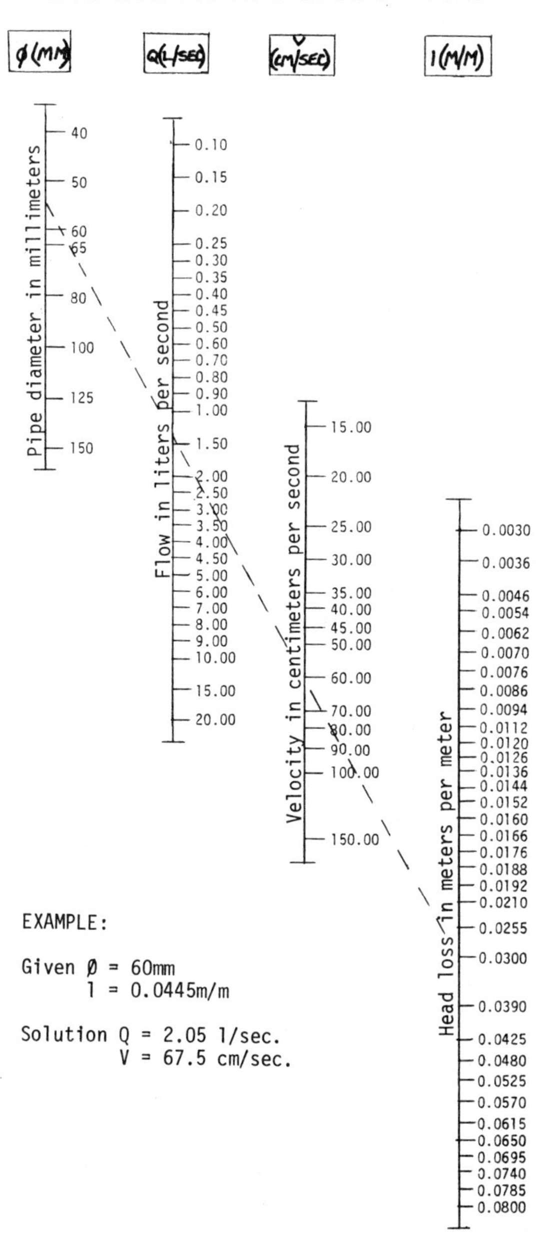


EXAMPLE:

Given Ø = 60mm
I = 0.0445m/m

Solution Q = 2.05 l/sec.
V = 67.5 cm/sec.

# WATER LIFTING

## Pump Specifications: Choosing or Evaluating a Pump

The form given in Figure 1, the "Pump Application Fact Sheet," is a check list for collecting the information needed to get help in choosing a pump for a particular situation. If you have a pump on hand, you can also use the form to estimate its capabilities. The form is an adaptation of a standard pump specification sheet used by engineers.

Fill out the form and send it off to a manufacturer or a technical assistance organization like VITA to get help in choosing a pump. If you are doubtful about how much information to give, it is better to give too much information than to risk not giving enough. When seeking advice on how to solve a pumping problem or when asking pump manufacturers to specify the best pump for your service, give complete information on what its use will be and how it will be installed. If the experts are not given all the details, the pump chosen may give you trouble.

The "Pump Application Fact Sheet" is shown filled in for a typical situation. For your own use, make a copy of the form. The following comments on each numbered item on the fact sheet will help you to complete the form adequately.

1. Give the exact composition of the liquid to be pumped: Fresh or salt water, oil, gasoline, acid, alkali, etc.

2. Weight percent of solids can be found by getting a representative sample in a pail. Let the solids settle to the bottom and decant the liquid (or filter the liquid through a cloth so that the liquid coming through is clear). Weigh the solids and the liquid, and give the weight percent of solids.

   If this is not possible, measure the volume of the sample (in liters, U.S. gallons, etc.) and the volume of solids (in cubic centimeters, teaspoons, etc.) and send these figures. Describe the solid material completely and send a small sample if possible. This is important; if the correct pump is not selected, the solids will erode and/or break moving parts.

   Weight percent of solids =

$$\frac{100 \text{ x weight of solids in liquid sample}}{\text{weight of liquid sample}}$$

3. If you do not have a thermometer to measure temperature, guess at it, making sure you guess on the high side. Pumping troubles are often caused when liquid temperatures at the intake are too high.

4. Gas bubbles or boiling cause special problems, and must always be mentioned.

FIGURE 1

PUMP APPLICATION FACT SHEET

NAME John Doe
ADDRESS P.O. Box 393
Xanadu
Gondwanaland

DATE July 24, 1981

1. Liquid to be handled: Fresh Water

2. Erosive effect of liquid:
   (a) Weight percent of solids: 1-2 percent
   (b) Type of solids: Sand
   (c) Size of solids: largest particle -1mm

3. Maximum temperature of liquid entering pump: 35°C

4. Special situations (explain):
   (a) Gases in liquid: no
   (b) Liquid boiling: no

5. Capacity required: ______ liters per minute
   or: 1200 kilograms per hour - made up of
   or: 600 Kg per hour from lower outlet and 600 Kg per hour from upper outlet

6. Power source available:
   (a) Electrical: 110 volts
   AC: single phase or: DC: ______ volts
   50 cycles per second
   (b) Fuel: ______
   (c) Other: ______

7. Differential head and suction head: See sketch

8. Pipe material: Suction: Galvanized Iron (see sketch for pipe size)
   Discharge: Galvanized Iron (see sketch for pipe size)

9. Pump connections required: Standard pipe thread
   Pipe size (inside diameter): Inlet: 5.25 cm* Outlet: 5.25 cm*

10. Sketch of piping (all fittings and valves shown) attached

11. Other comments:

Figure 1. Pump Specification Fact Sheet. Make a copy of this form for your own use.

NOTE: For advice on pump selection or application, send the completed form (keeping a copy for your own information)to a local university, a pump manufacturer or to VITA.

* Actually this piping is the same as 2" U.S. Schedule 40.

5. Give the capacity (the rate at which you want to move the liquid) in any convenient units (liters per minute, U.S. gallons per minute) by giving the total of the maximum capacity needed for each outlet.

6. Give complete details on the power source.

   A. If you are buying an electric motor for the pump, be sure to give your voltage. If the power is A.C. (Alternating Current) give the frequency (in cycles per second) and the number of phases. Usually this will be single phase for most small motors. Do you want a pressure switch or other special means to start the motor automatically?

   B. If you want to buy an engine driven pump, describe the type and cost of fuel, the altitude, maximum air temperature, and say whether the air is unusually wet or dusty.

   C. If you already have an electric motor or engine, give as much information about it as you can. Give the speed and sketch the machine, being especially careful to show the power shaft diameter and where it is with respect to the mounting. Describe the size and type of pulley if you intend to use a belt drive. Finally, you must estimate the power. The best thing is to copy the name plate data completely. If possible give the number of cylinders in your engine, their size, and the stroke.

7. The "head" or pressure to be overcome by the pump and the capacity (or required flow of water) determine the pump size and power. The entry "Determining Pump Capacity and Horsepower Requirements," explains the calculation of simple head situations. The best approach is to explain the heads by drawing an accurate piping sketch (see Item 10 in the "Pump Application Fact Sheet"). Be sure to give the suction lift and piping separately from the discharge lift and piping. An accurate description of the piping is essential for calculating the friction head. See Figure 2.

8. The piping material, inside diameter, and thickness are necessary for making the head calculations and to check whether pipes are strong enough to withstand the pressure. See "Water Lifting and Transport–Overview" for comments on specifying pipe diameter.

9. Connections to commercial pumps are normally flanged or threaded with standard pipe thread.

10. In the sketch be sure to show the following:

    (a) Pipe sizes; show where sizes are changed by indicating reducing fittings.

(b) All pipe fittings–elbows, tees, valves (show valve type), etc.

(c) Length of each pipe run in a given direction. Length of each size pipe and vertical lift are the most important dimensions.

11. Give information on how the pipe will be used. Comment on such points as:

- o Indoor or outdoor installation?
- o Continuous or intermittent service?
- o Space or weight limitations?

**Source:**

Benjamin P. Coe, VITA Volunteer, Schenectady, New York.

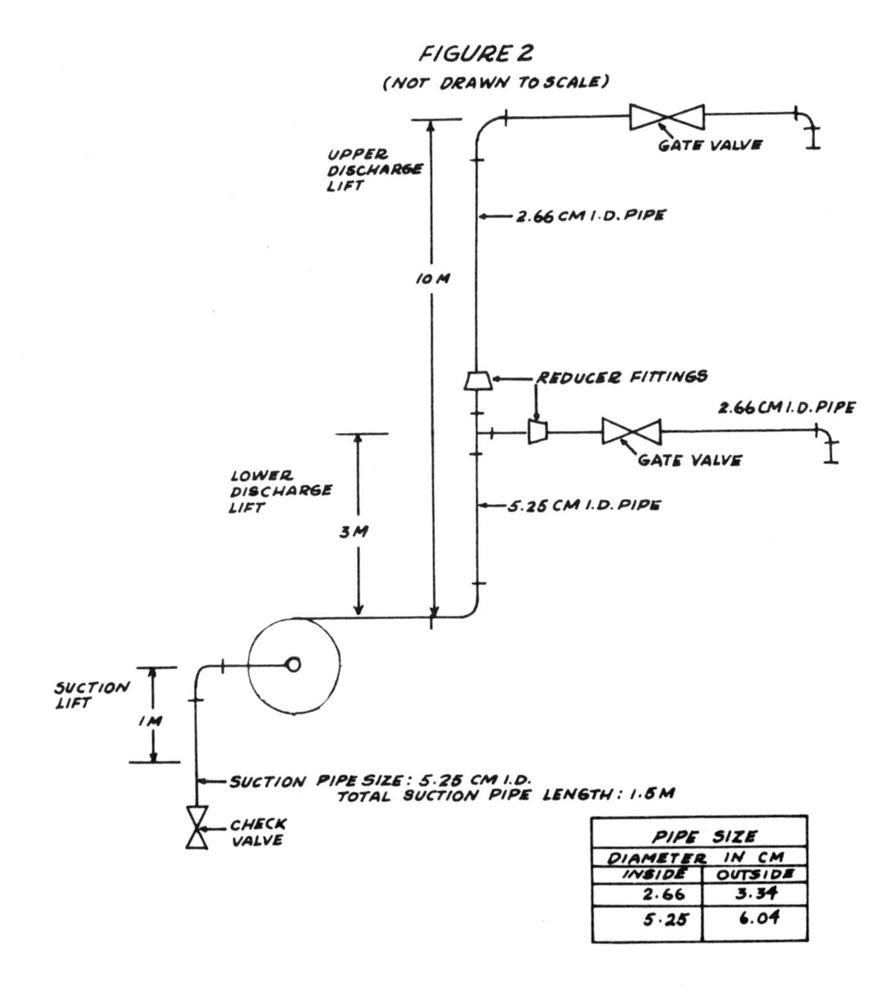


| PIPE SIZE | |
|---|---|
| DIAMETER IN CM | |
| INSIDE | OUTSIDE |
| 2.66 | 3.34 |
| 5.25 | 6.04 |

# Determining Pump Capacity and Horsepower Requirements

With the alignment chart in Figure 1, you can determine the necessary pump size (diameter or discharge outlet) and the amount of horsepower needed to power the pump. The power can be supplied by people or by motors.

An average healthy person can generate about 0.1 horsepower (HP) for a reasonably long period and 0.4HP for short bursts. Motors are designed for varying amounts of horsepower.

To get the approximate pump size needed for lifting liquid to a known height through simple piping, follow these steps:

1. Determine the quantity of flow desired in liters per minute.

2. Measure the height of the lift required (from the point where the water enters the pump suction piping to where it discharges).

3. Using the entry "Determining Pipe Size or Velocity of Water in Pipes," page 74, choose a pipe size that will give a water velocity of about 1.8 meters per second (6' per second). This velocity is chosen because it will generally give the most economical combination of pump and piping; Step 5 explains how to convert for higher or lower water velocities.

4. Estimate the pipe friction-loss head (a 3-meter head represents the pressure at the bottom of a 2-meter-high column of water) for the **total equivalent pipe length**, including suction and discharge piping and equivalent pipe lengths for valves and fittings, using the following equation:

   $$\text{Friction-loss head} = \frac{\text{F x total equivalent pipe length}}{100}$$

   where F equals approximate friction head (in meters) per 100 meters of pipe. To get the value of F, see the table below. For an explanation of **total equivalent pipe length**, see preceding sections.

5. To find F (approximate friction head in meters per 100m of pipe) when water velocity is higher or lower than 1.8 meters per second, use the following equation:

   $$F = \frac{F_{\text{at } 1.8\text{m/sec}} \times V^2}{1.8\text{m/sec}^2}$$

   where V = higher or lower velocity

FIGURE I

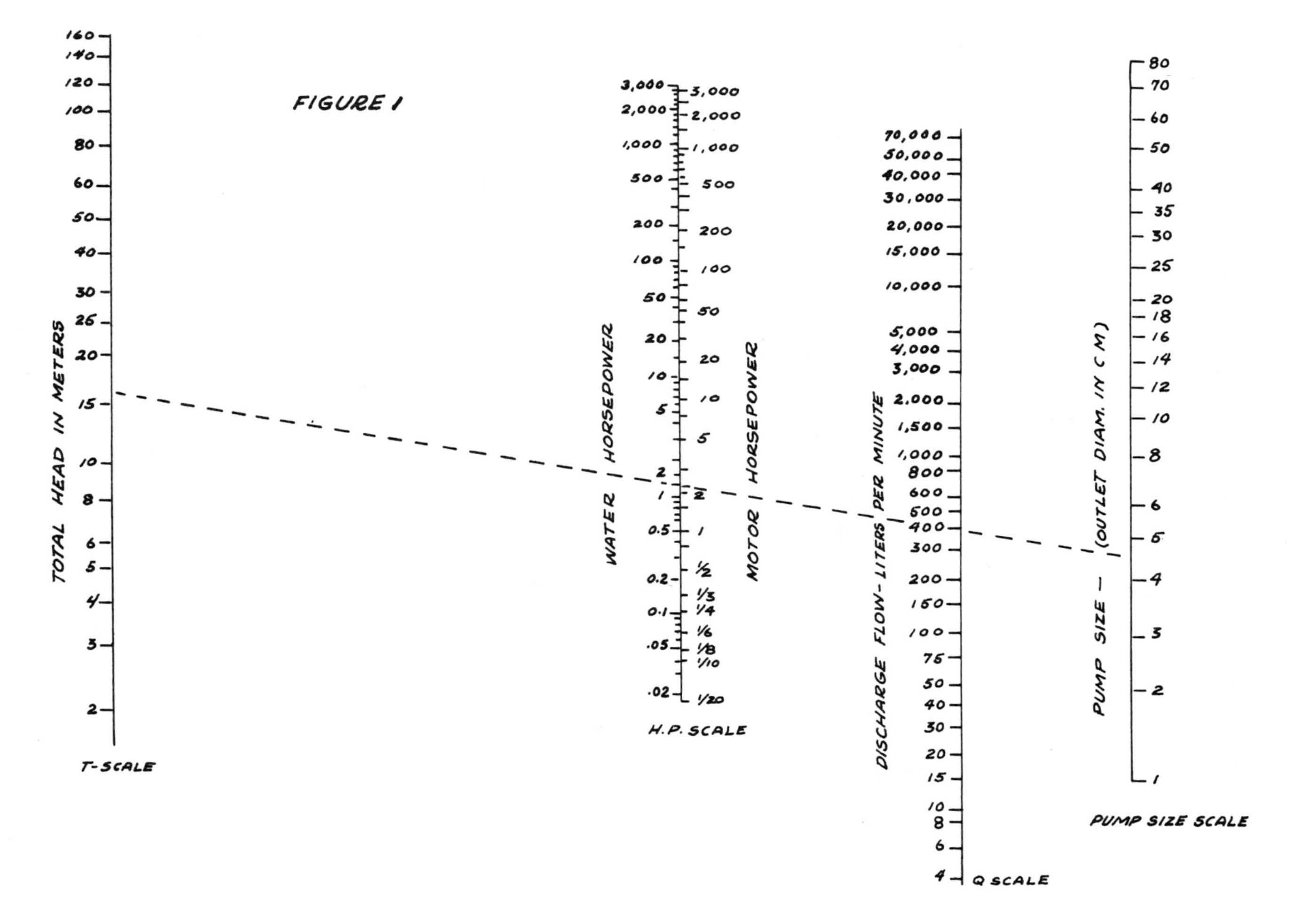
TOTAL HEAD IN METERS
T-SCALE
WATER HORSEPOWER
MOTOR HORSEPOWER
H.P. SCALE
DISCHARGE FLOW-LITERS PER MINUTE
Q SCALE
PUMP SIZE — (OUTLET DIAM. IN C M)
PUMP SIZE SCALE

Example:

If the water velocity is 3.6m per second and F at 1.8m/sec is 16, then:

$$F = \frac{16 \times 3.6^2}{1.8^2} = \frac{16 \times 13}{3.24} = 64$$

6. Obtain "Total Head" as follows:

Total Head = Height of Lift + Friction-loss Head

**Average friction loss in meters for fresh water flowing through steel pipe velocity is 1.8 meters (6 feet) per second**

| Pipe inside diameter: cm | 2.5 | 5.1 | 7.6 | 10.2 | 15.2 | 20.4 | 30.6 | 61.2 |
|---|---|---|---|---|---|---|---|---|
| inches* | 1" | 2" | 3" | 4" | 6" | 8" | 12" | 24" |
| F (approximate friction loss in meters per 100 meters of pipe) | 16 | 7 | 5 | 3 | 2 | 1.5 | 1 | 0.5 |

*For the degree of accuracy of this method, either actual inside diameter in inches, or nominal pipe size, U.S. Schedule 40, can be used.

7. Using a straightedge, connect the proper point on the T-scale with the proper point on the Q-scale; read motor horsepower and pump size on the other two scales.

Example:

Desired flow: 400 liters per minute
Height of lift: 16 meters, No fittings
Pipe size: 5cm
Friction-loss head: about 1 meter
Total head: 17 meters

Solution:

Pump size: 5cm
Motor horsepower: 3HP

Note that water horsepower is less than motor horsepower (see HP-scale, Figure 1). This is because of friction losses in the pump and motor. The alignment chart should be used for rough estimate only. For an exact determination, give all information on flow and piping to a pump manufacturer or an independent expert.

He has the exact data on pumps for various applications. Pump specifications can be tricky especially if suction piping is long and the suction lift is great.

For conversion to metric horsepower given the limits of accuracy of this method, metric horsepower can be considered roughly equal to the horsepower indicated by the alignment chart (Figure 1). Actual metric horsepower can be obtained by multiplying horsepower by 1.014.

**Source:**

Kulman, C.A. *Nomographic Charts.* New York: McGraw-Hill Book Co., 1951.

## Determining Lift Pump Capability

The height that a lift pump can raise water depends on altitude and, to a lesser extent, on water temperature. The graph in Figure 1 will help you to find out what a lift pump can do at various altitudes and water temperatures. To use it, you will need a measuring tape and a thermometer.

If you know your altitude and the temperature of your water, Figure 1 will tell you the maximum allowable distance between the pump cylinder and the lowest water level expected. If the graph shows that lift pumps are marginal or will not work, then a force pump should be used. This involves putting the cylinder down in the well, close enough to the lowest expected water level to be certain of proper functioning.

The graph shows normal lifts. Maximum possible lifts under favorable conditions would be about 1.2 meters higher, but this would require slower pumping and would probably give much difficulty in "losing the prime."

Check predictions from the graph by measuring lifts in nearby wells or by experimentation.

Example:

> Suppose your elevation is 2,000 meters and the water temperature is 25°C. The graph shows that the normal lift would be four meters.

**Source:**

Baumeister, Theodore. *Mechanical Engineer's Handbook,* 6th edition. New York: McGraw-Hill Book Co., 1958.

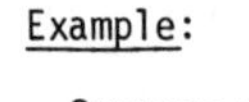
Example:

Suppose your elevation is 2000 meters and the water temperature is 25C. The graph shows that the normal lift would be 4 meters.

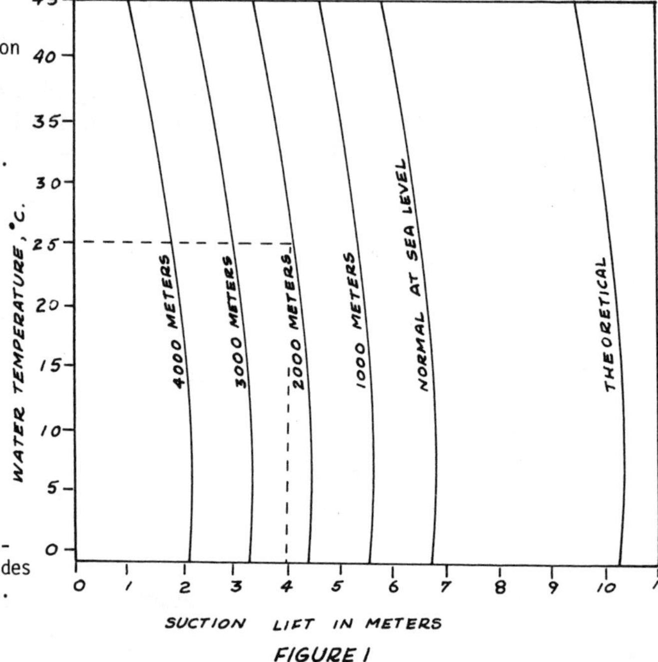


Figure 1. Graph showing lift pump capabilities at various altitudes and water temperatures. Broken lines indicate example given in text.

# SIMPLE PUMPS

## Chain Pump for Irrigation

The chain pump, which can be powered by hand or animal, is primarily a shallow-well pump to lift water for irrigation (see Figure 1). It works best when the lift is less than 6 meters (20'). The water source must have a depth of about 5 chain links.

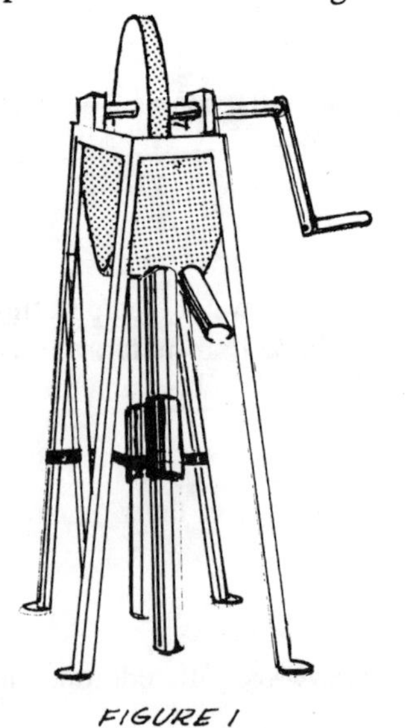

FIGURE 1

Both the pump capacity and the power requirement for any lift are proportional to the square of the diameter of the tube. Figure 2 shows what can be expected from a 10cm (4") diameter tube operated by four people working in two shifts.

The pump is intended for use as an irrigation pump because it is difficult to seal for use as a sanitary pump.

FIGURE 2

| LIFT | QUANTITY |
|---|---|
| 6 METERS (18 FEET) | 11 CUBIC METERS/HOUR (2906 GALLONS/HOUR) |
| 3 METERS (9 FEET) | 20 CUBIC METERS/HOUR (5284 GALLONS/HOUR |
| 1.5 TO 2 METERS (4.5 TO 6 FEET) | 25-30 CUBIC METERS/HOUR (6605 TO 7926 GALLONS/HOUR |

**Tools and Materials**

Welding or brazing equipment
Metal-cutting equipment
Woodworking tools
Pipe: 10cm (4") outside diameter, length as needed
5cm (2") outside diameter, length as needed
Chain with links about 8mm (5/16") in diameter, length as needed
Sheet steel, 3mm (1/8") thick
Sheet steel, 6mm (1/4") thick
Steel rod, 8mm (5/16") in diameter
Steel rod, 12.7mm (1/2") in diameter
Leather or rubber for washers

The entire chain pump is shown in Figure 3. Details of this pump can be changed to fit materials available and structure of the well.

The piston links (see Figures 4, 5, 6 and 7) are made from three parts:

1. a leather or rubber washer (see Figure 4) with an outside diameter about two thicknesses of a washer larger than the inside diameter of the pipe.

2. a piston disk (see Figure 5).

3. a retaining plate (see Figure 6).

The piston link is made as shown in Figure 7. Center all three parts and clamp them together temporarily. Drill a hole about 6mm (1/4") in diameter through all three parts and fasten them together with a bolt or rivet.

The winch is built as shown in Figure 3. Two steel disks 6mm (1/4") thick are welded to the pipe shaft.

Twelve steel rods, 12.7mm (1/2") thick, are spaced at equal distances, at or near the outside diameter, and are welded in place. The rods may be laid on the outside of the disks, if desired.

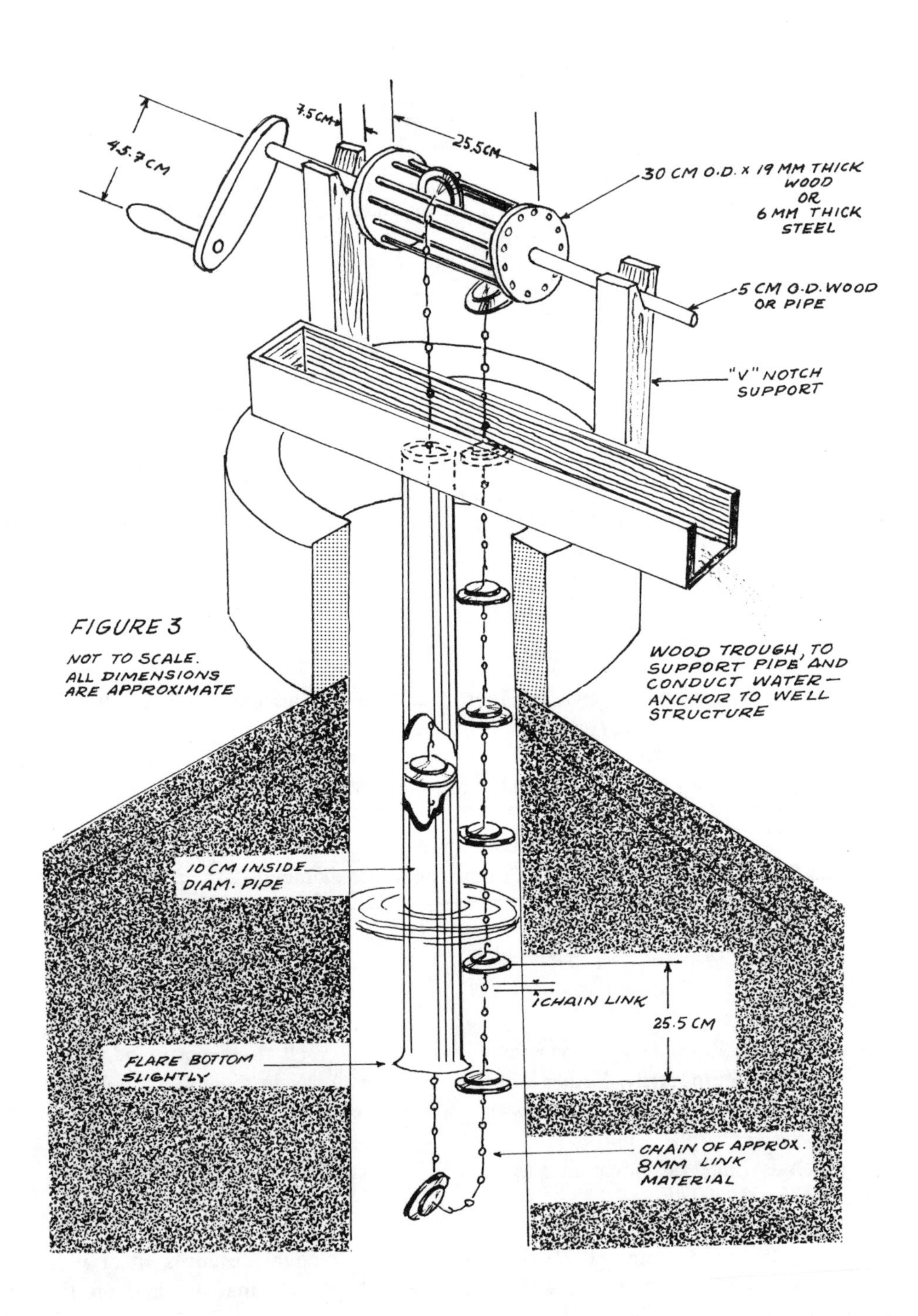


FIGURE 3

NOT TO SCALE.
ALL DIMENSIONS ARE APPROXIMATE

A crank and handle of wood or metal is then welded or bolted to the winch shaft.

The supports for the winch shaft (see Figure 3) can be V-notched to hold the shaft, which will gradually wear its own groove. A strap or block can be added across the top, if necessary, to hold the shaft in place.

The pipe can be supported by threading or welding a flange to its upper end (see Figure 8). The flange should be 8mm to 10mm (5/16" to 3/8") thick. The pipe passes through a hole in the bottom of the trough and hangs from the trough into the well.

**Sources:**


Robert G. Young, VITA Volunteer, New Holland, Pennsylvania

Molenaar, Aldert. *Water Lifting Devices for Irrigation.* Rome: Food and Agriculture Organization, 1956.


FIGURE 4
LEATHER WASHER

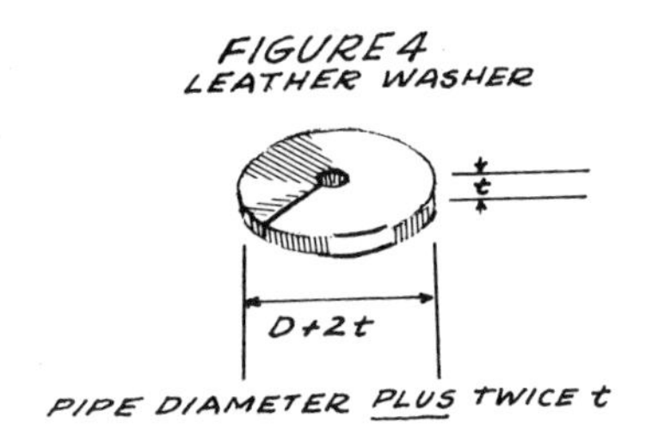


FIGURE 5

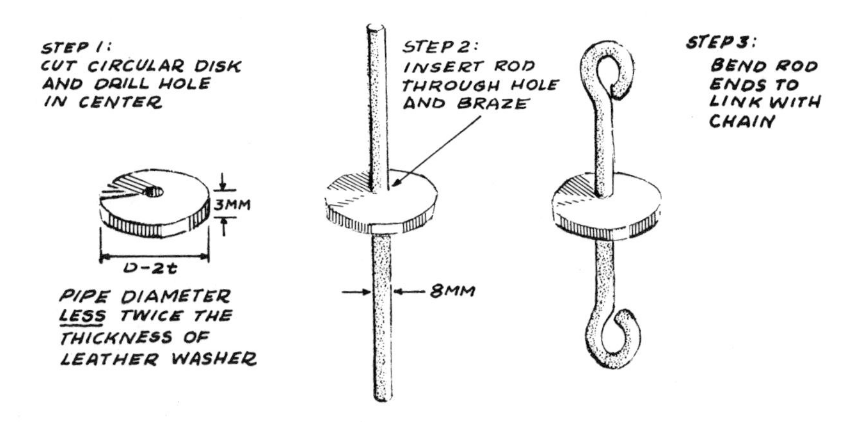

FIGURE 6
RETAINING PLATE

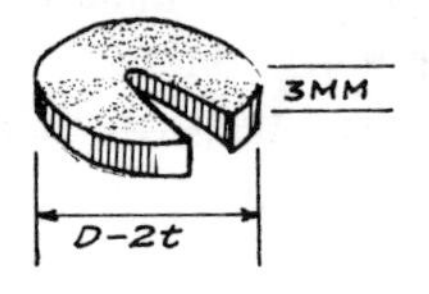


FIGURE 7
PISTON LINK
ASSEMBLED

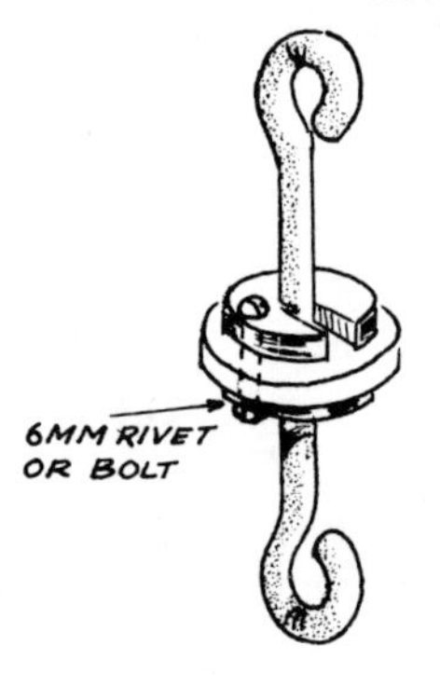


FIGURE 8 PIPE SUPPORT

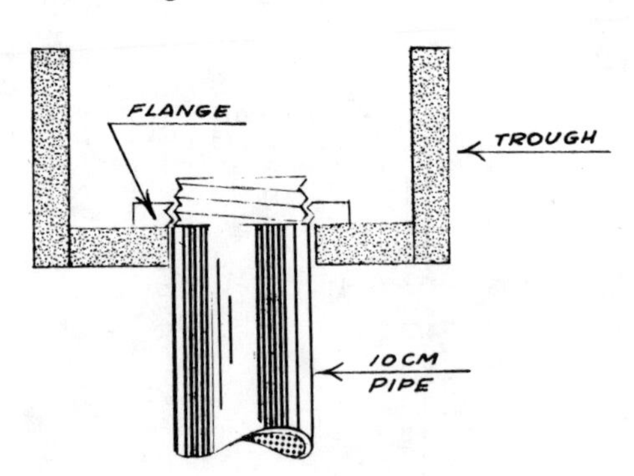

# Inertia Hand Pump

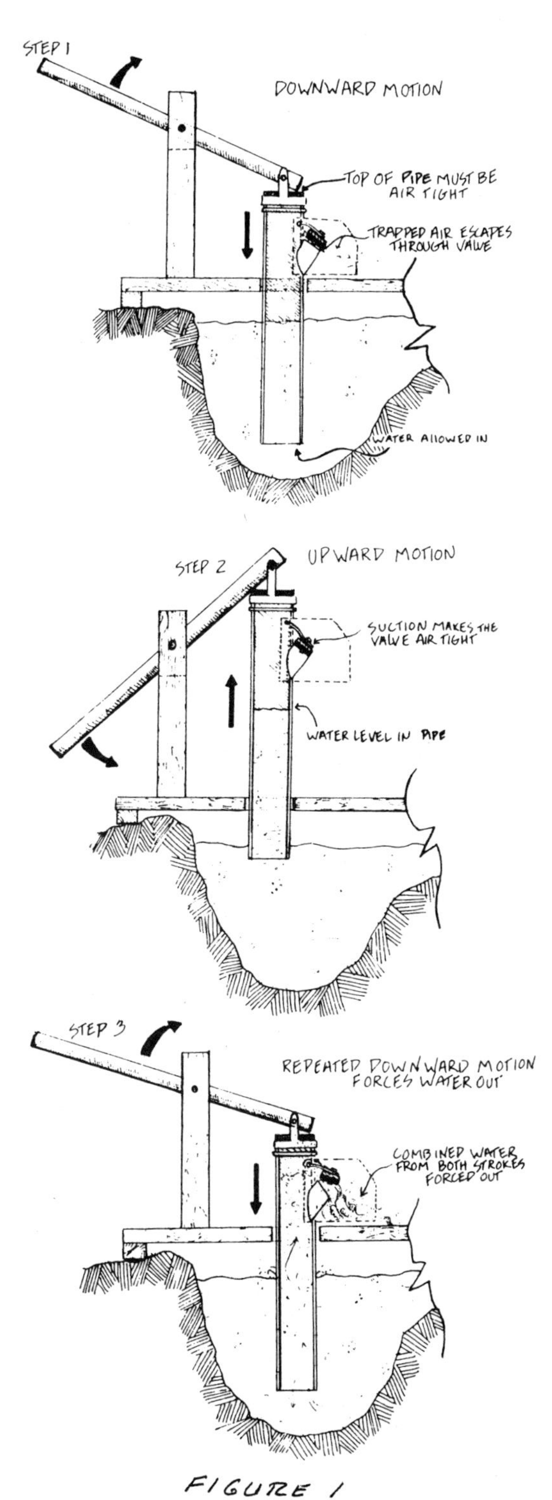

FIGURE 1

The inertia hand pump described here (Figure 1) is a very efficient pump for lifting water short distances. It lifts water 4 meters (13') at the rate of 75 to 114 liters (20 to 30 U.S. gallons) per minute. It lifts water 1 meter (3.3') at the rate of 227 to 284 liters (60 to 75 gallons) per minute. Delivery depends on the number of persons pumping and their strength.

The pump is easily built by a tinsmith. Its three moving parts require almost no maintenance. The pump has been built in three different sizes for different water levels.

The pump is made from galvanized sheet metal of the heaviest weight obtainable that can be easily worked by a tinsmith (24- to 28-gauge sheets have been used successfully). The pipe is formed and made air tight by soldering all joints and seams. The valve is made from the metal of discarded barrels and a piece of truck inner tube rubber. The bracket for attaching the handle is also made from barrel metal.

Figure 1 shows the pump in operation. Figure 2 gives the dimensions of parts for pumps in three sizes and Figure 3 shows the capacity of each size. Figures 4, 5, and 6 are construction drawings.

**Tools and Materials**
**(for 1-meter (3.3') pump)**

Soldering equipment
Drill and bits or punch
Hammer, saws, tinsnips
Anvil (railroad rail or iron pipe)
Galvanized iron (24 to 28 gauge):
Shield: 61cm x 32cm, 1 piece (24" x 12 5/8")
Shield cover: 21cm x 22cm, 1 piece (8 1/4" x 8 5/8")
Pipe: 140cm x 49cm, 1 piece (55 1/8" x 19 1/4")
Top of pipe: 15cm x 15cm, 1 piece (6" x 6")
"Y" pipe: 49cm x 30cm, 1 piece (19 1/4" x 12")
Barrel metal:
  Bracket: 15cm x 45cm, 1 piece (6" x 21 1/4")
  Valve-bottom: 12cm (4 3/4") in diameter, 1 piece
  Valve-top: 18cm (7 1/8") in diameter, 1 piece
Wire:
  Hinge: 4mm (5/32") in diameter, 32cm (12 5/8") long

This pump can also be made from plastic pipe or bamboo.

There are two points to be remembered concerning this pump. One is that the distance from the top of the pipe to the top of the hole where the short section of pipe is connected must be 20cm (8"). See Figure 4. The air that stays in the pipe above this junction serves as a cushion (to prevent "hammering") and regulates the number of strokes pumped per minute. The second point is to remember to operate the pump with short strokes, 15 to 20cm (6" to 8"), and at a rate of about 80 strokes per minute. There is a definite speed at which the pump works best and the operators will soon get the "feel" of their own pumps.

In building the two larger size pumps it is sometimes necessary to strengthen the pipe to keep it from collapsing if it hits the side of the well. It can be strengthened by forming "ribs" about every 30cm (12") below the valve or banding with bands made from barrel metal and attached with 6mm (1/4") bolts.

The handle is attached to the pump and post with a bolt 10mm (3/8") in diameter, or a large nail or rod of similar size.

**Source:**
Dale Fritz, VITA Volunteer, Schenectady, New York.

FIGURE 2

| PART | MATERIAL | 8 CM PIPE | 10 CM PIPE | 15 CM PIPE |
|---|---|---|---|---|
| HANDLE BRACKET | BARREL METAL | | | |
| A | | 34 CM | 40 CM | 54 CM |
| B | | 24 | 30 | 44 |
| C | | 3½ | 5 | 8½ |
| D | | 7 | 10 | 17 |
| SHIELD | GALVANIZED TIN | | | |
| E | | 43 | 49 | 61 |
| F | | 14 | 16 | 20 |
| G | | 14 | 16 | 20 |
| H | | 3 | 3 | 2½ |
| I | | 8 | 10 | 15 |
| J | | 4 | 4 | 4 |
| K | | 30 | 30 | 32 |
| SHIELD COVER | GALVANIZED TIN | | | |
| L | | 15 | 17 | 21 |
| M | | 20 | 20 | 22 |
| N | BARREL METAL | 6 | 8 | 12 |
| O | INNER TUBE RUBBER | 11 | 13 | 18 |
| P | BARREL METAL | 11 | 13 | 18 |
| Q | WIRE (4 MM) | 16 | 18 | 22 |
| HANDLE | WOOD POLE | | | |
| POST | WOOD POST | | | |

FIGURE 3

| DIAMETER OF PIPE | LENGTH OF PIPE | HEIGHT OF LIFT | LITERS PER MINUTE AT 1830 METERS ELEVATION |
|---|---|---|---|
| 8 CM | 450 CM | 2 TO 4 METERS | 75 TO 114 |
| 10 CM | 270 CM | 1 TO 2 METERS | 114 TO 152 |
| 15 CM | 140 CM | 1 METER | 227 TO 284 |

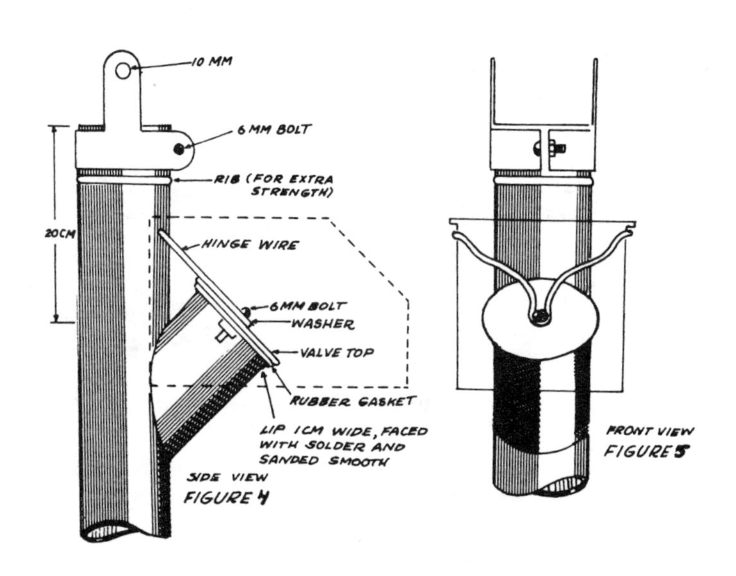


SIDE VIEW
FIGURE 4

FRONT VIEW
FIGURE 5

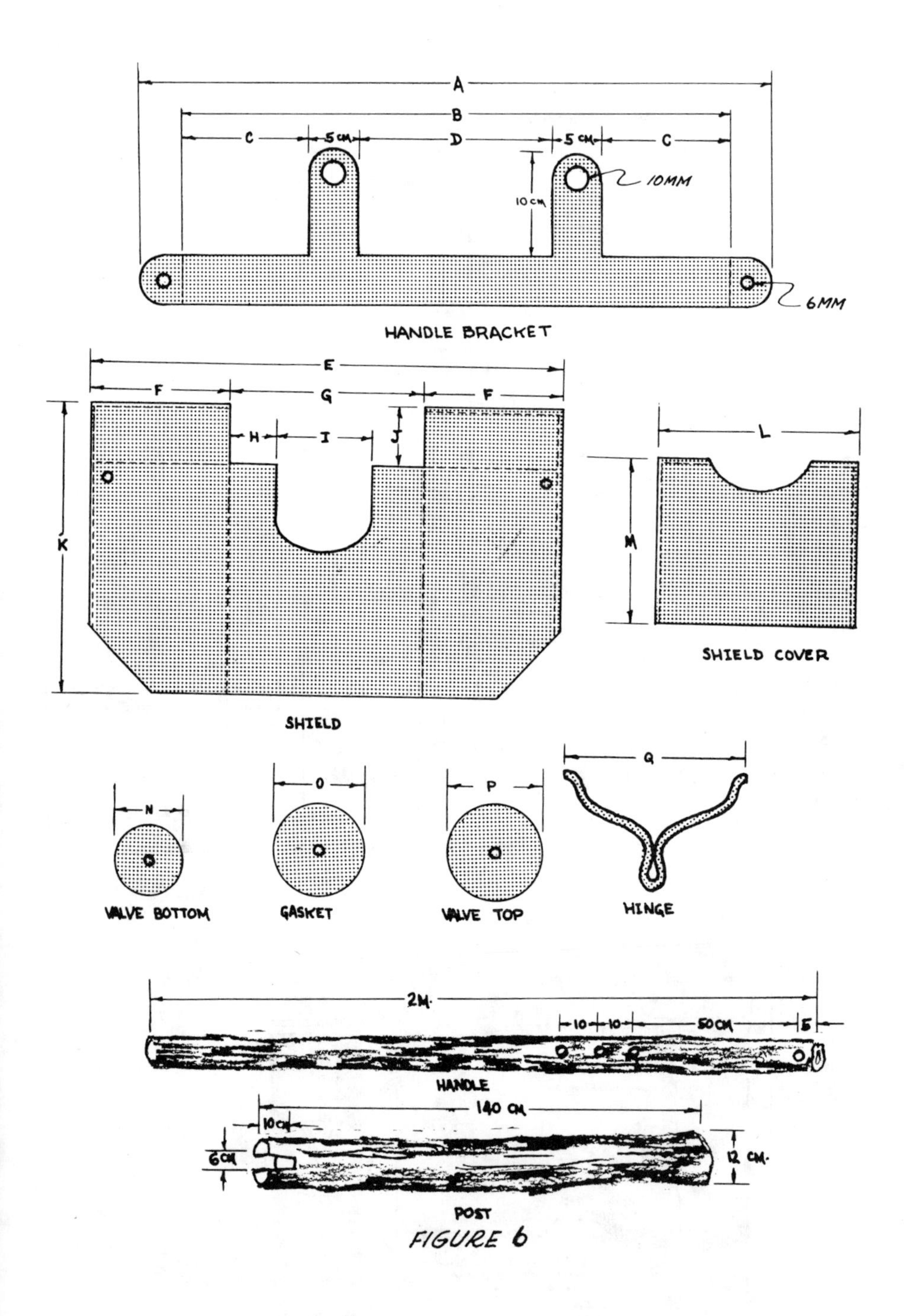
A
B
C
5 CM
D
5 CM
C
10 CM
10MM
6MM
HANDLE BRACKET
E
F
G
F
H
I
J
K
L
M
SHIELD COVER
SHIELD
N
O
P
Q
VALVE BOTTOM
GASKET
VALVE TOP
HINGE
2M.
10
10
50 CM
5
HANDLE
140 CM
10 CM
6 CM
12 CM.
POST

FIGURE 6

# Handle Mechanism for Hand Pumps

The wearing parts of this durable handpump handle mechanism are wooden (see Figure 1). They can be easily replaced by a village carpenter. This handle has been designed to replace pump handle mechanisms which are difficult to maintain. Some have been in use for several years in India with only simple, infrequent repairs.

The mechanism shown in Figure 1 is bolted to the top flange of your pump. The mounting holes A and C in the block should be spaced to fit your pump (see Figure 6). Figure 2 shows a pump with this handle mechanism that is manufactured by F. Humane and Bros., 28 Strand Road, Calcutta, India.

**Tools and Materials**

Saw
Drill
Bits
Tap: 12.5mm (1/2")
Tap: 10mm (3/8")
Chisel
Drawknife, spokeshave or lathe
Hardwood86.4cm x 6.4cm x 6.4cm
(34" x 2 1/2" x 2 1/2")
Mild steel rod: 10mm (3/4") in diameter
and 46.5cm (16") long
Strap iron, 2 pieces: 26.7cm x 38mm x 6mm
(10 1/2" x 1 1/2" x 1/4")

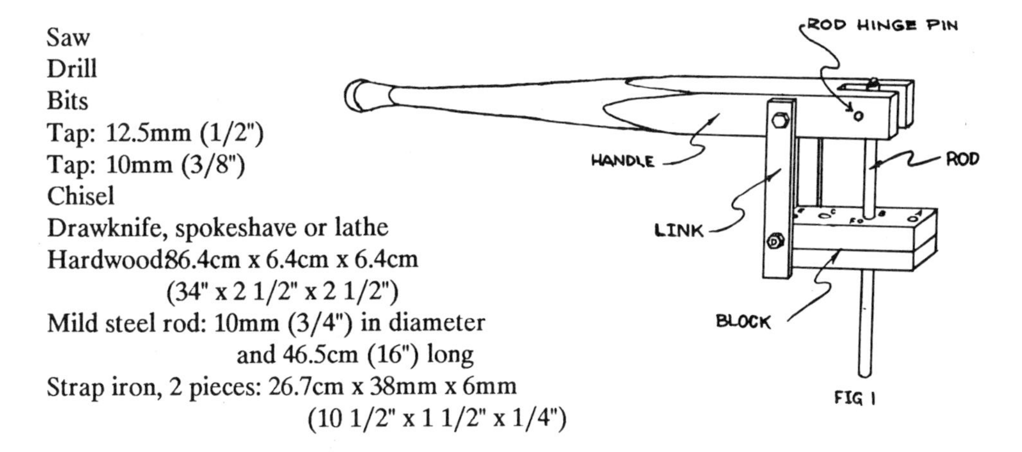


FIG 1

BOLT HARDWARE

| Number of bolts needed | Dia. mm | Length mm | Number of nuts needed | Number of lock-washers | Number of plain washers | Purpose–fastens: |
|---|---|---|---|---|---|---|
| 1 | 10 | 38 | 0 | 0 | 0 | 76mm bolt to rod |
| 1 | 10 | 76 | 0 | 0 | 2 | Rod to handle |
| 2 | 12.5 | 89 | 2 | 4 | 4 | Link to handle |
| | | | | | | Link to block |
| 2 | 12.5 | ? | 2 | 2 | 2 | Block to pump |
| 1 | 12.5 | ? | 1 | 1 | 0 | Rod to piston |

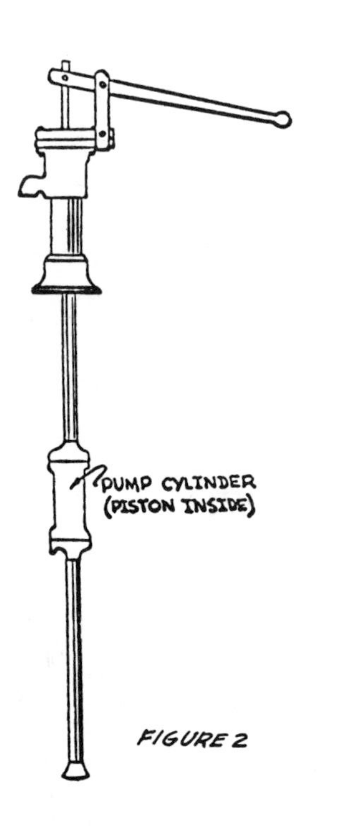

FIGURE 2

## Handle

Make the handle of tough hardwood, shaped on a lathe or by hand shaving. The slot should be cut wide enough to accommodate the rod with two plain washers on either side. See Figure 3.

## Rod

The rod is made of mild steel as shown in Figure 4. A 10mm (3/8") diameter machine bolt 38mm (1 1/2") long screws into the end of the rod to lock the rod hinge pin in place. The rod hinge pin is a 10mm (3/8") diameter machine bolt that connects the rod to the handle (see Figure 1). The end of the rod can be bolted directly to the pump piston with a 12.5mm bolt. If the pump cylinder is too far down for this, a threaded 12.5mm (1/2") rod should be used instead.

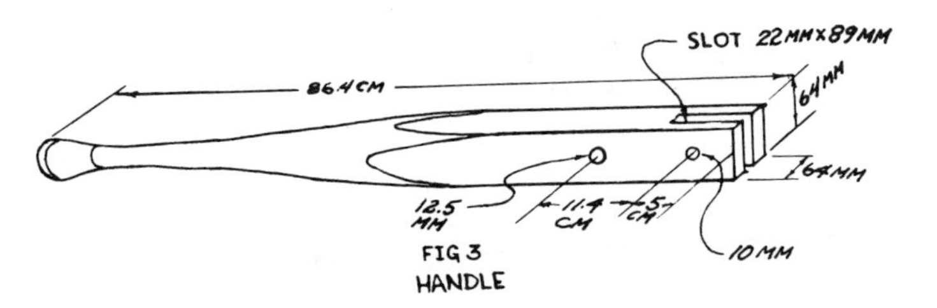

FIG 3
HANDLE

## Links

The links are two pieces of flat steel strap iron. Clamp them together for drilling to make the hole spacing equal. See Figure 5.

## Block

The block forms the base of the lever mechanism, serves as a lubricated guide hole for the rod, and provides a means for fastening the mechanism to the pump barrel. If the block is accurately made of seasoned tough hardwood without knots, the mechanism will function well for many years. Carefully square the block to 22.9cm x 6.4cm x 6.4cm (9" x 1 1/2" x 1 1/2"). Next holes, A, B, C, and D are drilled perpendicular to the block as shown in Figure 6. The spacing of the

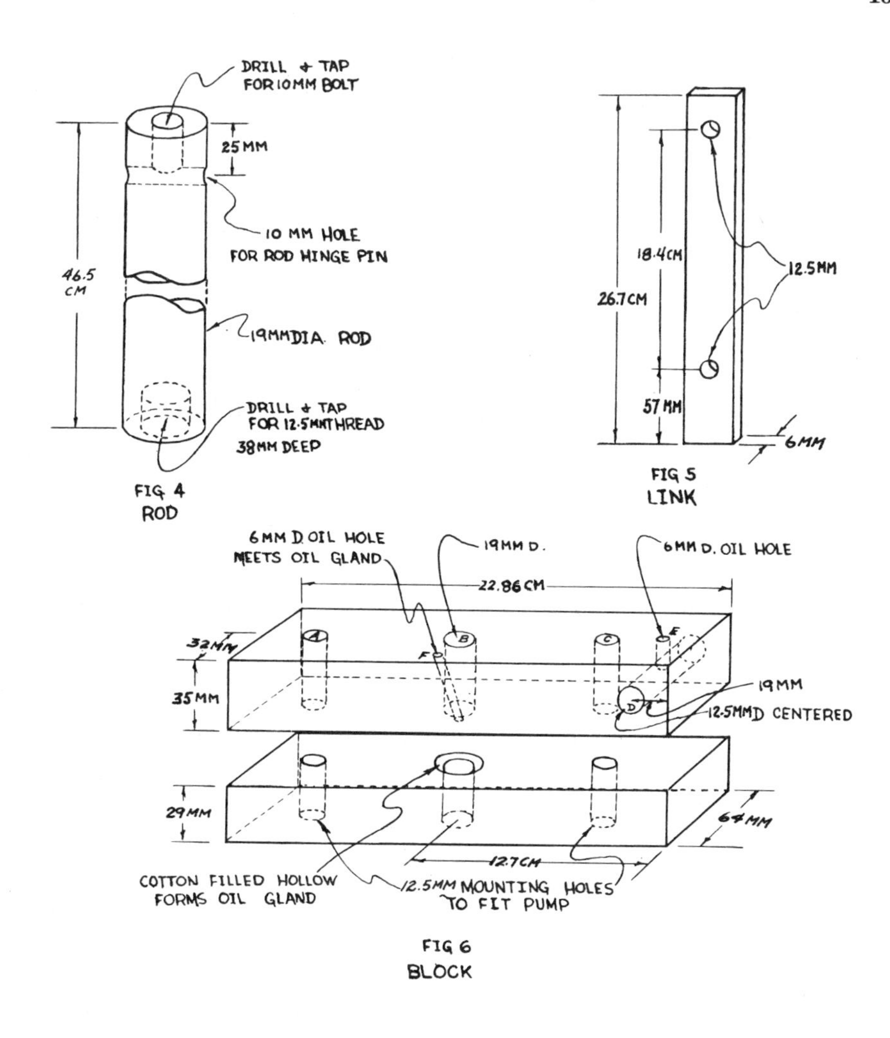


mounting holes A and C from hole B is determined by the spacing of the bolt holes in the barrel flange of your pump. Next saw the block in half in a plane 3.5cm (1 3/8") down from the top side. Enlarge hole B at the top of the lower section with a chisel to form an oil well around the rod. This well is filled with cotton. A 6mm (1/4") hole, F, is drilled at an angle from the oil well to the surface of the block. A second oil duct hole E is drilled in the upper section of the block to meet hole D. Use lockwashers under the head and nut of the link bolts to lock the bolts and links together. Use plain washers between the links and the wooden parts.

**Source:**

Abbott, Dr. Edwin. *A Pump Designed for Village Use.* Philadelphia: American Friends Service Committee, 1955.

# Hydraulic Ram

A hydraulic ram is a self-powered pump that uses the energy of falling water to lift some of the water to a level above the original source. This entry explains the use of commercial hydraulic rams, which are available in some countries. Plans for building your own hydraulic ram are also available from VITA and elsewhere.

## *Use of the Hydraulic Ram*

A hydraulic ram can be used wherever a spring or stream of water flows with at least a 91.5cm (3') fall in altitude. The source must be a flow of at least 11.4 liters (3 gallons) a minute. Water can be lifted about 7.6 meters (25') for each 30.5cm (12") of fall in altitude. It can be lifted as high as 152 meters (500'), but a more common lift is 45 meters (150').

The pumping cycle (see Figure 1) is:

- o Water flows through the drive pipe (D) and out the outside valve (F).
- o The drag of the moving water closes the valve (F).
- o The momentum of water in the drive pipe (D) drives some water into the air chamber (A) and out the delivery pipe (I).
- o The flow stops.
- o The check valve (B) closes
- o The outside valve (F) opens to start the next cycle.

This cycle is repeated 25 to 100 times a minute; the frequency is regulated by moving the adjustment weight (C).

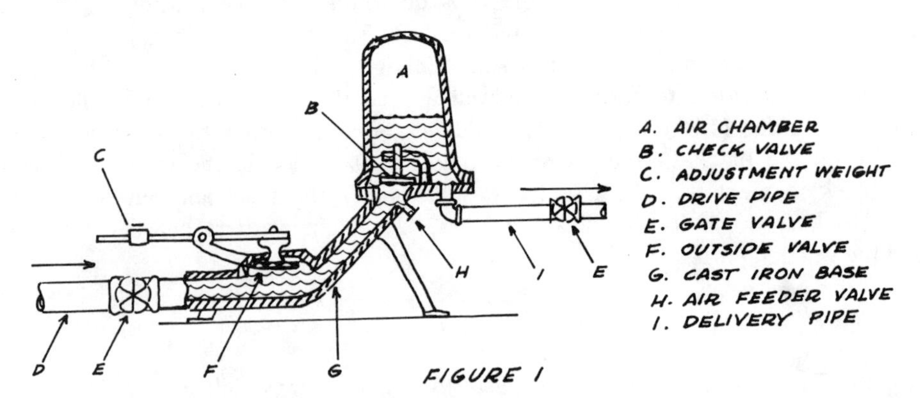


FIGURE 1

The length of the drive pipe must be between five and ten times the length of the fall (see Figure 2). If the distance from the source to the ram is greater than ten times the length of the fall, the length of the drive pipe can be adjusted by installing a stand pipe between the source and the ram (see B in Figure 2).

Once the ram is installed there is little need for maintenance and no need for skilled labor. The cost of a hydraulic ram system must include the cost of the pipe and installation as well as the ram. Although the cost may seem high, it must be remembered that there is no further power cost and a ram will last for 30 years or more. A ram used in freezing climates must be insulated.

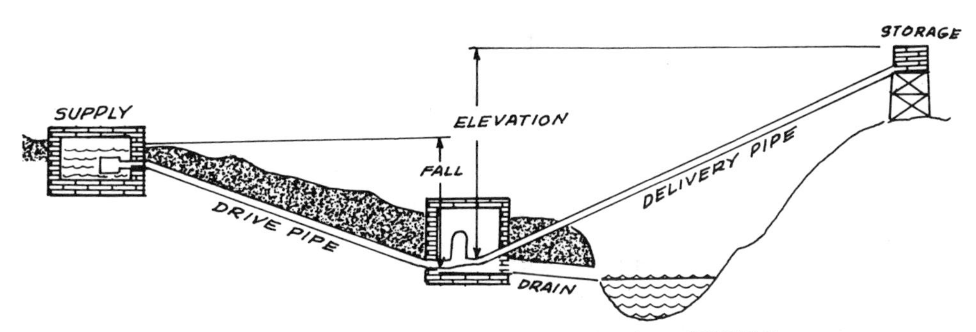


A. COMMON ARRANGEMENT OF DRIVE PIPE, RAM AND STORAGE

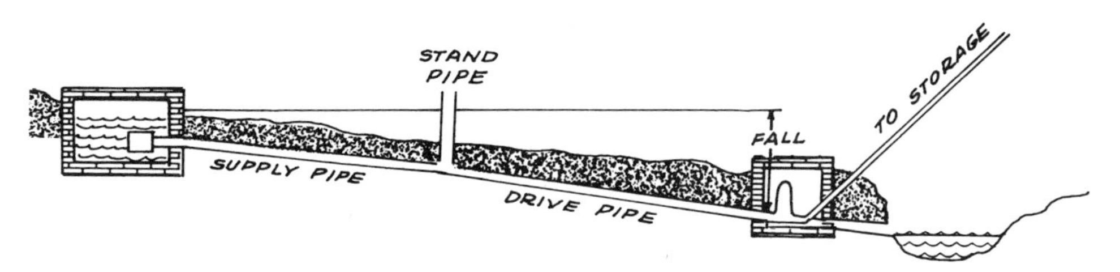


B. ARRANGEMENT OF DRIVE PIPE FOR A DISTANT WATER SUPPLY

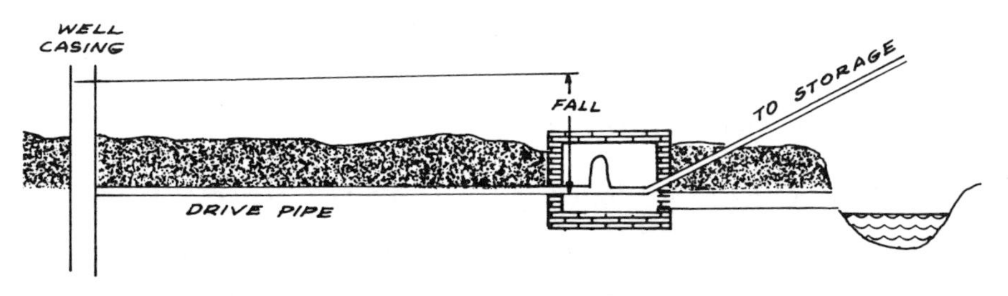


C. ARTESIAN WELL OPERATING A RAM

FIGURE 2

A double-acting ram will use an impure water supply to pump two-thirds of the pure water from a spring or similar source. A third of the pure water mixes with the impure water. A supplier should be consulted for this special application.

To calculate the approximate pumping rate, use the following equation:

$$\text{Capacity (gallons per hour)} = \frac{V \times F \times 40}{E}$$

V = gallons per minute from source
F = fall in feet
E = height the water is to be raised in feet

## *Data Needed for Ordering a Hydraulic Ram*

1. Quantity of water available at the source of supply in liters (or gallons) per minute

2. Vertical fall in meters (or feet) from supply to ram

3. Height to which the water must be raised above the ram

4. Quantity of water required per day

5. Distance from the source of supply to the ram

6. Distance from the ram to the storage tank

**Sources:**

Loren G. Sadler, New Holland, Pennsylvania

Rife Hydraulic Engine Manufacturing Company, Millburn, New Jersey

Sheldon, W.H. *The Hydraulic Ram.* Extension Bulletin 171, July 1943, Michigan State College of Agriculture and Applied Science.

"Country Workshop." *Australian Country.* September 1961, pages 32-33.

"Hydraulic Ram Forces Water to Pump Itself." *Popular Science,* October 1948, pages 231-233.

"Hydraulic Ram." *The Home Craftsman,* March-April 1963, pages 20-22.

# RECIPROCATING WIRE POWER TRANSMISSION FOR WATER PUMP

A reciprocating wire can transmit power from a water wheel to a point up to 0.8km (1/2 mile) away where it is usually used to pump well water. These devices have been used for many years by the Amish people of Pennsylvania. If they are properly installed, they give long, trouble-free service.

The Amish people use this method to transmit mechanical power from small water wheels to the barnyard, where the reciprocating motion is used to pump well water for home and farm use. The water wheel is typically a small undershot wheel (with the water flowing under the wheel) one or two feet in diameter. The wheel shaft is fitted with a crank, which is attached to a triangular frame that pivots on a pole (see Figure 2). A wire is used to connect this frame to another identical unit located over the well. Counterweights keep the wire tight.

### Tools and Materials

Wire: galvanized smooth fence wire
Water wheel with eccentric crank to give a motion slightly less than largest stroke of farmyard pump
Galvanized pipe for triangle frames: 2cm (3/4") by 10 meters long (32.8')
Welding or brazing equipment to make frames
Concrete for counterweight
2 Poles: 12 to 25cm (6" to 10") in diameter.

FIGURE 1.

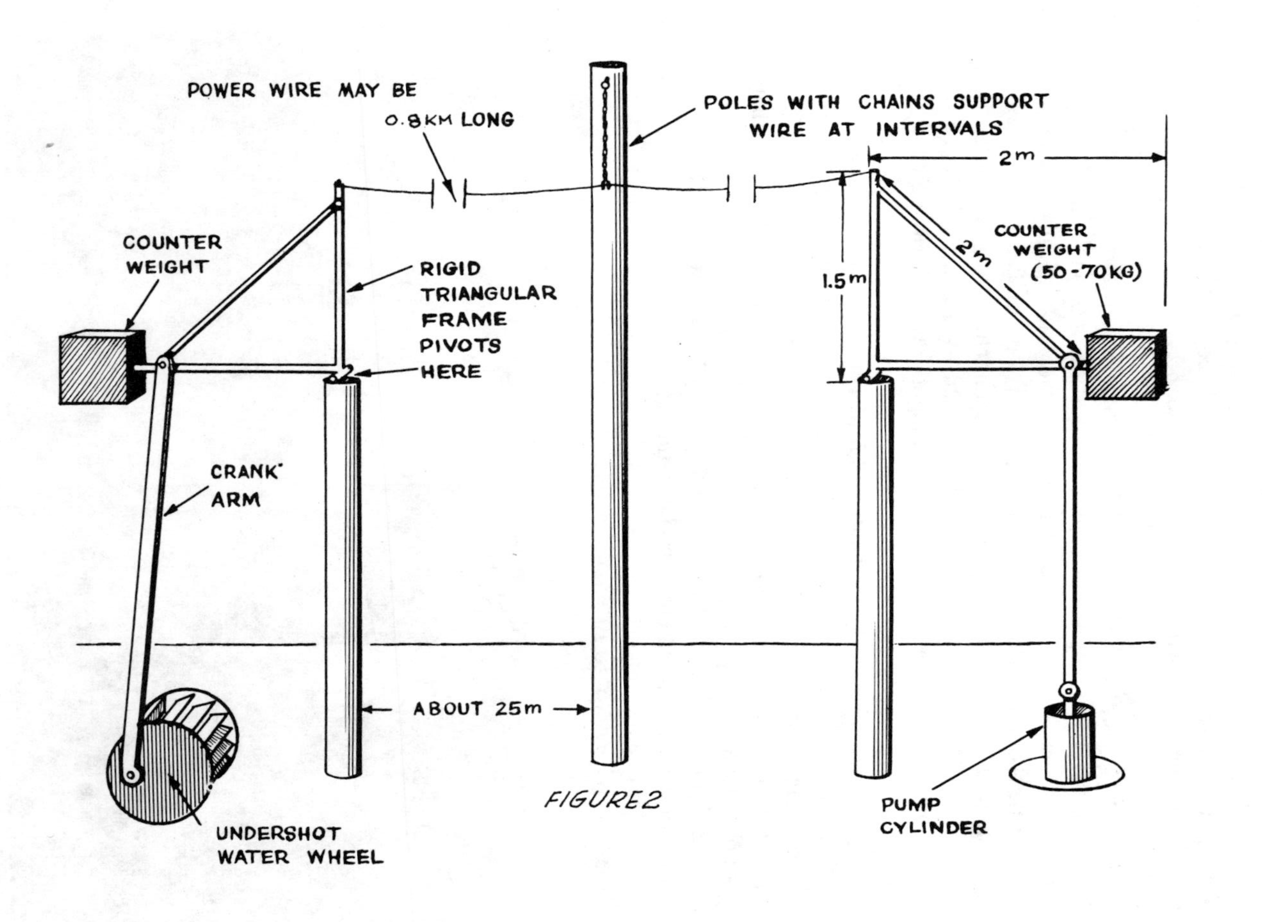
POWER WIRE MAY BE
0.8KM LONG
POLES WITH CHAINS SUPPORT
WIRE AT INTERVALS
2m
COUNTER
WEIGHT
RIGID
TRIANGULAR
FRAME
PIVOTS
HERE
1.5m
2m
COUNTER
WEIGHT
(50-70KG)
CRANK
ARM
ABOUT 25m
UNDERSHOT
WATER WHEEL
PUMP
CYLINDER

FIGURE 2

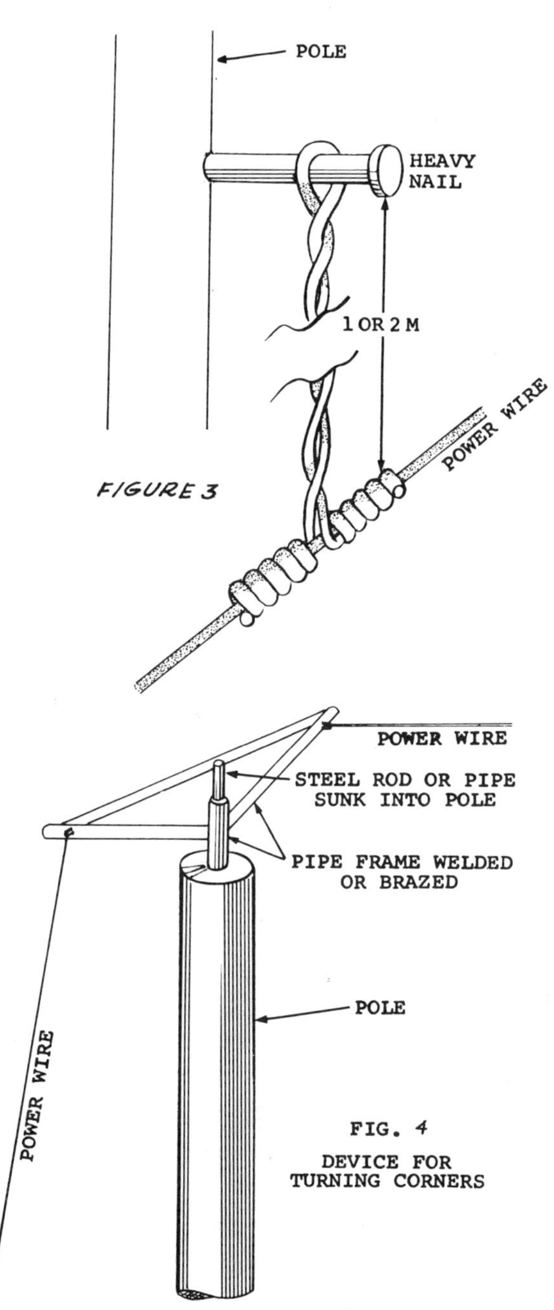


As the water wheel turns, the crank tips the triangular frame back and forth. This action pulls the wire back and forth. One typical complete back and forth cycle takes 3 to 4 seconds. Sometimes power for several transmission wires comes from one larger water wheel.

The wire is mounted up on poles to keep it overhead and out of the way. If the distance from stream to courtyard is far, extra poles will be needed to help support the wire. Amish folks use a loop of wire covered with a small piece of garden hose attached to the top of the pole. The reciprocating wire slides back and forth through this loop. If this is not possible, try making the pole 1-2 meters higher than the power wire. Drive a heavy nail near the pole top and attach a chain or wire from it to the power wire as shown in Figure 3.

Turns can be made in order to follow hedgerows by mounting a small triangular frame horizontally at the top of a pole as shown in Figure 4.

Figures 5, 6, and 7 show how to build and install a small water wheel made from wood and bamboo.

**Source:**

New Holland, Pennsylvania VITA Chapter.

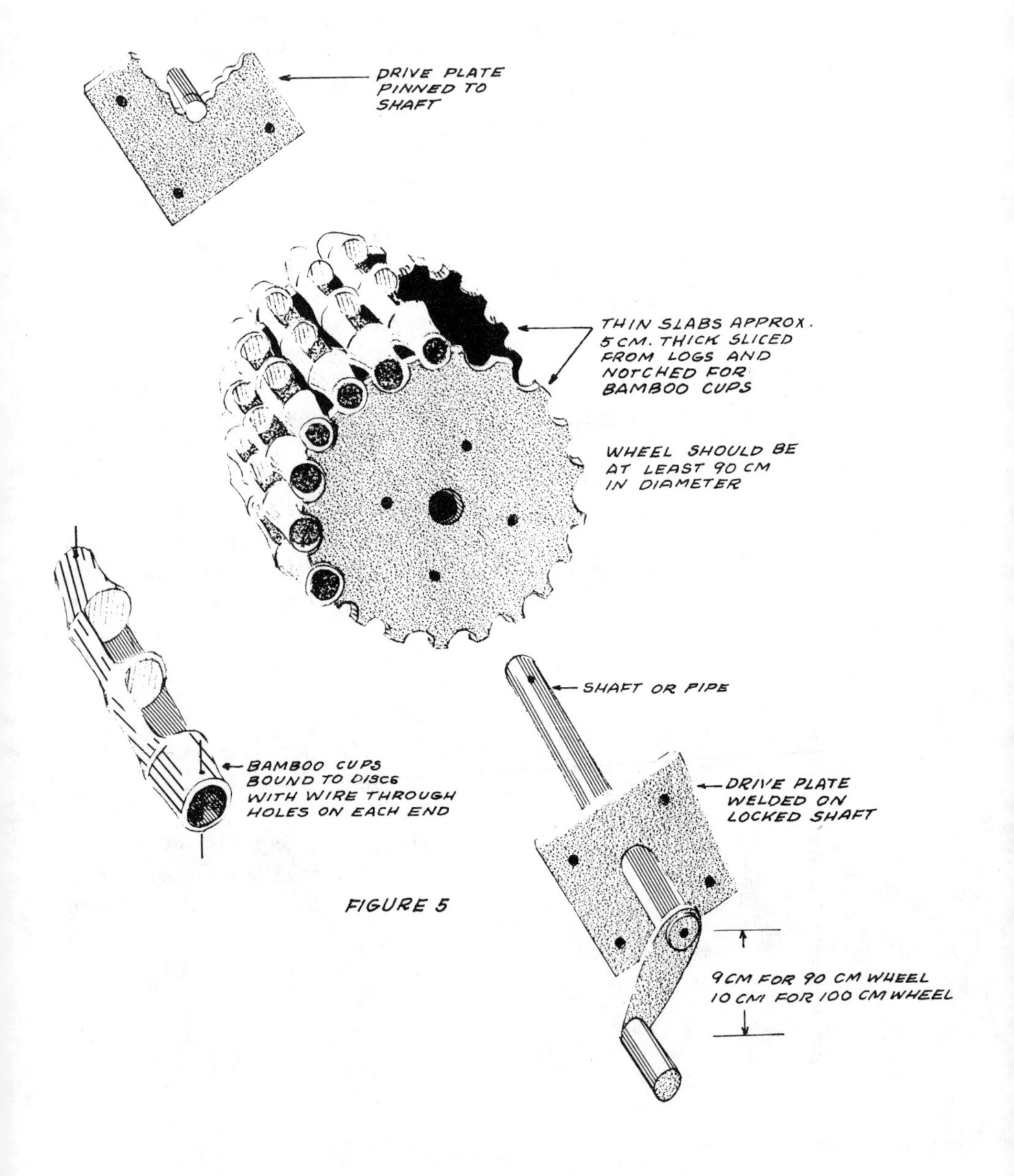
DRIVE PLATE
PINNED TO
SHAFT
THIN SLABS APPROX.
5 CM. THICK SLICED
FROM LOGS AND
NOTCHED FOR
BAMBOO CUPS
WHEEL SHOULD BE
AT LEAST 90 CM
IN DIAMETER
SHAFT OR PIPE
BAMBOO CUPS
BOUND TO DISCS
WITH WIRE THROUGH
HOLES ON EACH END
DRIVE PLATE
WELDED ON
LOCKED SHAFT
9 CM FOR 90 CM WHEEL
10 CM FOR 100 CM WHEEL

FIGURE 5

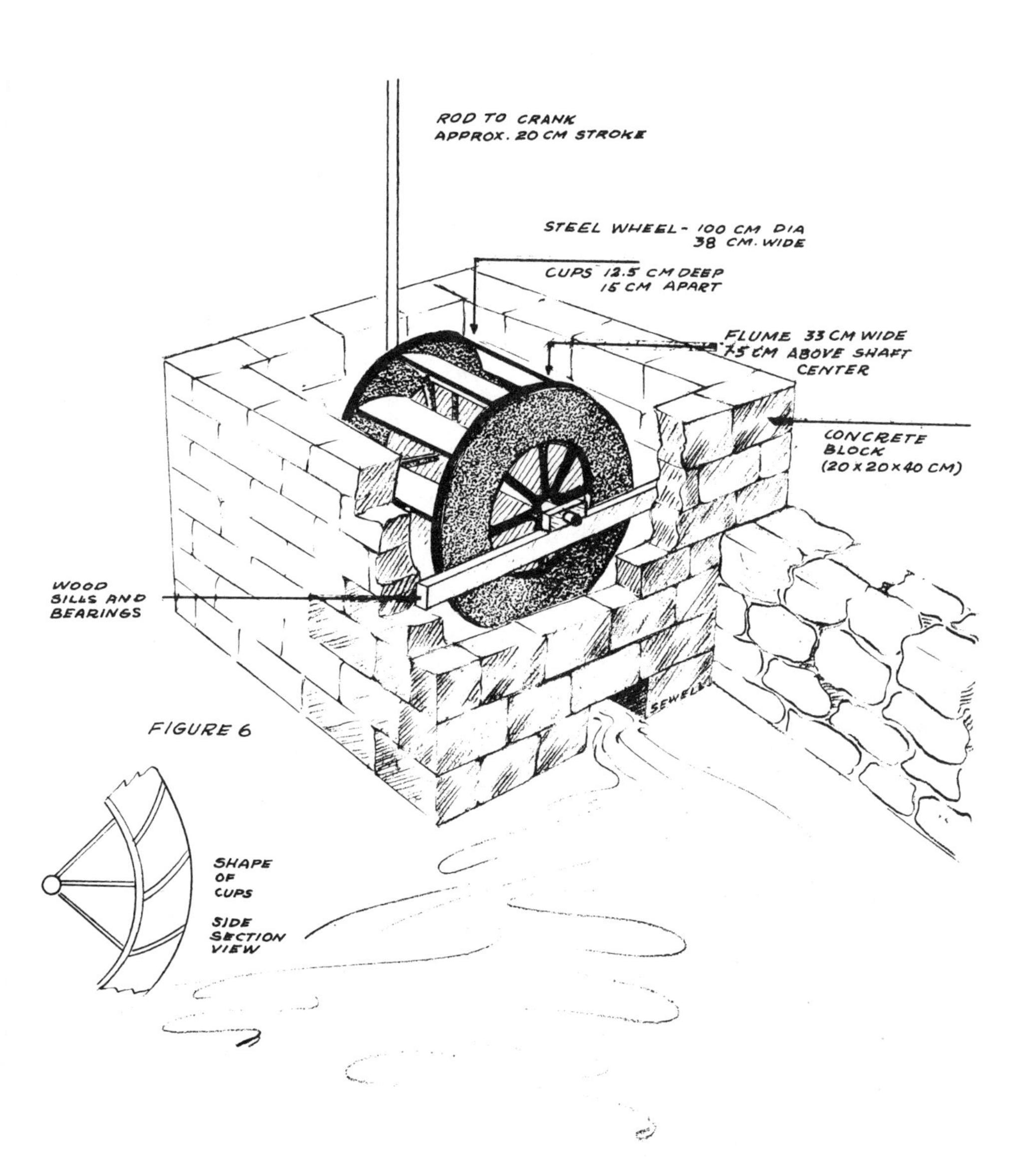
ROD TO CRANK
APPROX. 20 CM STROKE
STEEL WHEEL - 100 CM DIA
38 CM. WIDE
CUPS 12.5 CM DEEP
15 CM APART
FLUME 33 CM WIDE
7.5 CM ABOVE SHAFT
CENTER
CONCRETE
BLOCK
(20 x 20 x 40 CM)
WOOD
SILLS AND
BEARINGS
SEWELL
SHAPE
OF
CUPS
SIDE
SECTION
VIEW

FIGURE 6

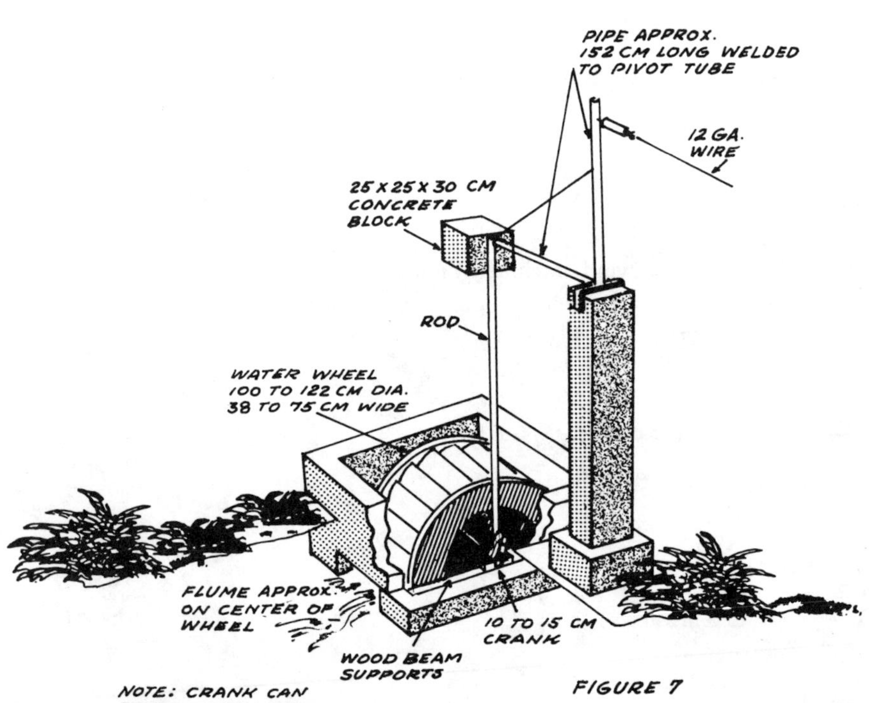
PIPE APPROX.
152 CM LONG WELDED
TO PIVOT TUBE
12 GA.
WIRE
25 X 25 X 30 CM
CONCRETE
BLOCK
ROD
WATER WHEEL
100 TO 122 CM DIA.
38 TO 75 CM WIDE
FLUME APPROX.
ON CENTER OF
WHEEL
10 TO 15 CM
CRANK
WOOD BEAM
SUPPORTS
NOTE: CRANK CAN
BE ADDED TO OTHER
END OF SHAFT AT 90°
ANGLE TO FIRST CRANK
TO ATTACH MORE POWER
TRANSMITTING WIRES

FIGURE 7

# WIND ENERGY FOR WATER PUMPING

## Overview

There are many places in the world where wind energy is a good alternative for pumping water. Specifically these include windy areas with limited access to other forms of power. In order to determine whether wind power is appropriate for a particular situation an assessment of its possibilities and the alternatives should be undertaken. The necessary steps include the following:

1. Identify the users of the water.
2. Assess the water requirement.
3. Find the pumping height and overall power requirements.
4. Evaluate the wind resources.
5. Estimate the size of the wind machine(s) needed.
6. Compare the wind machine output with the water requirement on a seasonal basis.
7. Select a type of wind machine and pump from the available options.
8. Identify possible suppliers of machines, spare parts, repair, etc.
9. Identify alternative sources for water.
10. Assess costs of various systems and perform economic analysis to find least cost alternative.
11. If wind energy is chosen, arrange for obtaining and installing the machines and for providing for their maintenance.

## Decision Making Process

The following summarizes the key aspects of those suggested steps.

### *1. Identify the Users*

This step seems quite obvious, but should not be ignored. By paying attention to who will use the wind machine and its water it will be possible to develop a project that can have continuing success. Questions to consider are whether they are villagers, farmers, or ranchers; what their educational level is; whether they have had experience with similar types of technology in the past; whether they have access to or experience with metal working shops. Who will be paying for the projects? Who will be owning the equipment; who will be responsible for keeping it running; and who will be benefitting most? Another important question is how many pumps are planned. A large project to supply many pumps may well be different than one looking to supply a single site.

## 2. *Assess the Water Requirements*

There are four main types of uses for water pumps in areas where wind energy is likely to be used. These are: 1) domestic use, 2) livestock watering, 3) irrigation, 4) drainage.

Domestic use will depend a great deal on the amenities available. A typical villager may use from 15 - 30 liters per day (4-8 gallons per day). When indoor plumbing is used, water consumption may increase substantially. For example, a flush toilet consumes 25 liters (6 1/2 gallons) with each use and a shower may take 230 (60 gallons.) When estimating water requirements, one must also consider population growth. For example, if the growth rate is 3 percent, water use would increase by nearly 60 percent at the end of 15 years, a reasonable lifetime for a water pump.

Basic livestock requirements range from about 0.2 liters (0.2 quart) a day for chickens or rabbits to 135 liters (36 gallons) a day for a milking cow. A single cattle dip might use 7500 liters (2000 gallons) a day.

Estimation of irrigation requirements is more complex and depends on a variety of meteorological factors as well as the types of crops involved. The amount of irrigation water needed is approximately equal to the difference between that needed by the plants and that provided by rainfall. Various techniques may be used to estimate evaporation rates, due for example to wind and sun. These may then be related to plant requirements at different stages during their growing cycle. By way of example, in one semi-arid region irrigation requirements varied from 35,000 liters (9,275 gallons) per day per hectare (2.47 acres) for fruits and vegetables to 100,000 liters (26,500 gallons) per day per hectare for cotton.

Drainage requirements are very site dependent. Typical daily values might range from 10,000 to 50,000 liters (2,650 to 13,250 gallons) per hectare.

In order to make the estimate for the water demand, each user's consumption is identified, and summed up to find the total. As will become apparent later. It is desirable to do this on a monthly basis so that the demand can be related to the wind resource.

## 3. Find Pumping Height and Total Power Requirement

If wells are already available their depth can be measured directly. If new wells are to be dug, depth must be estimated by reference to other wells and knowledge of ground water characteristics in the area. The total elevation, or head, that the pump must work against, however, is always greater than the static well depth. Other contributors are the well draw down (the lowering of the water table in the vicinity of the well while pumping is underway), the height above ground to which the water will be pumped (such as to a storage tank), and frictional losses in the piping. In a properly designed system the well depth and height above ground of the outlet are the most important determinants of pumping head.

The power required to pump water is proportional to its mass per unit volume, or density (1000 $kg/m^3$), the acceleration of gravity (g= 9.8 $m/s^2$, the total pumping head (m), and the volume flow rate of water ($m^3/s$). Power is also inversely proportional to the pump efficiency. Note that 1 cubic meter equals 1000 liters. Expressed as a formula,

$$\text{Power} = \text{Density x Gravity x Head x Flow rate}$$

Example:

To pump 50 $m^3$ in one day (0.000579 $m^3/s$) up a total head of 15 m would require:

$$\text{Power} = (1000\ kg/m^3)\ (9.8 m/s^2)\ (15m)\ (.000579 m^3/s) = 85 \text{ watts.}$$

Actual power required would be more because of the less than perfect efficiency of the pump.

Sometimes needed pumped power is described in terms of daily hydraulic requirement, which is often given in the units of $m^3\cdot$ m/day. For example, in the above example the hydraulic requirement is 750 $m^3\cdot$m /day.

## 4. Evaluate Wind Resource

It is well known that the power in the wind varies with the cube of the wind speed. Thus if the wind speed doubles, the available power increases by a factor of eight. Hence it is very important to have a good understanding of the wind speed patterns at a given site in order to evaluate the possible use of a wind pump there. It is sometimes recommended that a site should have an average wind speed at the height of a wind rotor of at least 2.5 m/s in order to have potential for water pumping. That is a good rule of thumb, but by no means the whole story. First of all, one seldom knows the wind speed at any height at a prospective windmill site, except by estimate and correlation. Second, mean wind speeds generally vary with the time of day and year and it makes an enormous difference if the winds occur when the water is needed.

The best way to evaluate the wind at a prospective site is to monitor it for at least a year. Data should be summarized at least monthly. This is often impossible, but there should be some monitoring done if a large wind project is envisioned. The most practical approach may be to obtain wind data from the nearest weather station (for reference) and try to correlate it with that at the proposed wind pump site. If at all possible the station should be visited to ascertain the placement of the measuring instrument (anemometer) and its calibration. Many times anemometers are placed too near the ground or are obscured by vegetation and so greatly underestimate the wind speed. The correlation with the proposed site is best done by placing an anemometer there for a relatively short time (at least a few weeks) and comparing resulting data with that taken simultaneously at the reference site. A scaling factor for the long-term data can be deduced and used to predict wind speed at the desired location.

Of course, possible locations for wind machines are limited by the placement of the wells, but a few basic observations should be kept in mind. The entire rotor should be well above the surrounding vegetation, which should be kept as low as possible for a distance of at least ten times the rotor diameter in all directions. Wind speed increases with elevation above ground, usually by 15-20 percent with every doubling of height (in the height range of most wind pumps). Because of the cubic relationship between wind speed and power, the effect on the latter is even more dramatic.

## *5. Estimate Wind Machines Size*

A typical wind pump is shown in Figure 1. Most wind pumps have a horizontal axis (that is, the rotating shaft is parallel to the ground). Vertical axis machines, such as the Savonius rotor, have usually been less successful in practice.

In order to estimate wind machine's size it is first necessary to have some idea how it will perform in real winds. As previously mentioned, the power in wind varies with the cube of the wind speed. It is also proportional to the density of the air. Atmospheric density is 1.293 $kg/m^3$ at sea level at standard conditions but is affected by temperature and pressure. The power that a wind machine produces, in addition, depends on the swept area of its rotor and the aerodynamic characteristics of its blades. Under ideal conditions the rotational speed of the rotor varies in direct relation to the wind speed. In this case the efficiency of the rotor remains constant and power varies as the cube of the wind speed (and rotational speed).

With wind pumps, however, the situation is more complicated. The majority use piston pumps, whose power requirements vary directly with the speed of the pump. At high wind speeds the rotor can produce more power than the pump can

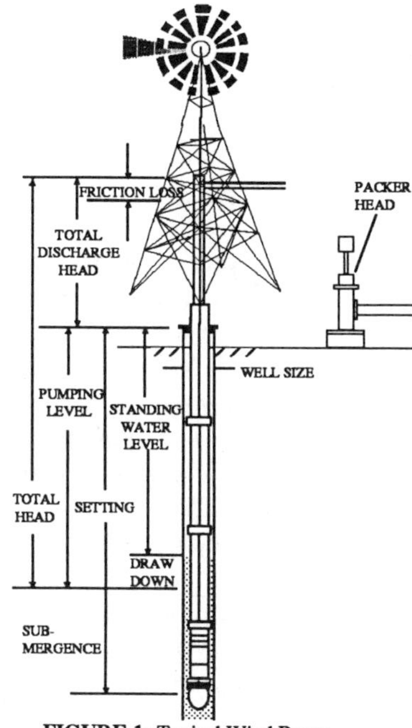


**FIGURE 1:** Typical Wind Pump

use. The rotor speeds up, causing its efficiency to drop, so it produces less power. The pump, coupled to the rotor, also moves more rapidly so it absorbs more power. At a certain point the power from the rotor equals the power used by the pump, and the rotational speed remains constant until the wind speed changes.

The net effect of all this is that the whole system behaves rather differently than an ideal wind turbine. Its actual performance is best described by a measured characteristic curve (Figure 2), which relates actual water flow at given pumping heads to the wind speed. This curve also reflects other important information such as the wind speeds at which the machine starts and stops pumping (low wind) and when it begins to turn away in high winds (furling).

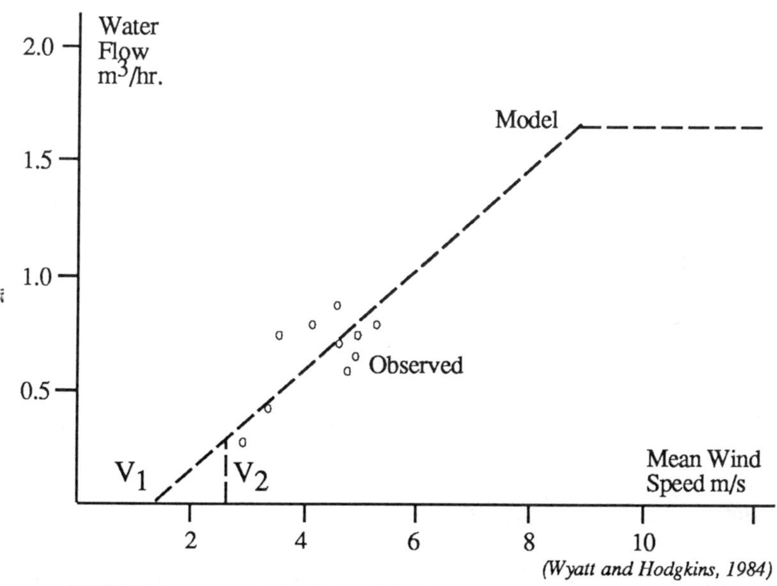


**FIGURE 2:** Multiblade Windmill Performance, Observed And Model Results, One Minute Average Readings

Most commercial machines and those developed and tested more recently have such curves and these should be used if possible in predicting wind machine output. On the other hand, it should be noted that some manufacturers provide incomplete or overly optimistic estimates of what their machines can do. Sales literature should be examined carefully.

In addition to the characteristic curve of the wind machine, one must also know the pattern of the wind in order accurately to estimate productivity. For example, suppose it is known how many hours (frequency) the average wind speed was between 0-1 m/s, 1-2 m/s, 2-3 m/s, etc., in a given month. By referring to the characteristic curve, one could determine how much water was pumped in each of the groups of hours corresponding to those wind speed ranges. The sum of water from all groups would be the monthly total. Usually such detailed information on the wind is not known. However, a variety of statistical techniques are available from which the frequencies can be predicted fairly accurately, using only the long-term mean wind speed and, when available, a measure of its variability (standard deviation). See Lysen, 1983, and Wyatt and Hodgkin, 1984.

Many times there is little information known about a possible machine or it is just desired to know very approximately what size machine would be appropriate. Under these conditions the following simplified formula can be used:

$$\text{Power} = \text{Area} \times 0.1 \times (\text{Vmean})^3$$

where

Power = useful power delivered in pumping the water, watts

Area = swept area of rotor (3.14 x Radius squared), $m^2$

Vmean = mean wind speed, m/s

By rearranging the above equation, an approximate diameter of the wind rotor can be found. Returning to the earlier example, to pump 50 $m^3$/day, 15 m would require an average of 85 watts. Suppose the mean wind speed was 4 m/s. Then the diameter (twice the radius) would be:

$$\text{Diameter} = 2\,[\text{Power}/(3.14) \times 0.1 \times \text{Vmean}^3)]$$

or

$$\text{Diameter} = 2 \times [85/(3.14 \times 0.1 \times 4^3)] = 4.1 \text{ m}$$

## *6. Compare Seasonal Water Production to Requirement*

This procedure is usually done on a monthly basis. It consists of comparing the amount of water that could be pumped with that actually needed. In this way it can be told if the machine is large enough and conversely if some of the time there will be excess water. This information is needed to perform a realistic economic analysis. The results may suggest a change in the size of machines to be used.

Comparison of water supply and requirement will also aid in determining the necessary storage size. In general storage should be equal to about one or two days of usage.

## *7. Select Type of Wind Machine and Pump*

There is a variety of types of wind machines that could be considered. The most common use relatively slow speed rotors with many blades, coupled to a reciprocating piston pump.

Rotor speed is described in terms of the tip speed ratio, which is the ratio between the actual speed of the blade tips and the free wind speed. Traditional wind pumps operate with highest efficiency when the tip speed ratio is about 1.0. Some of the more recently developed machines, with less blade area relative to their swept area, perform best at higher tip speed ratios (such as 2.0).

A primary consideration in selecting a machine is its intended application. Generally speaking, wind pumps for domestic use or livestock supply are designed for unattended operation. They should be quite reliable and may have a relatively high cost. Machines for irrigation are used seasonally and may be designed to be manually operated. Hence they can be more simply constructed and less expensive.

For most wind pump applications, there are four possible types or sources of equipment. These are: 1) Commercially available machines of the sort developed for the American West in the late 1800s; 2) Refurbished machines of the first types that have been abandoned; 3) Intermediate technology machines, developed over the last 20 years for production and use in developing countries; and 4) Low technology machines, built of local materials.

The traditional, American "fan mill," is a well developed technology with very high reliability. It incorporates a step down transmission, so that pumping rate is a quarter to a third of the rotational speed of the rotor. This design is particularly suitable for relatively deep wells (greater than 30m--100'). The main problem with these machines is their high weight and cost relative to their pumping capacity. Production of these machines in developing countries is often difficult because of the need for casting gears.

Refurbushing abandoned traditional pumps may have more potential than might at first appear likely. In many windy parts of the world a substantial number of these machines were installed early in this century, but were later abandoned when other forms of power became available. Often these machines can be made operational for much less cost than purchasing a new one. In many cases parts from newer machines are interchangeable with the older ones. By coupling refurbishing with a training program, a maintenance and repair infrastructure can be created at the same time that machines are being restored. Development of this infrastructure will facilitate the successful introduction of newer machines in the future.

For heads of less than 30m, the intermediate technology machines may be most appropriate. Some of the groups working on such designs are listed at the end of this entry. These machines typically use a higher speed rotor and have no gear box. On the other hand they may need an air chamber to compensate for adverse acceleration effects due to the rapidly moving piston. The machines are made of steel, and require no casting and minimal welding. Their design is such that they can be readily made in machine shops in developing countries. Many of these wind pumps have undergone substantial analysis and field testing and can be considered reliable.

Low technology machines are intended to be built with locally available materials and simple tools. Their fabrication and maintenance, on the other hand, are very labor intensive. In a number of cases projects using these designs have been less successful than had been hoped. If such a design is desired, it should first be verified that machines of that type have actually been built and operated successfully. For a sobering appraisal of some of the problems encountered in building wind machines locally, see *Wind Energy Development in Kenya* (see Sources).

Although most wind machines use piston pumps, other types include mono pumps (rotating), centrifugal pumps (rotating at high speed), oscillating vanes, compressed air pumps, and electric pumps driven by a wind electric generator. Diaphragm pumps are sometimes used for low head irrigation (5-10 m or 16-32'). No matter what type of rotor is used, the pump must be sized appropriately. A large pump will pump more water at high wind speeds than will a small one. On the other hand, it will not pump at all at lower wind speeds. Since the power required in pumping the water is proportional to the head and the flow rate, as the head increases the volume pumped will have to decrease accordingly. The piston travel, or stroke, is generally constant (with some exceptions) for a given windmill. Hence, piston area should be decreased in proportion to the pumping head to maintain optimum performance.

Selecting the correct piston pump for a particular application involves consideration of two types of factors: 1) the characteristics of the rotor and the rest of the machine, and 2) the site conditions. The important machine characteristics are: 1) the rotor size (diameter); 2) the design tip speed ratio; 3) the gear ratio; and 4) the stroke length. The first two have been discussed earlier. The gear ratio reflects the fact that most wind pumps are geared down by a factor of 3 to 4. Stroke length increases with rotor size. The choice is affected by structural considerations. Typical values for a machine geared down 3.5:1 range from 10 cm (4") for a rotor diameter of 1.8 m (6') to 40 cm (15")for a diameter of 5 m (16').

Note that it is the size of the crank driven by the rotor (via the gearing) that determines the stroke of the pump.

The key site conditions are: 1) mean wind speed and 2) well depth. These site factors can be combined with the machine parameters to find the pump diameter with the use of the following equation. This equation assumes that the pump is selected so that the machine performs best at the mean wind speed.

$$DP = \sqrt{\frac{(0.1)(\pi)(DIAMR)^3 (VMEAN)^2 (GEAR)}{(DENSW)(G)(HEIGHT)(TSR)(STROKE)}}$$

where:
DP = Diameter of piston, m
$\pi = 3.1416$
DIAMR = Diameter of the rotor, m
VMEAN = Mean wind speed, m/s
GEAR = Gear down ratio
DENSW = Density of water, 1000 $kg/m^3$
G = Acceleration of gravity, 9.8 $m/s^2$
HEIGHT = Total pumping head, m
TSR = Design tip speed ratio
STROKE = Piston stroke length, m

Example:

Suppose the wind machine of the previous examples has a gear down ratio of 3.5:1, a design tip speed ratio of 1.0 and a stroke of 30 cm. Then the diameter of the piston would be:

$$DP = \sqrt{\frac{(0.1)(3.14)(4.1)^3 (4.0)^2 (3.5)}{(1000)(9.8)(15)(1.0)(0.3)}} = .166m$$

## 8. *Identify Suppliers of Machinery*

Once a type of machine has been selected, suppliers of the equipment or the designs should be contacted for information about availability of equipment and spare parts in the region in question, references, cost, etc. If the machine is to be built locally, sources of material, such as sheet steel, angle iron, bearings, etc. will have to be identified. Possible machine shops should be visited and their work on similar kinds of fabrication should be examined.

## 9. *Identify Alternative Power Sources for Water Pumping*

There are usually a number of alternatives in any given situation. What might be a good option depends on the specific conditions. Some of the possibilities include pumps using human power (hand pumps), animal power (Persian wheels, chain pumps), internal combustion engines (gasoline, diesel, or biogas), external combustion engines (steam, Stirling cycle), hydropower (hydraulic rams, norias), and solar power (thermodynamic cycles, photovoltaics).

## *10. Evaluate Economics*

For all the realistic options the likely costs should be assessed and a life cycle economic analysis performed. The costs include the first cost (purchase or manufacturing price), shipping, installation, operation (including fuel where applicable), maintenance, spare parts, etc. For each system being evaluated the total useful delivered water must also be determined (as described in Step 6). The life cycle analysis takes account of costs and benefits that accrue over the life of the project and puts them on a comparable basis. The result is frequently expressed in an average cost per cubic meter of water (Figure 3).

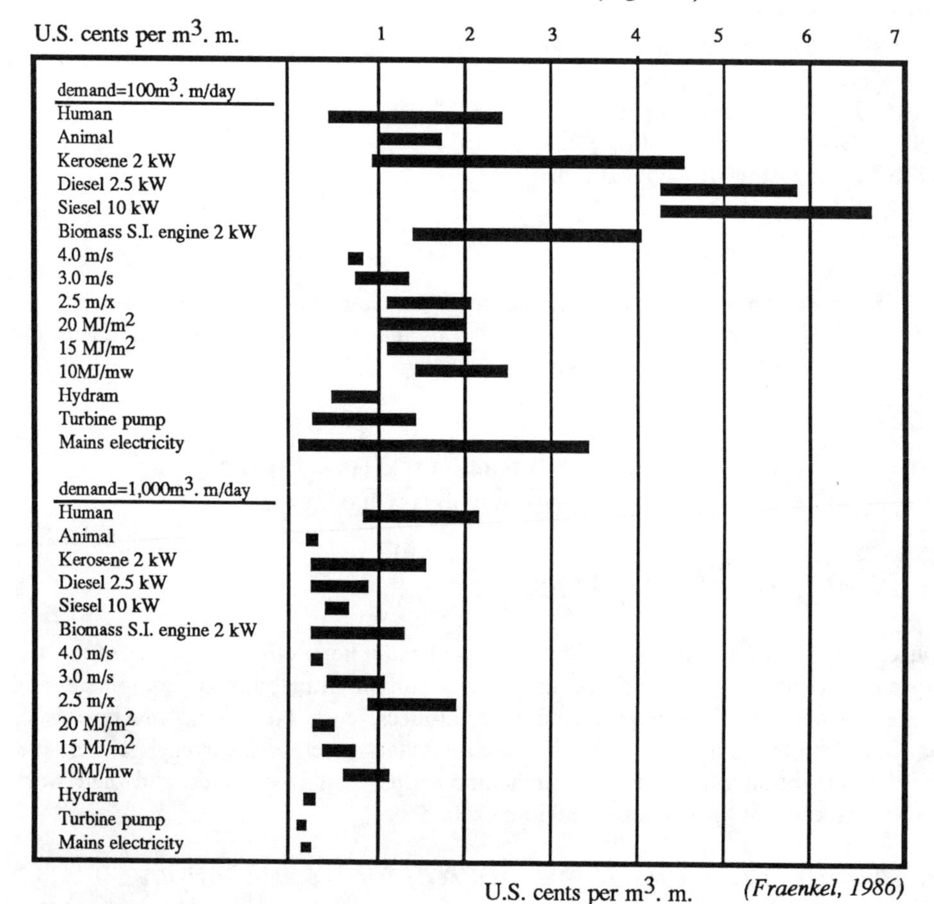


*(Fraenkel, 1986)*

**FIGURE 3:** Expected range of unit energy costs for two levels of demand, 100 and 1000 $m^3$.m/day, for different types of prime mover.

It should be noted that the most economic option is strongly affected by the size of the project. In general, wind energy is seldom competitive when mean winds are less than 2.5 m/s, but it is the least cost alternative for a wide range of conditions when the mean wind speed is greater than 4.0 m/s.

## *11. Install the Machines*

Once wind energy has been selected, arrangements should be made for the purchase or construction of the equipment. The site must be prepared and the materials all brought there. A crew for assembly and erection must be secured, and instructed. Someone must be in charge of overseeing the installation to ensure that it is done properly and to check the machine out when it is up. Regular maintenance must be arranged for.

With proper planning, organization, design, construction, and maintenance, the wind machines may have a very useful and productive life.

**Source:**

James F. Manwell, VITA Volunteer, University of Massachusetts.

**References:**

Fraenkel, Peter. *Water-Pumping Devices: A Handbook for Users and Choosers.* London: Intermediate Technology Publications, 1986.

Johnson, Garry. *Wind Energy Systems.* Englewood Cliffs, New Jersey: Prentice Hall, Inc.

Lierop, W.E. and van Veldheizen, L.R. *Wind Energy Development in Kenya,* Main Report, Vol. 1: Past and Present Wind Energy Activities, SWD 82-3/Vol. 1 Amersfoort, The Netherlands: Consultancy for Wind Energy in Developing Countries, 1982.

Lysen, E.H. *Introduction to Wind Energy.* SWD 82-1 Amersfoort, The Netherlands: Consultancy for Wind Energy in Developing Countries, 1983.

Manwell, J.F. and Cromack, D.E. *Understanding Wind Energy: An Overview.* Arlington, Virginia: Volunteers in Technical Assistance, 1984.

McKenzie, D.W. "Improved and New Water Pumping Windmills," Proceedings of Winter Meeting, American Society of Agricultural Engineers, New Orleans, December, 1984.

Vilsteren, A.V. *Aspects of Irrigation with Windmills.* Amersfoot, The Netherlands: Consultancy for Wind Energy in Developing Countries, 1981.

Wegley, H.L., et al. *A Siting Handbook for Small Wind Energy Conversion Systems.* Richland, Washington: Battelle Memorial Institute, 1978.

Wyatt, A.S. and Hodgkin, J., *A Performance Model for Multiblade Water Pumping Windmills.* Arlington, Virginia: VITA, 1984.

**Groups Involved with Wind Pumping in Developing Countries**

Consultancy for Wind Energy in Developing Countries, P.O. Box 85, 3800 AB, Amersfoort, The Netherlands

Intermediate Technology Development Group, Ltd., 9 King Street, Coven Garden, London, WC2E 8HW, UK

IPAT, Technical University of Berlin, Sekr. TH2, Lentzallee 86, D-1000 Berlin 33, West Germany

Renewable Energy Research Laboratory, Dept. of Mechanical Engineering, University of Massachusetts, Amherst, Massachusetts 01003, USA

SKAT, Varnbuelstr. 14, CH-9000 St. Gallen, Switzerland

The Danish Center for Renewable Energy, Asgaard, Sdr. Ydby, DK-7760 Hurup Thy, Denmark

Volunteers in Technical Assistance (VITA), 1815 N. Lynn Street, Suite 200, Arlington, Virginia 22209-2079 USA

**Manufacturers of Water Pumping Windmills**

Aermotor, P.O. Box 1364, Conway, Arkansas 72032, USA

Dempster Industries, Inc., Beatrice, Nebraska 68310, USA

Heller Aller Company, Perry & Oakwood St., Napoleon, Ohio 43545, USA

# Water Storage and Treatment

## CISTERNS

Cisterns for family use are most practical in areas of adequate rainfall and where ground water is difficult to obtain or where it contains too many minerals. A sealed well usually requires no filtration, no chemical disinfection, and little upkeep, while a cistern needs all of these. And cisterns generally cost more to build than wells. Cistern water has few minerals, however, and is ideal for washing clothes.

A cistern water supply has four basic parts: tank, catchment area, filter, and pump. (Pumps are discussed in the section on "Water Lifting.")

## Cistern Tank

The tank described here can be used for sanitary storage of rainwater for family use. It can be constructed of reinforced concrete sealed with asphalt sealing compound.

The cistern tank must be watertight to prevent surface contamination from polluting the supply. Reinforced concrete is the best material because it is strong, it has a long life and it can be made watertight.

A manhole and drain must be provided so that the tank can be cleaned. (See Figure 1.) A vent and a place through which chlorine can be added easily for disinfection are also necessary. (Note: Chlorine can be added through the vent by removing the U elbow. Lubricate the threads of the elbow to make removal easy.)

The size of the cistern depends on the family's daily needs and the length of time between rainy periods. If a family needs 94.6 liters (25 U.S. gallons) of water a day and there are 125 days between rainy periods, then the cistern must hold:

94.6 liters x 125 days = 11,835 liters

or

25 U.S. gallons x 125 days = 3,125 U.S. gallons

FIG. 1 CISTERN WITH SAND FILTER (PUMP INSTALLATION OPTIONAL)

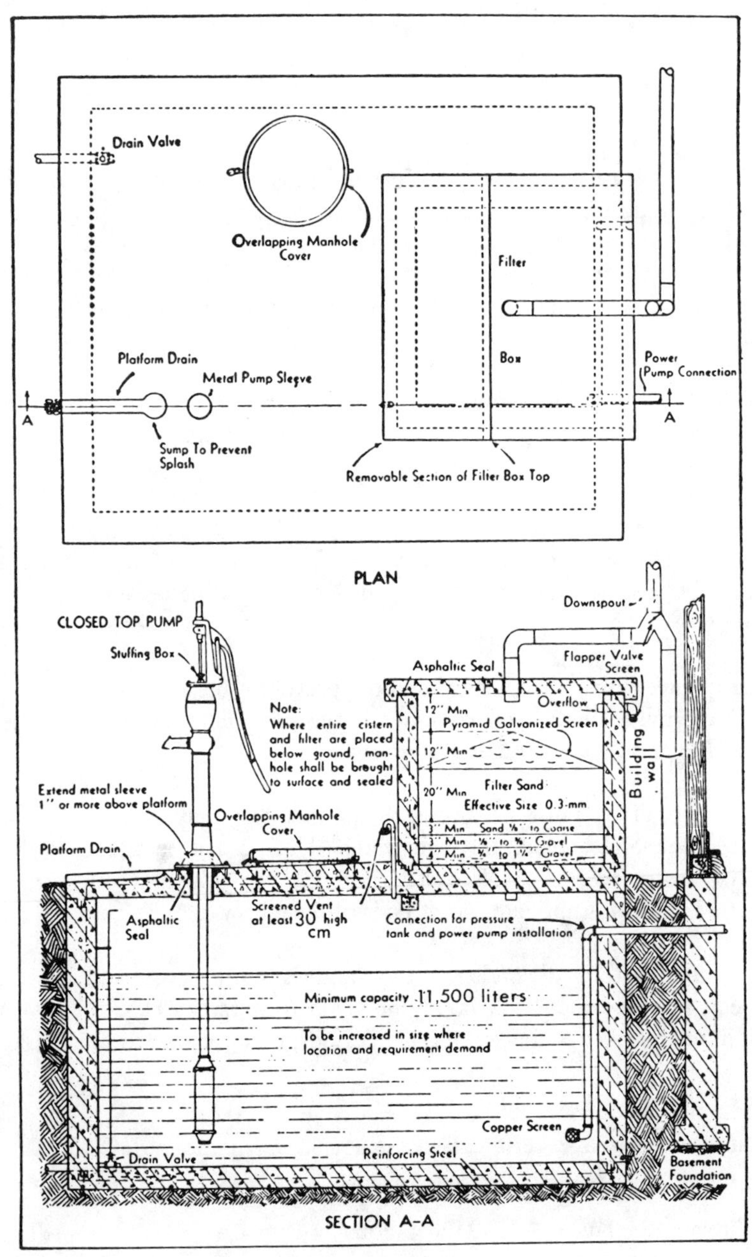


Reproduced from US Public Health Service, Joint Committee on Rural Sanitation (1950) *Individual water supply systems*, Washington, p. 32

A cistern with an inside size of 3 meters x 2 meters x 2 meters (7 1/2' x 7 1/2' x 7 1/2') holds 11,355 liters (3,000 U.S. gallons). The top surfaces of the cistern walls should be about 10cm above ground.

To be sure that the cistern is watertight, use about 28 liters of water per 50kg sack of cement (5 1/2 U.S. gallons per 94 pound or one cubic foot sack) when mixing the concrete. (See section on "Concrete Construction.") Tamp the concrete thoroughly and keep the surface damp for at least 10 days. If possible, pour the walls and floor at the same time. The manhole entrance must be 10cm (4") above the cistern surface and the cover should overlap by 5cm (2"). Slope the bottom of the cistern, making one part lower than the rest, so that water can be more easily siphoned or bailed out when the cistern is being cleaned. You can do this by scraping the bottom to the proper contour. Do not use fill dirt under the cistern because this may cause the cistern to settle unevenly and crack. A screened drain pipe and valve will make cleaning easier.

An overflow pipe is not needed if a roof-cleaning butterfly valve is properly used. If the overflow is installed, be sure to cover the outlet carefully with copper window screen. A screened vent is necessary if there is no overflow, to allow displaced air to leave the cistern. The hand pump must be securely mounted to bolts cast into the concrete cistern cover. The flanged base of the pump should be solid, with no holes for contamination to enter, and sealed to the pump cover, or the drop pipe must be sealed in with concrete and asphalt sealing compound.

A small pipe with a screw-on cap is needed to allow for measuring the water in the cistern and adding chlorine solution **after each rainfall.** The amount of water in the cistern is measured with a stick marked in thousands of liters (or thousands of gallons). To disinfect after each rainfall, add a 5 parts per million dosage of chlorine (see section on "Chlorination").

A newly built or repaired cistern should always be disinfected with a 50 parts per million chlorine solution. The cistern walls and the filter should be thoroughly washed with this strong solution and then rinsed. A small-pressure system can be disinfected readily by pumping this strong solution throughout the system and letting it stand overnight.

## Catchment Area

A catchment area of the proper size is a necessary part of a cistern water supply. Rainwater for a cistern can be collected from the roof of a house. The method given here for estimating catchment size should be checked against the actual size of nearby catchment installations.

The catchment or collecting area should be a smooth, watertight material, like a galvanized sheet-metal roof. Wood or thatch roofs may taint the water and retain dust, dirt and leaves; water from these roofs contains more organic matter and bacteria than water from smooth surfaces. Stone, concrete, and plastic film catchments are sometimes built on the ground. For family use, roofs are usually best because humans and animals cannot contaminate them.

To estimate your required catchment area, estimate the minimum yearly rainfall and the amount of water required by the family during one year. Sometimes the government meteorological section can give you the minimum rainfall expected. If they cannot, estimate the minimum rainfall at two-thirds of the yearly average. Take the average amount of water needed by the family for one day and multiply it by 365 to learn how much is needed for one year. Then use the chart to find how much roof space is needed (Figure 2). Add 10 percent to the area given by the chart to allow for water lost to evaporation and discarded at the beginning of each rainfall.

Example:

> With an average rainfall of 75cm a year, and a family needing 135 liters of water a day, then:
>
> 2/3 x 75 = minimum annual rainfall of 50cm
>
> 365 x 135 liters/day = 49,275 liters a year.
>
> Round this figure off to 50,000 liters a year. The example worked out on the chart (Figure 2) shows that a catchment area of about 115 square meters is needed. Add 10 percent to this area to allow for water loss, giving a total required catchment area of about 126.5 square meters.

A collecting trough and downspout are needed. Be sure there is a good pitch to the trough so that the water flows freely and does not hold small puddles that can attract mosquitoes and other insects. Troughs and downspouts need periodic inspection and cleaning. Extending the trough increases the catchment area.

## Cistern Filter

The sand filter described here will remove most organic matter from water but it will not produce safe drinking water by removing all harmful bacteria. Water collected in the cistern tank should be chlorinated after each rainfall. A catchment area always collects leaves, bird droppings, road dust, and insects. A cistern filter removes as much of this material as possible before the water enters the cistern (Figure 3).

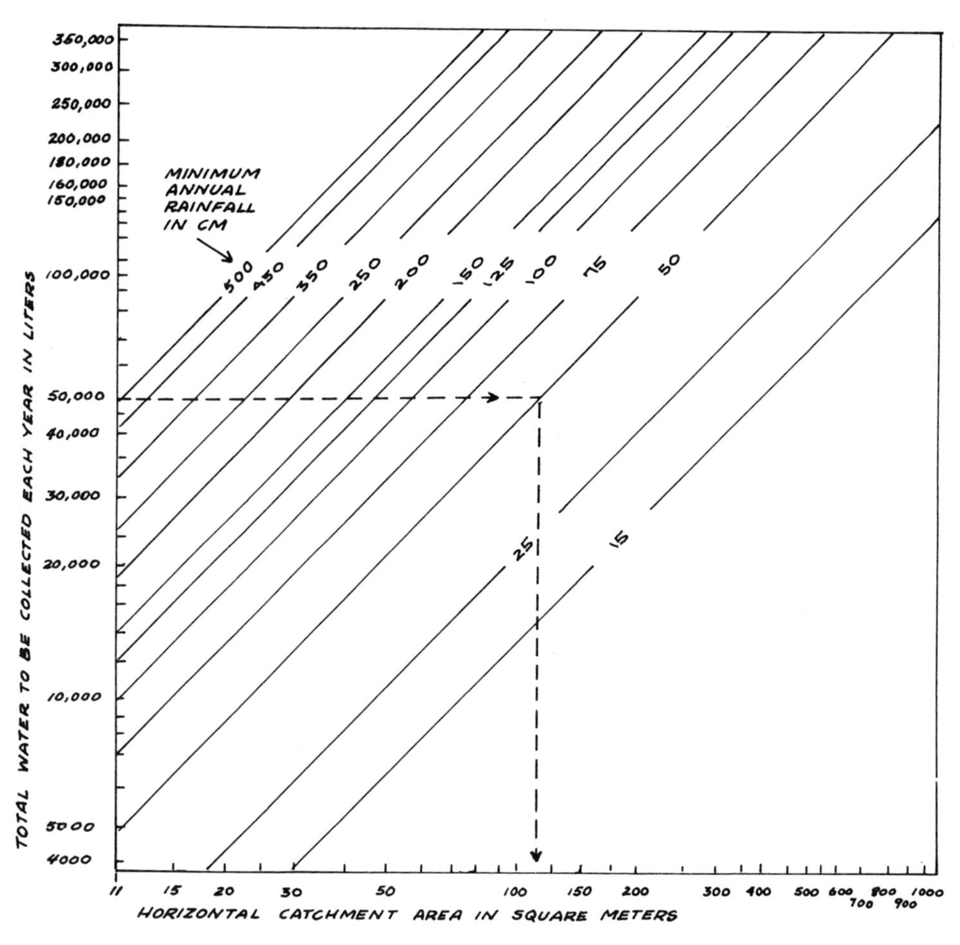

MINIMUM ANNUAL RAINFALL IN CM
500
450
350
250
200
150
125
100
75
50
25
15
TOTAL WATER TO BE COLLECTED EACH YEAR IN LITERS
350,000
300,000
250,000
200,000
180,000
160,000
150,000
100,000
50,000
40,000
30,000
20,000
10,000
5000
4000
HORIZONTAL CATCHMENT AREA IN SQUARE METERS
11
15
20
30
50
100
150
200
300
400
500
600
700
800
900
1000


FIGURE 3

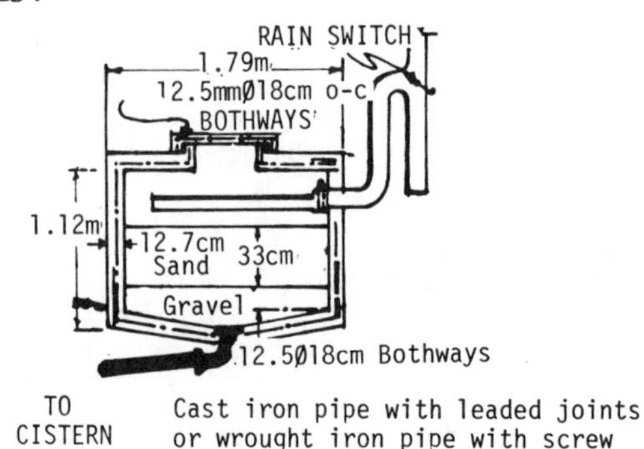


FIGURE 3

The sand filter is usually built at ground level and the filtered water runs into the cistern, which is mostly underground. The largest pieces, such as leaves, are caught in the splash plate. The splash plate also distributes the water over the surface of the filter, so that the water does not make holes in the sand. Several layers of copper window screen form the splash plate.

If a filter is made too small to handle the normal rush of water from rainstorms, the water will overflow the filter or dig a channel in the sand, ruining the filter. The filter area should be not less than one-tenth of the catchment area. A typical filter would be 122cm x 122cm (4' x 4') for a family-sized unit where rainfall intensity is average.

About every 6 months, remove the manhole cover and clean the filter. Remove all matter from the splash plate and scrape off and remove the top 1.25cm (1/2") of sand. When the sand is down to 30cm (12") in depth, rebuild it with clean sand to the original depth of 46cm (18").

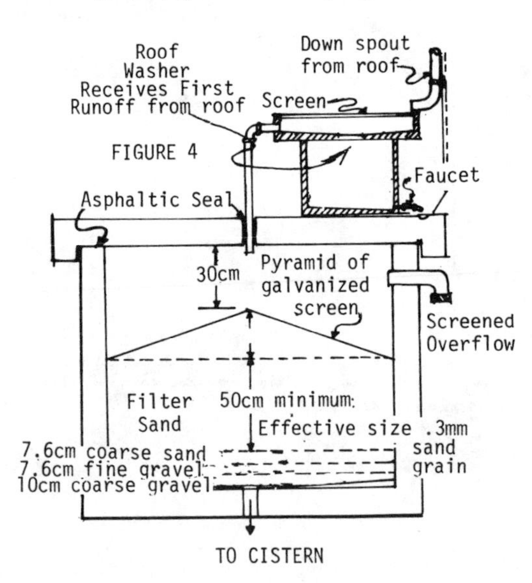


The first runoff from the roof, which usually contains a great deal of leaves and dirt, should be discarded. The simplest way to do this is to have a butterfly valve (like a damper in a stovepipe) in the downspout. After the rain has washed the roof, the valve is turned to let the runoff water enter the filter. A semi-automatic filter is shown in Figure 4.

In building the filter, it is important to use properly-sized sand and gravel and to make sure the filter can be cleaned easily. The filter must have a screened overflow.

**Sources:**

Wagner, E.G. and Lanoix, J.N. *Water Supply for Rural Areas and Small Communities.* Geneva: World Health Organization, 1959.

*Cisterns.* State of Illinois, Department of Public Health, Circular No. 833.

*Manual of Individual Water Supply Systems.* U.S. Department of Health, Education and Welfare, Public Health Service Publication No. 24.

# SELECTING A DAM SITE

A water reservoir can be formed by building a dam across a ravine. Building a dam takes time, labor, materials, and money. Furthermore, if a dam holding more than a few acre-feet of water breaks, a great deal of damage can be caused. Therefore, it is important to choose a dam site carefully, to guard against dam collapse, and to avoid excessive silting, porous soil, polluted water, and water shortages because the catchment area is too small. Careful selection of the dam site will save labor and material costs and help ensure a strong dam.

The preliminary evaluation described here will help to determine whether or not a particular site will be good for building a dam. **Remember that dams can have serious environmental consequences and an improperly constructed dam can be extremely dangerous. Consult an expert before starting to build.**

Six factors are important in site selection.

1. Enough water to meet your requirements and fill the reservoir.

2. Maximum water storage with the smallest dam.

3. A sound, leakproof foundation for the reservoir.

4. Reasonable freedom from pollution.

5. A storage site close to users.

6. Available materials for construction.

7. Provision for a simple spillway.

8. Authorization from local authorities to build the dam and use the water.

One acre-foot of water is equivalent to the amount required to cover an acre of land (30cm of water covering 0.4 hectares) to a depth of 1 foot. One acre-foot equals 1,233.49 cubic meters. The annual rainfall and type of catchment (or natural drainage) area will determine the amount of water the reservoir will collect.

## Catchment Area

A catchment area with steep slopes and rocky surfaces is very good. If the catchment area has porous soil on a leak-proof rock base, springs will develop and will carry water to the reservoir, but more slowly than rocky slopes. Trees with small leaves, such as conifers, will act as a windbreakers and reduce loss of water from evaporation.

Swamps, heavy vegetation, permeable ground, and slight slopes will decrease the yield of water from a catchment area.

## Rainfall

The average catchment area will, in a year, drain 5 acre-feet (6,167 cubic meters) into a reservoir for every inch (2.5cm) of annual rainfall falling on a square mile (2.59 square kilometers); that is, about 10 percent of the rainfall.

## Location

The best location for building a dam is where a broad valley narrows with steep sides and a firm base on which to build the dam (see Figure 1). Ground that contains large boulders, weathered or fissured bedrock, alluvial sands, or porous rock is not good. The best bases for building a dam are granite or basalt layers at or near the surface or a considerable depth of silty or sandy clay.

Location of a dam upstream from its point of use can lower pollution and may allow for gravity feed of the water to its point of use.

It is best if stone is nearby when building a masonry dam. When building an earth dam, rock will still be required for the spillway. The best soils for earth dams contain clay with some silt or sand. There should be enough of this soil close to the dam site for building the entire dam of reasonably uniform material.

**Source:**

Wagner, E.G. and Lanoix, J.N. *Water Supply for Rural Areas and Small Communities.* Geneva: World Health Organization, 1959.

FIGURE 1

# WATER PURIFICATION

The purification of unsafe water requires some trained supervision if it is to be done effectively. Such supervision is rarely available in the villages and the procedure tends to be neglected sooner or later. Under these circumstances **every effort must be made to obtain a source that provides naturally wholesome water and then to collect that water and protect it against pollution** by the methods already described. Thus, the necessity for treatment of the water may be avoided, and the practical importance of managing this can hardly be overemphasized.

Water treatment under rural conditions should be restricted by the responsible control agency to cases where such treatment is necessary and where proper plant operation and maintenance is assured.

If the water needs treatment, this should, if at all possible, be done for the whole community and certainly before, or on entry to the dwelling so that the water from all the taps in the house is safe. The practice, common in the Tropics, of sterilizing (by filtration and boiling) only the water to be used for drinking, teeth cleaning, etc., though efficient in itself (when carefully done) is frequently nullified by carelessness. Furthermore, children are likely to use water from any tap. Contrary to an all too common opinion, ordinary freezing of water, though it may retard the multiplication of bacteria, does not kill them, and ice from a household refrigerator is no safer than the water from which it was made.

The principal methods of purifying water on a small scale are boiling, chemical disinfection, and filtration. These methods may be used singly or in combination, but if more than filtration is needed the boiling or chemical disinfection should be done last. Each method is discussed briefly below. Following this general introduction are descriptions of a variety of water purification technologies: boiler for drinking water, chlorination of polluted water, water purification plant, and sand filter.

**Boiling** is the most satisfactory way of destroying disease-producing organisms in water. It is equally effective whether the water is clear or cloudy, whether it is relatively pure or heavily contaminated with organic matter. Boiling destroys all forms of disease-producing organisms usually encountered in water, whether they be bacteria, viruses, spores, cysts, or ova. To be safe the water must be brought to a good "rolling" boil (not just simmering) and kept there for 15-20 minutes. Boiling drives out the gases dissolved in the water and gives it a flat taste, but if the water is left for a few hours in a partly filled container, even though the mouth of the container is covered, it will absorb air and lose its flat, boiled taste. It is wise to store the water in the vessel in which it was boiled. Avoid pouring the water from one receptacle to another with the object of aerating or cooling it as that introduces a risk of re-contamination.

**Chlorine** is a good disinfectant for drinking water as it is effective against the bacteria associated with water-borne disease. In its usual doses, however, it is ineffective against the cysts of amoebic dysentery, ova of worms, cercariae which cause schistosomiasis, and organisms embedded in solid particles.

Chlorine is easiest to apply in the form of a solution and a useful solution in one which contains 1 percent available chlorine, for example, Milton Antiseptic. Dakin's solution contains 0.5 percent available chlorine, and bleaching powder holds 25 percent to 30 percent available chlorine. About 37cc (2 1/2 tablespoons) of bleaching powder dissolved in 0.95 liter (1 quart) of water will give a 1 percent chlorine solution. To chlorinate the water, add 3 drops of 1 percent solution to each 0.95 liter (1 quart) of water to be treated (2 tablespoonfuls to 38 gallons), mix thoroughly and allow it to stand for 20 minutes or longer before using the water.

Chlorine may be obtained in table form as "Sterotabs" (formerly known as "Halazone"), "Chlor-dechlor" and "Hydrochlorazone," which are obtainable on the market. Directions for use are on the packages.

**Iodine** is also a good disinfecting agent. Two drops of ordinary tincture of iodine are sufficient to treat 0.95 liter (1 quart) of water. Water that is cloudy or muddy, or water that has a noticeable color even when clear, is not suitable for disinfection by iodine. Filtering may render the water fit for treatment with iodine. If the water is heavily polluted, the dose should be doubled. Though the higher dosage is harmless it will give the water a medicinal taste. To remove any medicinal taste add 7 percent solution of sodium thiosulphate in a quantity equal to the amount of iodine added.

Iodine compounds for the disinfection of water have been put into table form, for example, "Potable Aqua Tablets," "Globaline" and "Individual Water Purification Tablets"; full directions for use are given on the packages. These tablets are among the most useful disinfection devices developed to date and they are effective against amoeba cysts, cercariae, leptospira, and some of the viruses.

**Source:**

*Small Water Supplies,* Bulletin No. 10 London: The Ross Institute, 1967.

**Other Useful References:**

Mann, H.T. and Williamson, P. *Water Treatment and Sanitation.* London: Intermediate Technology Publications, 1976.

*Iornech Disinfection System,* Iornech Ltd., 2063 Lakeshore Blvd. West Toronto, Ontario, Canada, (undated).

*Manual of Individual Water Supply Systems.* Public Health Service Publication No. 24, Washington, D.C. U.S. Department of Health amd Human Services, 1962.

Decade Watch newsletter. United Nations Development Program, Division of Information.

*International Reference Center for Community Water Supply and Sanitation,* newsletter. P.O. Box 93190, 2509 AD, The Hague, Netherlands.

# Boiler for Drinking Water

The boiler described here (Figure 1) will provide safe preparation and storage of drinking water in areas where pure water is not available and boiling is practical. When the unit was used in work camps in Mexico, a 208-liter (55-gallon) drum supplied 20 persons with water for a week.

### Tools and Materials

208-liter (55-gallon) drum
10mm (3/4") pipe nipple, 5cm (2") long
Bricks for two 30cm (1') layers to support drum
Sand and 1 sack of cement for mortar and base of fireplace
Large funnel and filter medium for filling drum
Metal plate to control draft in front of fireplace
19mm (3/4") valve, preferably all metal, such as a gate valve, that can withstand heat.

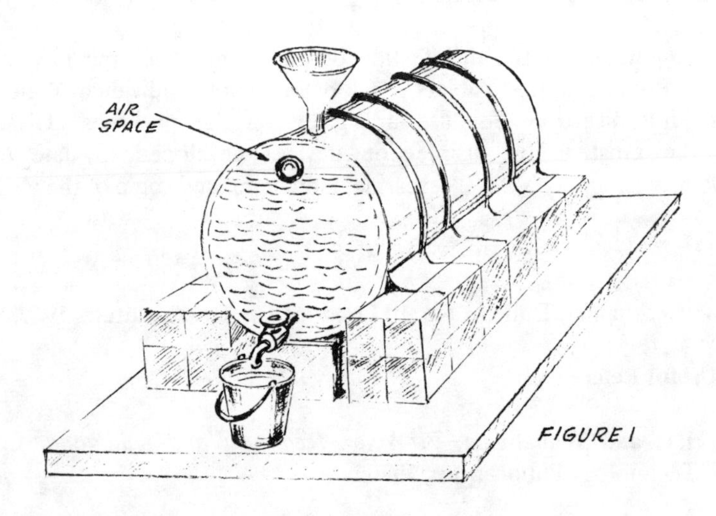


FIGURE 1

The fireplace for this unit (see Figure 2) is simple. It should be oriented so that the prevailing wind or draft goes between the bricks from the front to the back of the drum. A chimney can be provided, but it is not necessary.

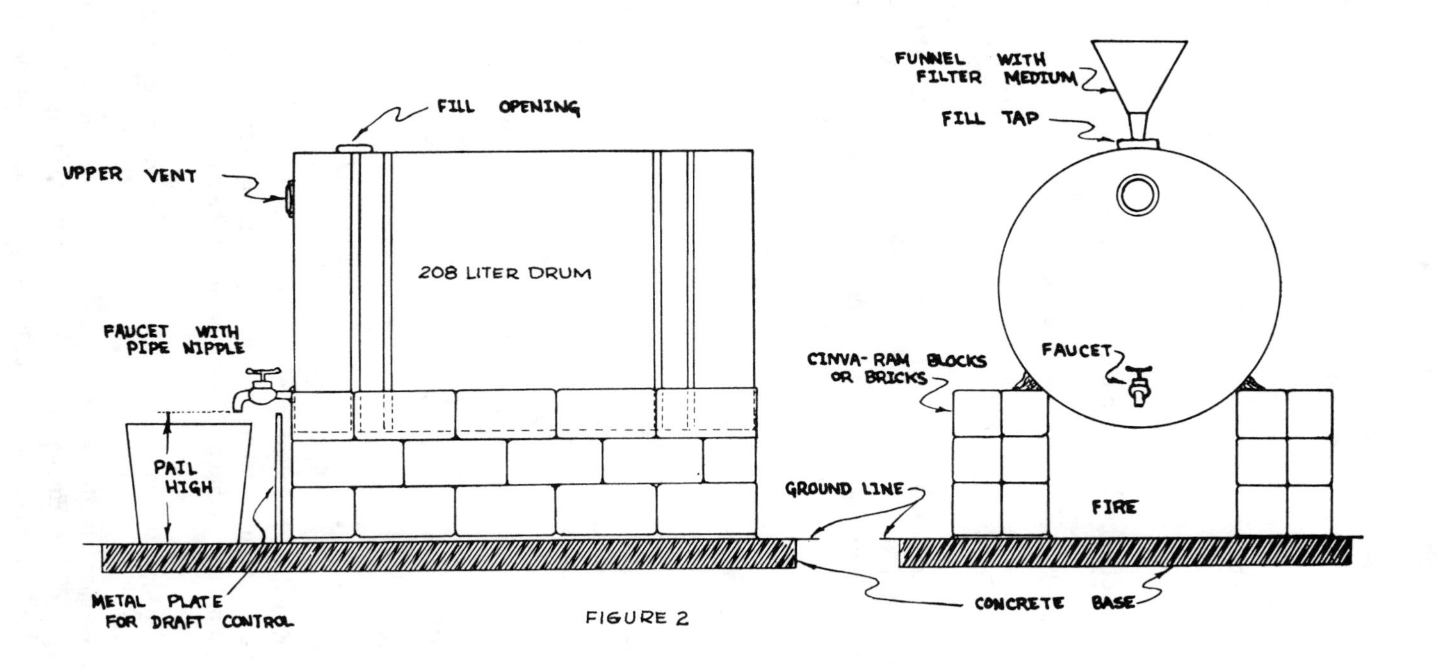
FILL OPENING
UPPER VENT
208 LITER DRUM
FAUCET WITH PIPE NIPPLE
PAIL HIGH
METAL PLATE FOR DRAFT CONTROL
FUNNEL WITH FILTER MEDIUM
FILL TAP
CINVA-RAM BLOCKS OR BRICKS
FAUCET
GROUND LINE
FIRE
CONCRETE BASE

FIGURE 2

When filling the drum, do not fill it completely, but leave an air space at the top as shown in Figure 1. Replace the funnel with a filler plug, but leave the plug completely loose.

Water must boil at least 15 minutes with steam escaping around the loose filler plug. Make sure that the water in the pipe nipple and valve reach boiling temperature by letting about 2 liters (2 quarts) of water out through the valve while the drum is at full boil.

**Source:**

Chris Ahrens, VITA Volunteer, Swannanoa, North Carolina

# Chlorinating Wells, Springs, and Cisterns

Chlorination, when properly applied, is a simple way to ensure and protect the purity of water. Guidelines given here include tables to give a rough indication of the amounts of chlorine-bearing chemical needed. Instructions are also given for super-chlorination for disinfecting newly built or repaired wells, spring encasements, or cisterns. Chlorine-bearing compounds, such as ordinary laundry bleach made with chlorine are used because pure chlorine is difficult and dangerous to use.

## *Determining the Proper Amount of Chlorine*

The amounts of chlorine suggested here will normally make water reasonably safe. A water-treatment system should be checked by an expert. In fact, the water should be tested periodically to make sure that it remains safe. Otherwise, the system itself could become a source of disease.

**Tools and Materials**

Container to mix chlorine
Chlorine in some form
Scale to weigh additive

The safest way to treat water for drinking is to boil it (see "Boiler for Drinking Water"). However, under controlled conditions, chlorination is a safe method; it is often more convenient and practical than boiling. Proper treatment of water with chlorine requires some knowledge of the process and its effects.

When chlorine is added to water, it attacks and combines with any suspended organic matter as well as some minerals such as iron. There is always a certain amount of dead organic matter in water, as well as live bacteria, viruses, and perhaps other types of life. Enough chlorine must be added to oxidize all of the organic matter, dead or alive, and to leave some excess uncombined or "free" chlorine. This residual free chlorine prevents recontamination. Too much residual chlorine, however, is harmful and extremely distasteful.

Some organisms are more resistant to chlorine than others. Two particularly resistant varieties are amoebic cysts (which cause amoebic dysentery) and the cercariae of schistosomes (which cause bilharziasis or schistosomiasis). These, among others, require much higher levels of residual free chlorine and longer contact periods than usual to be safe. Often special techniques are used to combat these and other specific diseases.

It always takes time for chlorine to work. Be sure that water is thoroughly mixed with an adequate dose of the dissolved chemical, and that it **stands for at least 30 minutes before consumption.**

Polluted water that contains large quantities of organic matter, or cloudy water, is not suitable for chlorination. It is best, and safest, to choose the clearest water available. A settling tank and simple filtration can help reduce the amount of suspended matter, especially particles large enough to see. Filtration that can be depended upon to remove all of the amoebic cysts, schistosomes, and other parthogens normally requires professionals to set up and operate.

**NEVER** depend on home-made filters alone to provide drinking water. However, a home-made slow sand filter is an excellent way to prepare water for chlorination.

Depending on the water to be treated, varying amounts of chlorine are needed for adequate protection. The best way to control the process is to measure the amount of free chlorine in the water after the 30 minute holding period. A simple chemical test, which uses a special organic indicator called **orthotolidine**, can be used. Orthotolidine testing kits available on the market come with instructions on their use.

When these kits are not available, the chart in Table 1 can be used as a rough guide to how strong a chlorine solution is necessary. The strength of the solution is measured in parts by weight of active chlorine per million parts by weight of water, or "parts per million" (ppm).

The chart in Table 2 gives the amount of chlorine-compound to add to 1,000 liters or to 1,000 gallons of water to get the solutions recommended in Table 1.

Usually it is convenient to make up a solution of 500 ppm strength that can then be further diluted to give the chlorine concentration needed. The 500 ppm solution must be stored in a sealed container in a cool dark place, and should be used as quickly as possible since it does lose strength. Modern chlorination plants use bottled chlorine gas, but this can only be used with expensive machinery by trained experts.

## TABLE 1

### INITIAL CHLORINE DOSE TO SAFEGUARD DRINKING WATER SUPPLY*

| Water Condition | Initial Chlorine Dose in Parts Per Million (ppm) | |
|---|---|---|
| | No hard-to-kill organism suspected | Hard-to-kill organisms present or suspected |
| Very Clear, few minerals | 5 ppm | Get expert advice; in an emergency boil and cool water first, then use 5 ppm to help prevent recontamination. If boiling is impossible, use 10 ppm. |
| A coin in the bottom of 1/4 liter (8 ounce) glass of the water looks hazy. | 10 ppm | Get expert advice; in an emergency boil and cool first. If boiling is impossible use15 ppm. |

* Parts per million (ppm) is the number of parts by weight of chlorine to a million parts by weight of water. It is equivalent to milligrams of per liter.

## TABLE 2

### AMOUNTS OF CHLORINE COMPOUND TO ADD TO DRINKING WATER

| Chlorine Compound | Percent by Weight Active Chlorine | Quantity to add to 1000 U.S. gallons of water required strength | | | Quantity to add to 1000 liters to get required strength | | |
|---|---|---|---|---|---|---|---|
| | | 5 PPM | 10 PPM | 15 PPM | 5 PPM | 10 PPM | 15 PPM |
| High test Calcium Hypochlorite $Ca(OCl)_2$ | 70% | 1 oz | 2 oz | 3 oz | 8 gms | 15 gms | 23 gms |
| Chlorinated Lime | 25% | 2 1/2 oz | 5 oz | 7 1/2 oz | 20 gms | 40 gms | 60 gms |
| Sodium hypochlorite NaOCl | 14% | 5 oz | 10 oz | 15 oz | 38 gms | 75 gms | 113 gms |
| Sodium hypochlorite | 10% | 7oz | 13 oz | 20 oz | 48 gms | 95 gms | 143 gms |
| Bleach—A Solution of Chlorine in water | usually 5.25% | 13 oz | 26 oz | 39 oz | 95 gms | 190 gms | 285 gms |

## *Super-Chlorination*

Super-chlorination means applying a dose of chlorine that is much stronger than the dosage needed to disinfect water. It is used to disinfect new or repaired wells, spring encasements, and cisterns. Table 3 gives recommended doses.

**TABLE 3**

**RECOMMENDED DOSES FOR SUPER-CHLORINATION***

| Application | Recommended Dose | Procedure |
|---|---|---|
| New or repaired well | 50 ppm | 1. Wash casing, pump exterior and drip pipe with solution.<br>2. Add dosage to water in well.<br>3. Pump until water coming from pump has strong chlorine odor for deep wells, repeat this a few times at 1 hour intervals.)<br>4. Leave solution in well at least 24 hours.<br>5. Flush all chlorine from well. |
| Spring encasements | 50 ppm | Same as above. |
| Cisterns | 100 ppm | 1. Flush with water to remove any sediment.<br>2. Fill with dosage.<br>3. Let stand for 24 hours.<br>4. Test for residual chlorine. If there is none, repeat dosage.<br>5. Flush system with treated water. |

* To find the correct amounts of chlorine compound needed for the required dosage, multiply the amounts given under 10ppm in Tables 2 or 3 to get 50ppm and by 10 to get 100ppm.

Example 1:

A water-holding tank contains 8,000 U.S. gallons. The water comes from a rapidly moving mountain stream and is passed through a sand filter before storage. How much bleach should be added to make this water drinkable? How long should the water be mixed after adding?

Solution:

In this case 5 ppm are probably sufficient to safeguard the water. To do this with bleach requires 13 ounces per 1,000 gallons. Therefore the weight of bleach to be added is 13 x 8 or 104 ounces.

Always mix thoroughly, for at least a half hour. A good rule of thumb is to mix until you are certain that the chemical is completely dissolved and distributed and then ten minutes longer. In this case, with an 8,000-gallon tank, try to add the bleach to several different locations in the tank to make the mixing easier. After mixing, test the water by sampling different locations, if possible. Check the corners of tank especially.

Example 2:

A new cistern has been built to hold water between rainstorms. On its initial filling it is to be super-chlorinated. How much chlorinated lime should be added? The cistern is 2 meters in diameter and 3 meters high.

Solution:

First calculate the volume of water. For a cylinder, Volume is $\frac{D^2}{4} = H$ (D is diameter, H is height and is 3.14.)

Here D = 2 meters H = 3 meters.

$$V = \frac{3.14}{4} \times (2 \text{ meters}) \times (2 \text{ meters}) \times (3 \text{ meters})$$

V = 9.42 cubic meters = 9,420 liters (Each cubic meter contains 1,000 liters.)

From Table 3 we learn that a cistern should be super-chlorinated with 100 ppm of chlorine. From Table 2, we learn that it takes 40 grams of chlorinated lime to bring 1,000 liters of water to 10 ppm C1. To bring it to 100 ppm, then, will require ten times this amount, or 400 grams.

$$\frac{400 \text{ grams}}{\text{thousand liters}} \times 9.42 \text{ thousand liters} = 3{,}768 \text{ grams.}$$

**Source:**

Salvato, J.S. *Environmental Sanitation.* New York: John Wiley & Sons, Inc., 1958

*Field Water Supply,* TM 5-700.

# Water Purification Plant

The water purification plant described here uses laundry bleach as a source of chlorine. Although this manually-operated plant is not as reliable as a modern water system, it will provide safe drinking water if it is operated according to instructions.

Many factors in this system require operating experience. When starting to use the system, it is safest to have the assistance of an engineer experienced in water supplies.

**Tools and Materials**

3 barrels, concrete tanks, or 208 liter (55-gallon) drums
20cm (8") funnel, or sheet metal to make a funnel
2 tanks, about 20 liters (5 gallons) in size
4 shut-off valves
Throttle or needle valve (clamps can be used instead of valves if hose is used)
Pipe or hose with fittings
Hypochlorite of lime or sodium hypo-chlorite (laundry bleach)

The water purification plant is made as in Figure 3. The two tanks at the top of the structure are for diluting the bleach. (The system can be simplified by eliminating the concentrate tank; the bleach is then added directly to the mixing tank.)

The two smaller tanks on the shelf below are for holding equal amounts of diluted bleach solution and water at a constant pressure; this makes the solution and the water flow at the same speed into the hoses that lead to the mixing point. The mix, which can be seen through the open funnel, is further controlled by the valves. If a needle or throttle valve is not available a throttle action can be obtained by installing another shut-off valve in series with Valve #4.

Placing the two barrels at a height of less than 1.8 meters (6') above the float valve causes a pressure of less than 0.35kg per square centimeter (5 pounds per square inch). Thus, the plumbing does not have to be of high quality except for Valve #1 and the float valve of the water hold-up tank, if the water supply is under higher pressure.

A trial and error process is necessary to learn how much concentrate should be put in the concentrate tank, how much concentrate should flow into the mixing tank, and how much solution should be allowed past the funnel. A suggested starting mixture is 1/4 liter (1/2 pint) of concentrated bleach for a mix tank capacity of 190 liters (50 gallons) to treat 1,900 liters (500 gallons) of water.

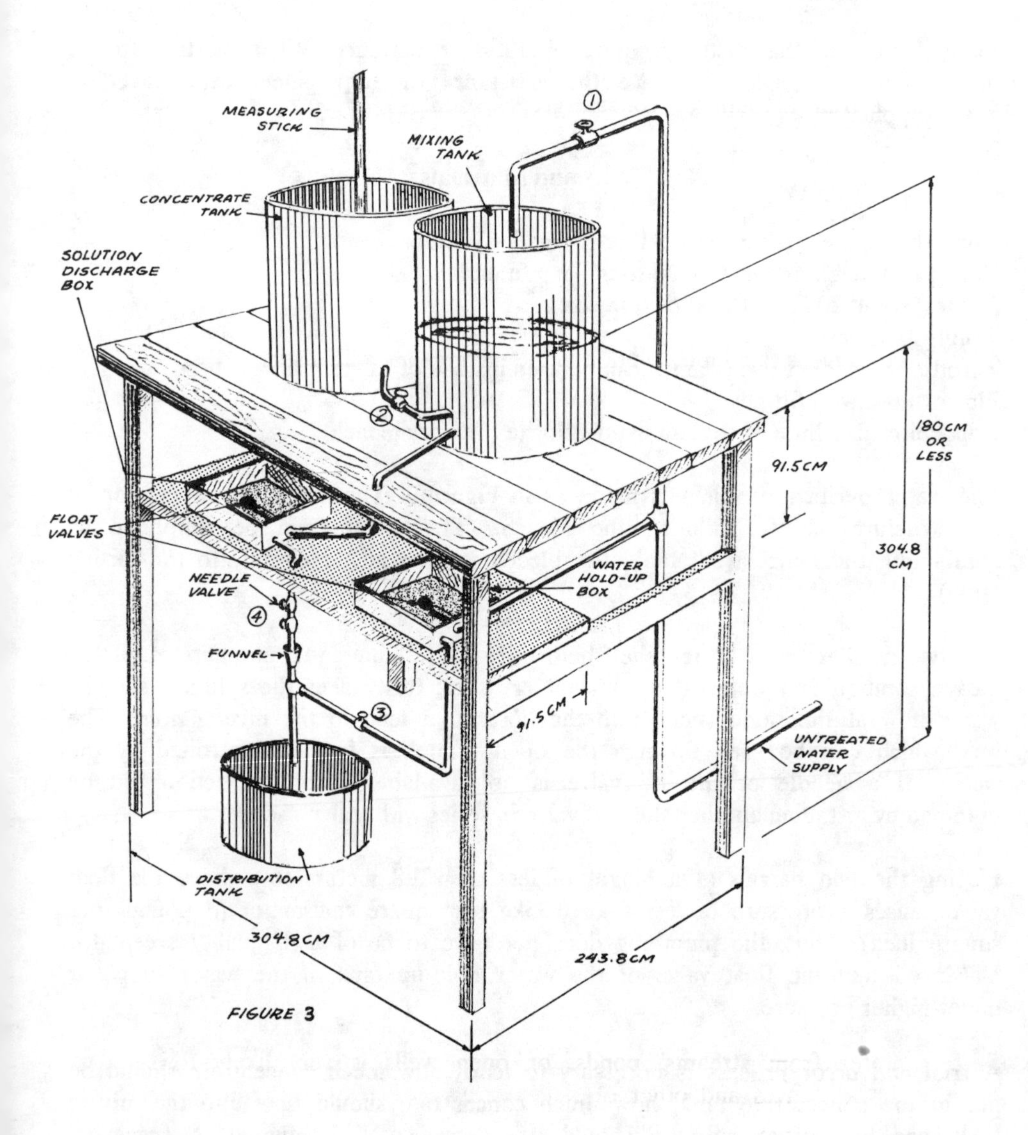
MEASURING STICK
MIXING TANK
CONCENTRATE TANK
SOLUTION DISCHARGE BOX
FLOAT VALVES
NEEDLE VALVE
FUNNEL
WATER HOLD-UP BOX
91.5 CM
180 CM OR LESS
304.8 CM
91.5 CM
UNTREATED WATER SUPPLY
DISTRIBUTION TANK
304.8 CM
243.8 CM

FIGURE 3

The water in the distribution tank should have a noticeable chlorine taste. The amount of bleach solution required depends on how dirty the water is.

1. Mix concentrated bleach with water in the concentrate tank with all valves closed. The mixing tank should be empty.

2. Fill the pipe from the mixing tank to the solution tank with water after having propped the float valve in a closed position.

3. Let a trial amount of concentrate flow into the mixing tank by opening Valve #2.

4. Use a measuring stick to see how much concentrate was used.

5. Close Valve #2 and open Valve #1 so that untreated water enters the mixing tank.

6. Close Valve #1 and mix solution in the mixing tank with a stick.

7. Remove the prop from the float valve of the solution tank so that it will operate properly.

8. Open wide the needle valve and Value #4 to clean the system. Let 4 liters (1 gallon) drain through the system, if the pipe mentioned in the second step is not permitted to empty before recharging the mixing tank.)

9. Close down to needle valve until only a stream of drops enter the funnel.

10. Open valve #3.

The flow into the funnel and the taste of the water in the distribution tank should be checked regularly to ensure proper treatment.

**Source:**

Chris Ahrens, VITA Volunteer, Swannanoa, North Carolina

# Sand Filter

Surface water from streams, ponds, or open wells is very likely to be contaminated with leaves and other organic matter. A gravity sand filter can remove most of this suspended organic material, but it will always let rivus and some bacteria pass through. For this reason, it is necessary to boil or chlorinate water after it has been filtered.

By removing most of the organic matter, the filter:

- o Removes large worm eggs, cysts, and cercariae, which are difficult to kill with chlorine.
- o Allows the use of smaller and fixed doses of chlorine for disinfection, which results in drinkable water with less taste of chlorine.
- o Makes the water look cleaner.
- o Reduces the amount of organic matter, including living organisms and their food, and the possibility of recontamination of the water.

Although sand filtration **does not** make polluted water safe for drinking, a properly built and maintained filter will make chlorination more effective. Sand filters must be cleaned periodically.

The household sand filter described here should deliver 1 liter (1 quart) per minute of clear water, ready for boiling or chlorinating.

**Tools and Materials**

Steel drum: at least 60cm wide by 75cm (2' x 29 1/2")
Sheet metal, for cover: 75cm (29 1/2") square
Wood: 5cm x 10cm (2" x 4"), 3 meters (9.8') long
Sand: 0.2 cubic meter (7 cubic feet)
Gravel
Blocks and nails
Pipe, to attach to water supply
Optional: valve and asphalt roofing compound to treat drum

The gravity sand filter is the easiest type of sand filter to understand and set up. It uses sand to strain suspended matter from the water, although this does not always stop small particles or bacteria.

Over a period of time, a biological growth forms in the top 7.5cm (3") of sand. This film increases the filtering action. It slows the flow of water through the sand, but it traps more particles and up to 95 percent of the bacteria. The water level must always be kept above the sand to protect this film.

Sand filters can get partially clogged with organic matter; under some conditions this can cause bacterial growth in the filter. If the sand filter is not operated and maintained correctly, it can actually add bacteria to the water.

The drum for the sand filter shown in Figure 4 should be of heavy steel. It can be coated with asphalt material to make it last longer.

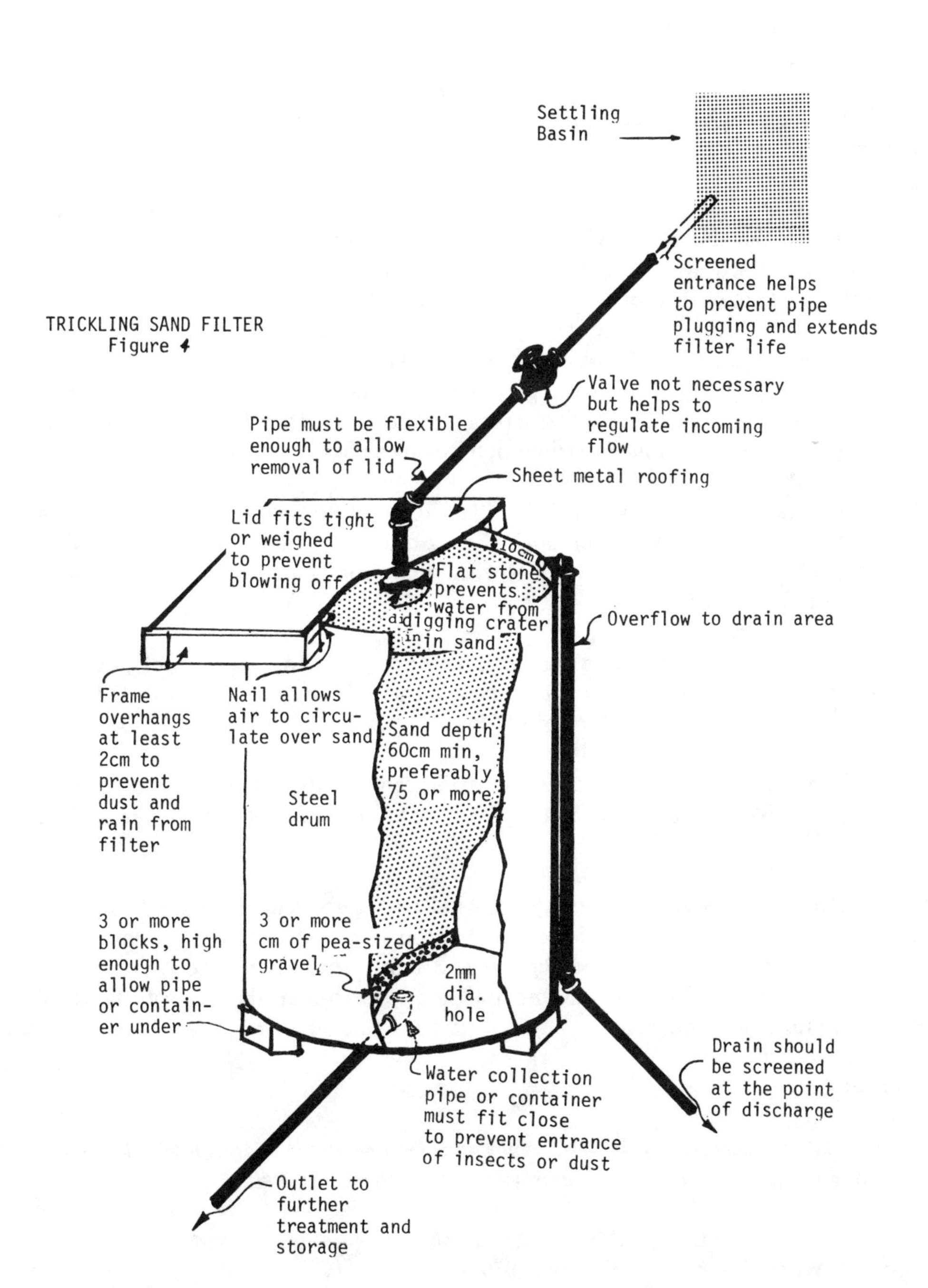

Settling Basin
Screened entrance helps to prevent pipe plugging and extends filter life
TRICKLING SAND FILTER
Figure 4
Valve not necessary but helps to regulate incoming flow
Pipe must be flexible enough to allow removal of lid
Sheet metal roofing
Lid fits tight or weighed to prevent blowing off
10cm
Flat stone prevents water from digging crater in sand
Overflow to drain area
Frame overhangs at least 2cm to prevent dust and rain from filter
Nail allows air to circulate over sand
Sand depth 60cm min, preferably 75 or more
Steel drum
3 or more blocks, high enough to allow pipe or container under
3 or more cm of pea-sized gravel
2mm dia. hole
Water collection pipe or container must fit close to prevent entrance of insects or dust
Drain should be screened at the point of discharge
Outlet to further treatment and storage

The 2mm (3/32") hole at the bottom regulates the flow: it **must not** be made larger.

The sand used should be fine enough to pass through a window screen. It should also be clean; it is best to wash it.

The following points are very important in making sure that a sand filter operates properly.

- Keep a continuous flow of water passing through the filter. Do not let the sand dry out, because this will destroy the film of microorganisms that forms on the surface layer of sand. The best way to ensure a continuing flow is to set the intake so that there is always a small overflow.

- Screen the intake and provide a settling basin to remove as many particles as possible before the water goes into the filter. This will keep the pipes from becoming plugged and stopping the flow of water. It will also help the filter to operate for longer periods between cleanings.

- Never let the filter run faster than 3.6 liters per square meter per minute (4 gallons per square foot per hour) because a faster flow will make the filter less efficient by keeping the biological film from building up at the top of the sand.

- Keep the filter covered so that it is perfectly dark to prevent the growth of green algae on the surface of the sand. But let air circulate above the sand to help the growth of the biological film.

- When the flow becomes too slow to fill daily needs, clean the filter: Scrape off and discard the top 1/2cm (1/4") of sand and rake or scratch the surface lightly.

After several cleanings, the sand layer should be returned to its original thickness by adding clean sand. Before doing this, scrape the sand in the filter down to a clean level. The filter should not be cleaned more often than once every several weeks or even months, because the biological growth at the top of the sand makes the filter more efficient.

**Source:**

Hubbs, S.A. *Understanding Water Supply and Treatment for Individual and Small Community Systems.* Arlington, Virginia: VITA Publications, 1985.

Wagner, E.G. and Lanoix, J.N. *Water Supply for Rural Areas and Small Communities.* World Health Organization, 1959.

# Health And Sanitation

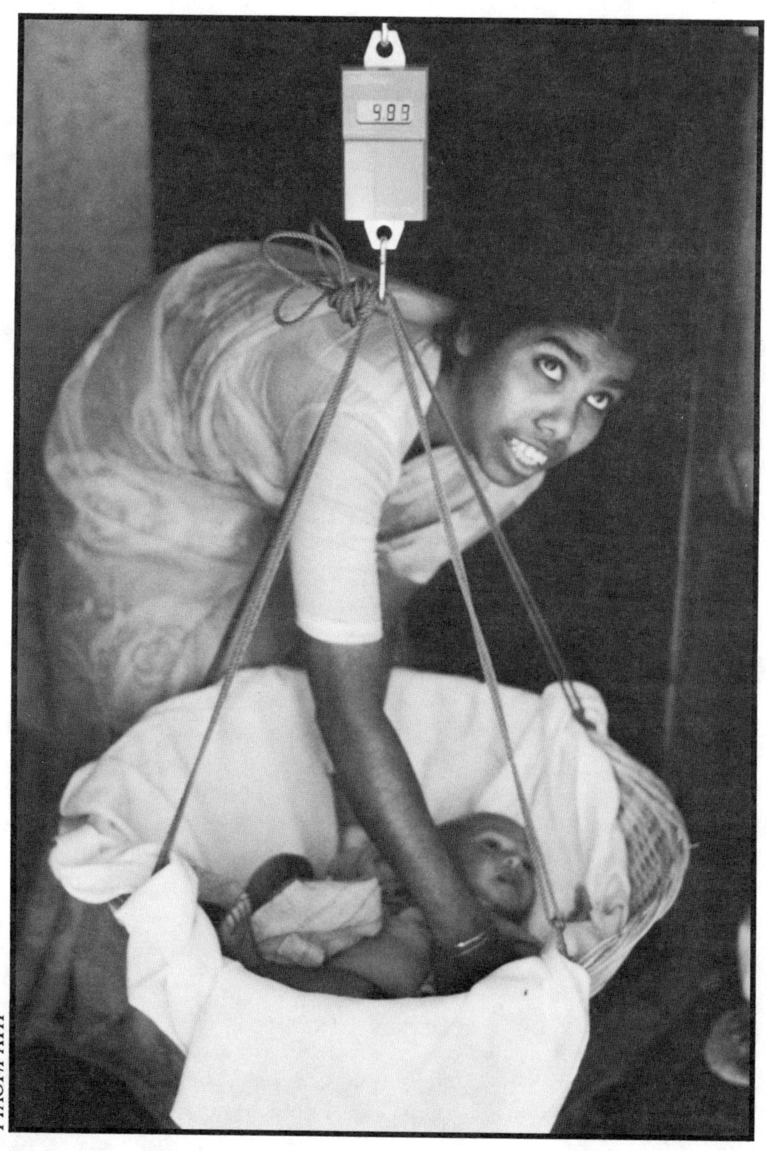


PIACT/PATH

# Sanitary Latrines

## OVERVIEW

The proper disposal of human waste (called night soil in many parts of the world) is one of the most pressing public health problems in many rural communities. The use of sanitary latrines or privies can be very effective in helping to control disease, which can be spread by water, soil, insects, or dirty hands. While it is necessary to have a sanitary water and food supply, sufficient medical service, and adequate diet to stop disease, the sanitary latrine breaks the disease cycle. Some sicknesses that can be controlled by widespread use of sanitary latrines are dysentery, cholera, typhoid, and worms. The human suffering and economic loss caused by these is staggering. It has been said that half of the food eaten by a person with intestinal parasites is consumed by the very worms that make the person sick.

Most countries that have actively participated in the 1980-90 U.N. Decade of Water Supply and Sanitation have developed latrine designs to meet the sanitary and cultural requirements of their people. Before building latrines the local health or development agency should be contacted for their advice and help. A latrine program must reach most or all of the people. This means a carefully planned, continuing long-range program with participation by government agencies, community leaders and most of all by the individual families. Proper latrine designs that fit the cultural pattern are economically possible and can satisfy the sanitary needs of a successful latrine program. Selected plans and designs for sanitary latrines are given in the entries that follow.

The recommended kinds of privies are:

- Pit privies: a simple hole in the ground, covered with a properly built floor and a shelter. It has two forms, the dry pit, which does not penetrate the water table, and the wet pit, which does. The addition of a ventilating pipe (see "The Ventilated Pit," page 156 helps reduce odors and fly problems.

- Water privies: where a watertight tank receives the nightsoil through a drop pipe or chute. An overflow pipe takes the digested material to an underground seepage pit or drainage area.

A water-seal slab may be used to cover either of these types of privies to provide a completely odorless privy.

Other types of simple latrines are not recommended for general use, because they usually fail to provide enough sanitary protection.

A good privy should fulfill the following conditions:

- o It should not contaminate the surface soil.
- o There should be no contamination of ground water that can enter springs or wells.
- o There should be no contamination of surface water.
- o Nightsoil should not be accessible to flies or animals.
- o There should be no handling of fresh nightsoil; if it is necessary, it should be handled as little as possible.
- o There should be no odors or unsightly conditions.
- o The latrine should be simple and inexpensive to build and use.

Other points to consider:

- o Superstructure can be made from any local building material that will give privacy and shelter from rain.
- o The privy can be squat or sit-down type.
- o The opening should be covered when not in use.
- o In water scarce areas, a standard pit latrine can be used. When pit is full after several years, latrine is moved to a new pit and old one is covered up and marked.
- o If space is limited to change the pit, a permanent location can be maintained with a double pit, as in the double septic tank (composting latrine) used in Vietnam. The urine is collected separately and diluted for use on crops. The composted material is used for fertilizer. One side is used until almost full, then it composts while the other side is used.
- o If water is readily available, a water-seal bore hole latrine can be used. When almost full, the latrine must be moved.
- o If a permanent location is desired, a double bore hole can be used as in India.

- o In most countries using water seal latrines the pan and trap are now available commercially or from a government agency for a nominal fee or for free.

- o Consider including a methane (biogas) generator when building new latrines.

**Source:**

Charles D. Spangler, VITA Volunteer, Bethesda, Maryland

Wagner, E.G. and Lanoix, J.N. *Excreta Disposal for Rural Areas and Small Communities.* Geneva: World Health Organization, 1958.

# Privy Location

Outhouses or privies should be close to the home, but they should be lower than water sources and far enough away from these sources that they will not pollute the water.

The information given here covers most normal situations, but it is always best to have a trained sanitary inspector or engineer review your installation or program.

A latrine site should be dry, well-drained, and above flood level.

If the bottom of a privy pit is in dry soil and at least 3 meters (10') above the highest water table, there is very little danger that it will contaminate water supplies. This is because the pollution will move downward no more than 3 meters with only 1 meter (3.3') of side movement. (See section on "Ground Water"). If the privy pit enters the water table or comes close to it when the water is at its highest level, pollution will spread to the ground water over a limited area and may endanger health.

Figure 1 shows the movement of pollution through the soil. It is particularly important to understand this movement when choosing a site for a privy or well. Put the privy downhill from a water source, or as far to one side as possible. On flat or gently sloping land, water moves toward the well as though it were going downhill. This is because when water is removed from a well, water from the surrounding soil flows toward it. Thus pollution from a nearby privy would move toward the well. If the land is flat or if the well is downhill from the privy, do not put the privy closer to the well than 10 meters (33'). In sandy soil, a distance of 7.5 meters (25') is sometimes enough because sand helps to stop bacterial pollution.

These rules do not apply in regions containing fissured rocks or limestone formation. Expert advice is necessary in these cases, because pollution can be carried great distances through solution channels to the drinking water supply.

MOVEMENT OF POLLUTION IN UNDERGROUND WATER

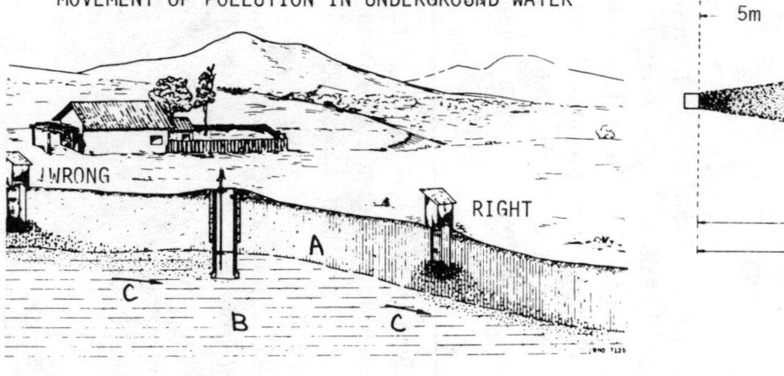


A = Top Soil
B = Water-bearing formation
C = Direction of ground-water flow

BACTERIAL AND CHEMICAL SOIL POLLUTION PATTERNS AND MAXIMUM MIGRATIONS*

Bacterial soil pollution pattern

5m
6m
2M
9m
25m
70m

Chemical soil pollution pattern

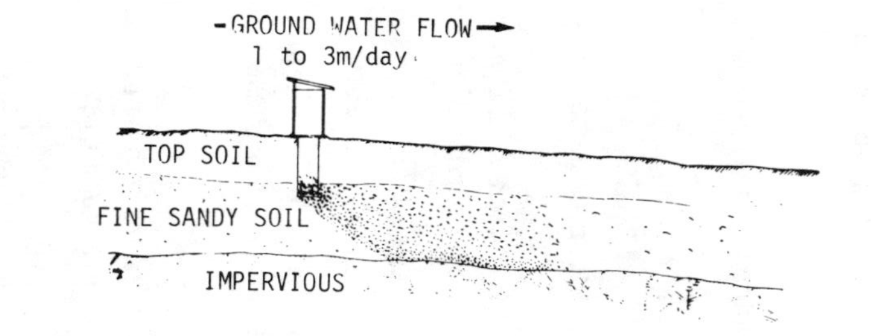


FIGURE 1

The source of contamination in these studies was human excreta placed in a hole which penetrated the ground-water table. Samples positive for coliform organisms were picked up quite soon between 4m and 6m (13ft and 19ft) from the source of contamination. The area of contamination widened out to a width of approximately 2m (7ft) at a point about 5m (16ft) from the privy and tapered off at about 11m (36ft). Contamination did not move "upstream" or against the direction of flow of the ground water. After a few months the soil around the privy became clogged, and positive samples could be picked up at only 2m to 3m (7ft to 10ft) from the pit.

The chemical pollution pattern is similar in shape to that of bacterial pollution but extends to much greater distances.

From the point of view of sanitation, the interest is in the maximum migrations and the fact that the direction of migration is always that of the flow of ground water. In locating wells, it must be remembered that the water within the circle of influence of the well flows towards the well. No part of the area of chemical or bacterial contamination may be within reach of the circle of influence of the well.

It is important to keep the latrine close to the house so that it will be used, but not too close. Putting the privy downhill also encourages use. People are more apt to keep a privy clean if it is close to the house.

Remember that all privies have to be closed up or moved when filled. This should be made easy or there will be a tendency to let them become overfull, which results in very unsanitary conditions and extra work to put the system in proper working order. A permanent location can have two pits that are used alternately. One pit is in use while the other composts before being emptied.

**Source:**

Wagner, E.G. and Lanoix, J.N. *Excreta Disposal for Rural Areas and Small Communities.* Geneva: World Health Organization, 1958.

## Privy Shelters

Several designs for privy shelters that have been found satisfactory in many parts of the world are shown in Figure 2.

The shelter should be built to suit the abilities and desires of the local people, because sanitary precautions are less important for the shelter than for the pit and slab. For a properly built shelter:

- Choose a standardized design for economy in building.
- Build the shelter to last as long as the pit, 8 to 15 years.
- Build the shelter to fit the floor slab. It should not be so large that people will be tempted to use any part of the floor when the area around the hole has been soiled by earlier users. The roof should be 2m (6 1/2') high at the entrance.
- Openings at the top of the shelter's walls, for airing the interior, should be 10cm to 15cm (4" to 6") wide.
- Some natural light should be let in, but the structure should give enough shade over an uncovered seat or holes that flies will not be attracted.
- The latrine should be kept neat and clean so that people will continue to use it. Paint or whitewash the shelter. Cut back nearby vegetation. The roof should have a large overhang to protect the walls and the mound from rain damage and to keep the privy area from getting muddy.

COMPLETED PRIVY, SHOWING PALM THATCH WALL AND ROOF COVERING

WATTLE HOUSE WITH PALM THATCH ROOF

HOUSE OF CUT LUMBER WITH CORRUGATED METAL OR ASBESTOS CEMENT ROOF

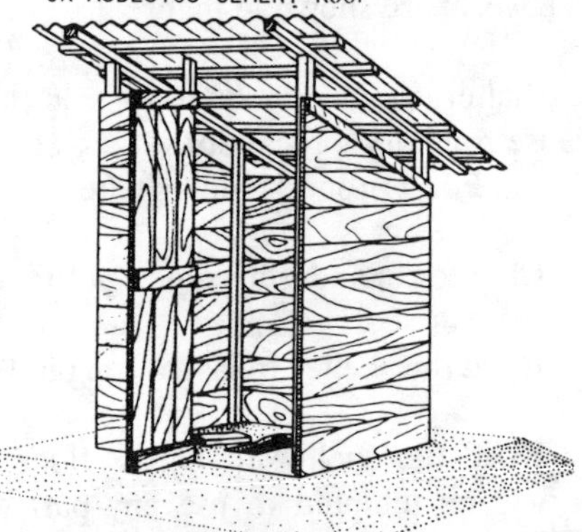

*FIGURE 2*

TYPE OF SUPERSTRUCTURE RECOMMENDED BY US PUBLIC HEALTH SERVICE

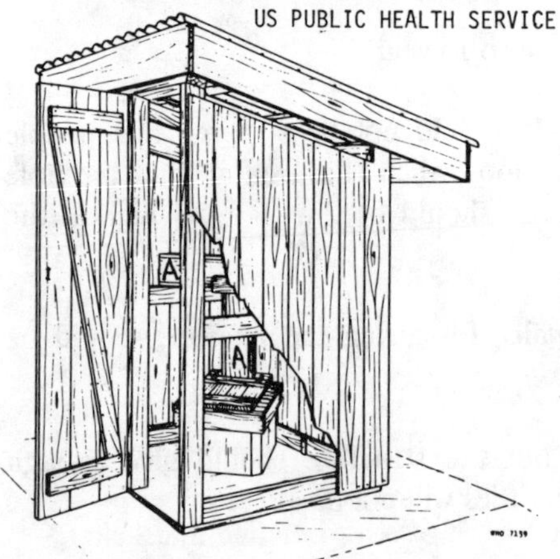


A = Vent pipe with lateral outlet

Adapted, by permission, from United States Public Health Service (1933) *The sanitary privy*, Washington, D.C. (Revised type No. IV of *Publ. Hlth Rep. (Wash.)*, Suppl.108).

HOUSE OF BRICK WITH TILE ROOF

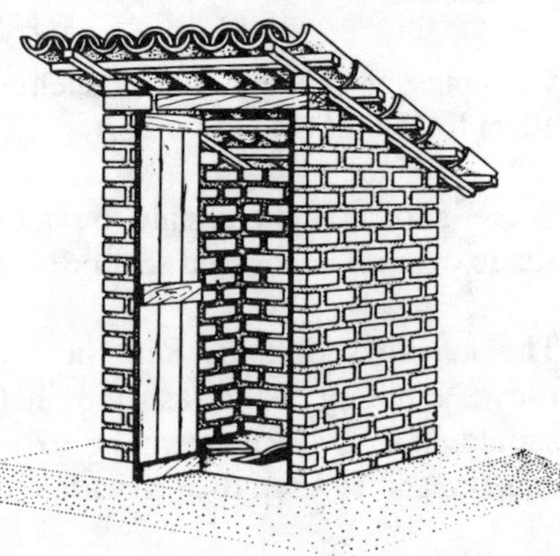

Here is a list of tools and materials needed to build one type of privy shelter:

**Tools and Materials**

Corrugated sheet metal roofing: 1.2m x 1.2m (4' x 4') or larger
Wooden posts: 5cm x 5cm (2" x 2") and 20m (66') long
Boards: 2cm (3/4") thick, 20cm (8") wide, 40m (132') long
Nails
Hand tools
Paint: 2 liters (2 quarts)

**Source:**

Wagner, E.G. and Lanoix, J.N. *Excreta Disposal for Rural Area and Small Communities.* Geneva: World Health Organization, 1958.

# PRIVY TYPES

## Pit Privy

FIG. 3 VARIOUS PARTS OF A SANITARY PRIVY

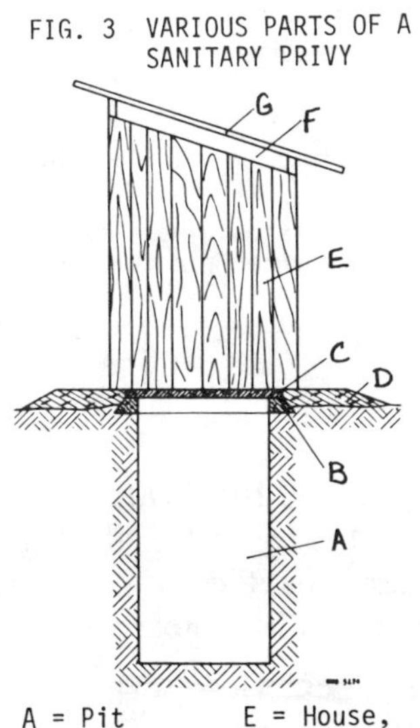


A = Pit
B = Base
C = Floor
D = Mound
E = House, including door
F = Ventilation
G = Roof

The pit privy is the simplest recommended latrine or privy. It consists of a hand-dug hole, a properly mounted slab, and a shelter (Figure 3). The addition of a ventilating pipe will help reduce odors and flies. Of the many existing designs for privies, the sanitary pit privy is the most widely applicable.

**Tools and Materials**

Materials for building the shelter

Hand tools for digging the pit, making concrete, and building the shelter

## The Pit

The pit is round or square, about 1m (3.3') in diameter or 1m (3.3') on each side, and usually from 1m (3.3') to 3m (10') deep. The pit may have to be lined with brick, wood, bamboo, or some other material to keep it from caving in, even in hard soil. The top 50cm (19 1/2") of the hole should be lined with mortar to make a solid base for the slab and the shelter.

The table in Figure 4 will help you to estimate the depth of hole to make.

| | | Estimated volume and depth* for hole with 1 square meter area — Personal Cleansing Material | | | |
|---|---|---|---|---|---|
| | | Water | | Solid (for example, grass or paper) | |
| Pit Type | Years of Service | Volume in cubic meters | Depth in meters | Volume in cubic meters | Depth in meters |
| Wet-Pit | 4 | 0.7 | 0.7 | 1.1 | 1.1 |
| | 8 | 1.5 | 1.5 | 2.3 | 2.3 |
| | 15 | 2.7 | 2.7 | 4.2 | 4.2 |
| Dry-Pit | 4 | 1.1 | 1.1 | 1.7 | 1.7 |
| | 8 | 2.3 | 2.3 | 3.4 | 3.4 |
| | 15 | 4.2 | 4.2 | - | - |

Figure 4. Privy capacities for a family of five. A wet-pit privy is one which penetrates the water table. A dry-pit privy does not.

*Add 50cm to the depth given in the table, because the pit should be closed and filled with earth when the waste comes to within this distance from the surface.

## The Ventilated Pit

The ventilated pit privy system was field tested during the late 1970s by the Blair Research Laboratories working with the Zimbabwe Ministry of Health (Figure 5). The idea was to reduce the health hazard caused by flies attracted to the standard pit privy. Thousands of the units are now in use in Zimbabwe, as well as in many other areas where water is scarce.

The Blair design depends on the aerodynamic properties of an efficient flue pipe, 150mm in diameter and about 2.5 meters high. The pipe is fitted onto the concrete latrine slab over a sealed tank or pit. The temperature difference between the inside and outside of the pipe causes a convection updraft, drawing the inside gases from the pit and thus causing a downdraft through the toilet opening.

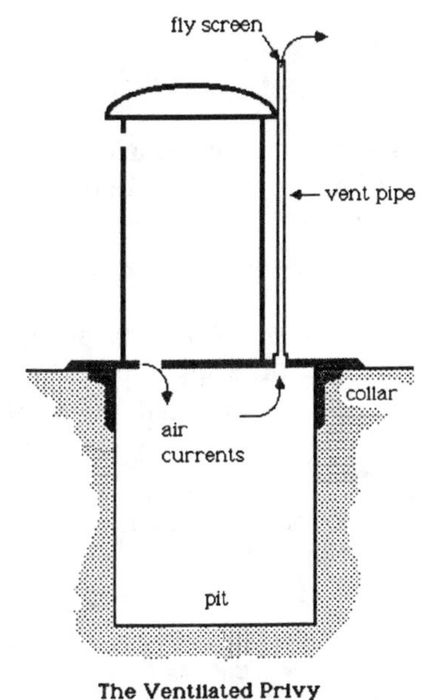


The Ventilated Privy

**FIGURE 5**

The toilet opening is kept closely covered between uses. Flies are then attracted to odors passing out the pipe rather than to the pit. Flies that do get into the pit travel up the pipe towards the light. There they are trapped by a screen over the pipe outlet.

It is essential that the pipe be large enough to enable the system to "breathe" efficiently and that it allow sufficient light to enter the pit to attract flies into the pipe. Efficiency is increased by painting the pipe black to increase the air flow and by facing it toward the Equator so it receives the most sunlight.

**Source:**

Cecelski, Elizabeth, "Appropriate Technology in Zimbabwe." *Energy Bulletin/VITA News,* July 1981.

## *The Base*

The base (see Figures 3, 6, and 7) serves as a solid, waterproof support for the floor. It also helps to prevent hookworm larvae from entering. If properly made of a hard, strong material, it helps to keep burrowing rodents and surface water out of the pit. The pit lining will in most cases serve as a base although it may need to be strengthened at the ground surface.

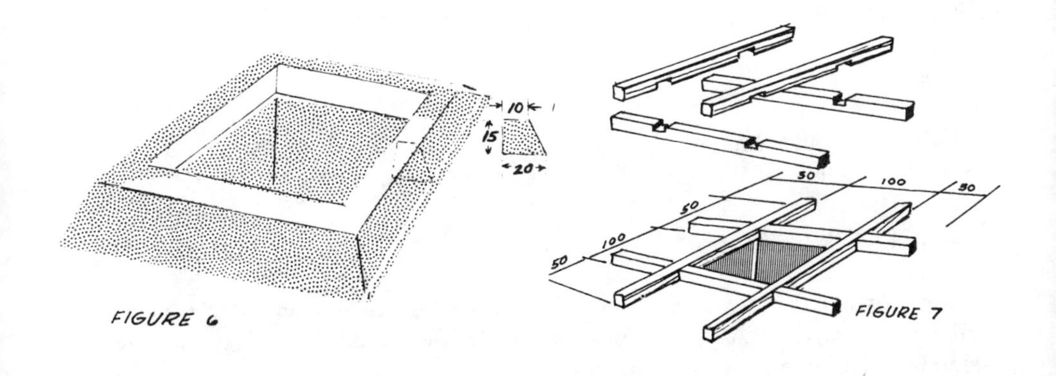


FIGURE 6

FIGURE 7

## The Slab

A concrete water-seal slab is best. It is inexpensive but it means added labor and construction. A concrete open-hole slab is the next best, while a wooden floor is adequate. A built-up floor of wood and compacted soil is sometimes used but it is difficult to keep clean; as it gets soiled, it is likely to spread hookworm.

The concrete should not be weaker than 1 part cement to 6 parts of aggregate with a minimum of water. It should be reinforced with strips of bamboo about 2.5cm (1") wide whose weaker fibers have been stripped away. Soak the bamboo in water overnight before use.

Slabs (see Figure 8) are cast upside down in one operation. The footrests are shaped by removing part of the wooden form so as to make two separate indentations in the wood. Sheet metal is placed around the form so that the metal extends above the wood to the thickness of the slab. Side walls of the hole and footrests are made with a slight slope so as to come out easily. The form for the open hole is removed when the concrete first sets. Slabs are removed from the forms in about 40 hours and should be stored under water for 10 days or more.

Round slabs can be rolled some distance when carrying is difficult. This is especially handy when the location of the privy has to be moved when the pit fills up.

## The Mound

The mound (see Figure 3) protects the pit and base from surface run-off that otherwise might enter and destroy the pit. It should be built up to the level of the floor and be very well tamped. It should extend 50cm (20") beyond the base on all sides. The mound may be built much higher than the ground in areas where protection is needed against floods and high tides. It will normally be built with earth removed from the pit or the surrounding area. A stone facing will help to keep it from being washed away by heavy rains. A masonry or brick step can be built in front of the entrance door to help keep the floor clean.

**Source:**

Wagner, E.G. and Lanoix, J.N. *Excreta Disposal for Rural Areas and Small Communities.* Geneva: World Health Organization, 1958.

# Water Privy

A water (or aqua) privy uses a watertight tank in which human excreta and urine partially decompose. A sewer pipe connects the tank's overflow pipe to an underground drain area or seepage pit.

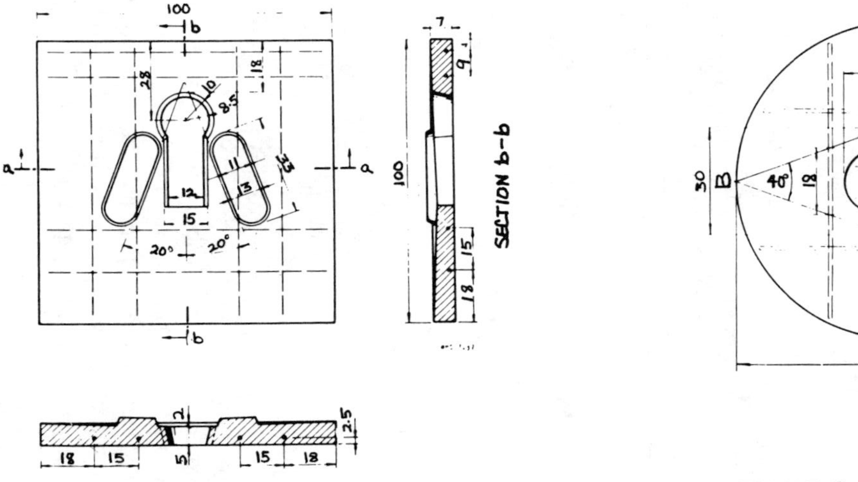


FIGURE 8

Measurements shown are in centimeters.

A = Centre open hole 2.5cm (1 in.) back of centre if slab is 80cm (31 in.) in diameter; centre open hole 8.0cm (3 in.) back of centre if slab is 90cm (35 in.) in diameter
B = Between back centre foot-rests
C = Reinforcement

Notes on construction of slab

Concrete for slabs should be not weaker than 1 part cement to 6 parts aggregate, with a minimum of water.

Slab is reinforced with strips of bamboo of timber quality. Reinforcing strips are about 2.5cm (1 in.) wide, have had inner, weaker fibres stripped away, and have been soaked in water overnight before use.

Slabs are cast upside down in one operation. Base of form is of wood with indentations for foot-rests. Base of form is encircled by sheet metal strip which makes outer wall of form. Side walls of hole form and foot-rests are made with slight slope so as to come out easily. Form for open hole is removed when concrete has taken initial set. Slabs are removed from form in about 40 hours and stored under water, preferably for 10 days or more. Since these slabs are round, they may be rolled some distance when conveyance is difficult.

This is a sanitary and permanent installation when it is properly built, used daily, and maintained properly. It can be placed near a building. The first cost of a water privy is high, but it is not expensive in the long run because it will be used for many years. It needs some water and cannot be used in freezing climates. And it is not practical in desert or water scarce areas. The water privy may not be successful in rural areas with no organized sanitation and health education services.

## The Process

The digesting or decomposing tank is usually made of watertight concrete (see Figures 9, 10, and 11). A drop-pipe, 10cm (4") in diameter, attached to the squatting plate or seat hangs down 10cm (4") below the surface of the liquid in the tank. This forms a water seal, which keeps bad odors from rising into the privy shelter.

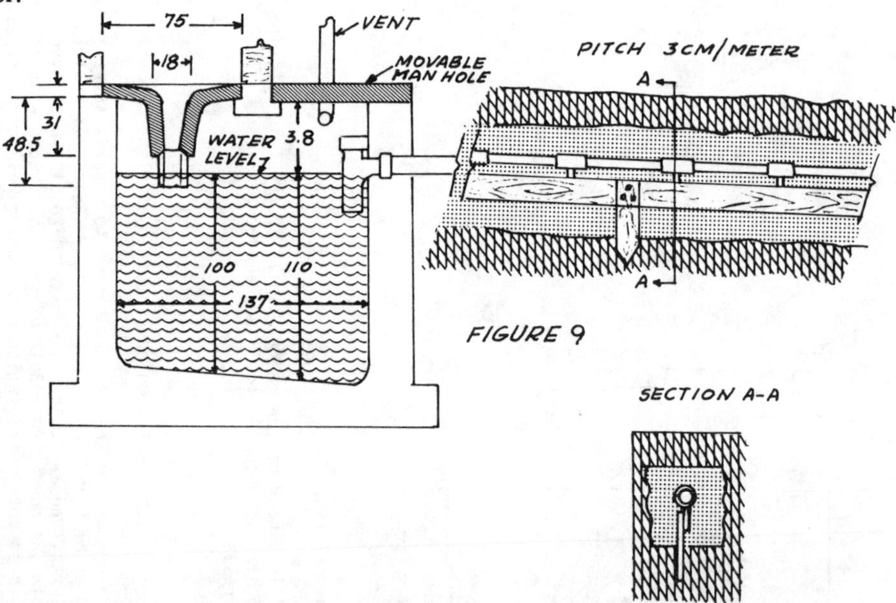


FIGURE 9

The decomposition process forms a sludge in the tank. The amount of sludge is only one-fourth the volume of the total waste deposited in the pit, because some of the solid matter breaks down into very small pieces, liquid, and gas. The liquid and the pieces of waste matter run out the overflow pipe to the drain field. The material that flows out is called *effluent.* The gas escapes through a vent pipe.

## The Tank

The tank must be watertight. If the tank leaks, the liquid level will fall below the drop pipe, odors will form, flies and mosquitos will breed, and the soil and ground water will be polluted. Tanks made from bricks or stone and mortar must be faced with a coat of rich cement plaster to make sure they are watertight.

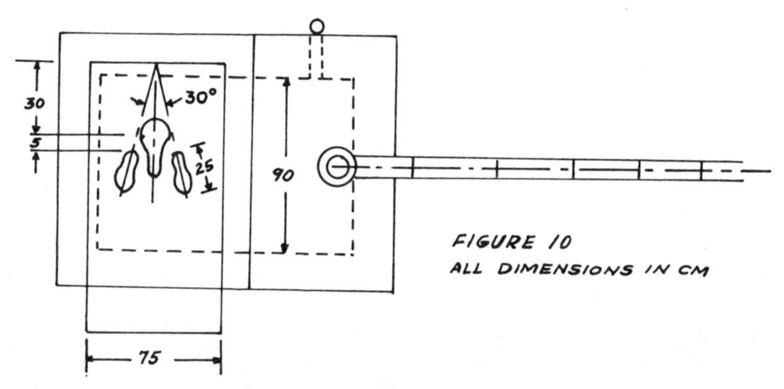


The tank can be made of plain concrete sewer pipes 90 or 120cm (36" to 47") in diameter and sealed at the bottom with concrete (see Figure 11).

Family-sized units should not be less than 1 cubic meter (35 cubic feet), which will usually allow 6 years or more between cleanings. Thus the family water privy need not be too deep, which is an advantage in rocky ground where the water table is high.

## *Drop-Pipe*

The 10cm drop-pipe with its end 10cm below the surface, prevents water from splashing and improves flushing. Nightsoil may stick in the pipe from time to time and must be flushed or poked down to stop odors and to keep flies from breeding. The pipe may be up to 20cm (8") in diameter and reach 20cm below the surface of the water in the pit, which will prevent sticking, but this size will release more odors and cause splashing, and the pipe may crust over.

## *Disposal of Effluent*

Disposal of effluent from a family unit is usually done in seepage pits or by below ground irrigation. The amount of effluent is equal to the amount of nightsoil and water put into the digesting pit. This averages 4.5 liters a person each day, but the drainage system should be designed to handle 9 liters a person each day. When a water tap is inside the privy, the effluent disposal system must be much larger. Too much water causes poor digestion of sludge.

The area of below ground irrigation ditches or seepage pits needed for a family of five will be from 1.4 square meters (10.7 square feet) in very light soil to 5 square meters (53 square feet) in soils that are hard to penetrate.

These methods are not practical in regions where the water table rises to within 1m (3') of the ground surface, or in clay soils or swampy land. Here some type of sand filter may help, but this requires help and approval from local health experts and continued maintenance.

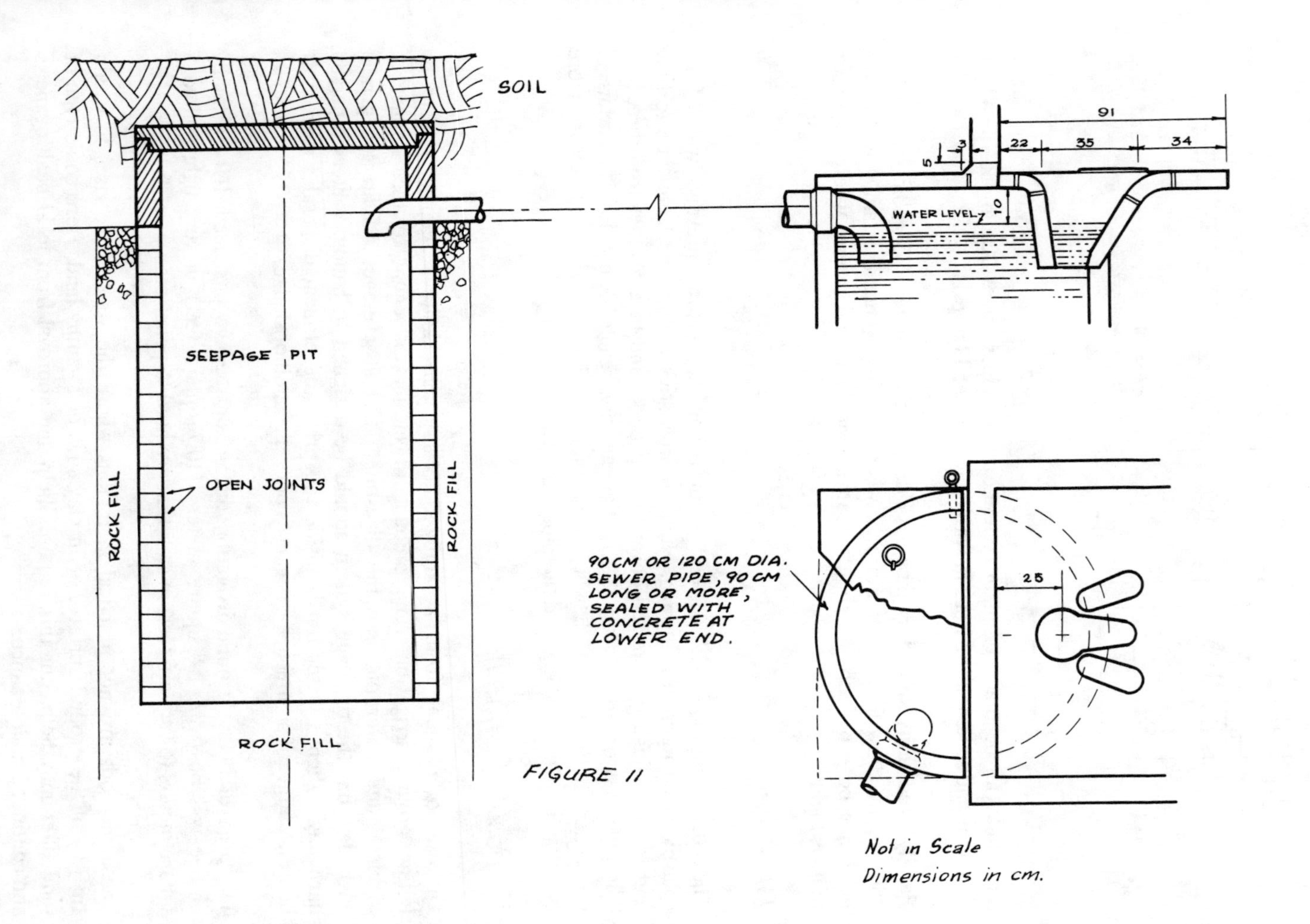
SOIL
SEEPAGE PIT
OPEN JOINTS
ROCK FILL
ROCK FILL
ROCK FILL
91
3
22
35
34
5
WATER LEVEL
10
90CM OR 120 CM DIA. SEWER PIPE, 90CM LONG OR MORE, SEALED WITH CONCRETE AT LOWER END.
25
Not in Scale
Dimensions in cm.

FIGURE 11

## Operation

The first step in putting a new water privy into operation is to fill the tank with water up to the overflow pipe. Digested sludge from another privy can be added to the tank; this will seed the water and start the decomposition process. If the tank is not seeded, it will take about 2 months for the process to get going efficiently. Once this level of operation is reached, the privy will keep the process going, provided it is used daily. Cleaning and flushing the slab and bowl daily with 25 to 40 liters (6 to 10 gallons) will give the tank the small amount of water it needs to keep the process going.

## Removing Sludge

The sludge that forms in the tank must be bailed out before the tank is half-full, about 6 to 8 years after the privy is put into operation. A manhole, often located outside the shelter, is made for this job.

Notice in Figure 9 that the tank floor slopes toward the manhole for easier cleaning. Both the vent and the drain are easily reached. The drain has a T-shaped section that helps to keep hard surface scum from entering and plugging the drain and whose shape makes it easy to clean. The overflow pipe in Figure 10 is an elbow.

Bury the sludge in shallow trenches about 40cm (16") deep.

**Source:**

Wagner E.G. and Lanoix, J.N. *Excreta Disposal for Rural Areas and Small Communities.* Geneva: World Health Organization, 1958.

# Philippine Water-Seal Latrine

A water-seal bowl improves a latrine by keeping flies out of the pit and preventing odors from escaping. The mold described here (see Figure 12) has been made and used successfully in sanitary latrine programs in the Philippines. The advantage of this mold over a concrete mold is that it requires no drying time.

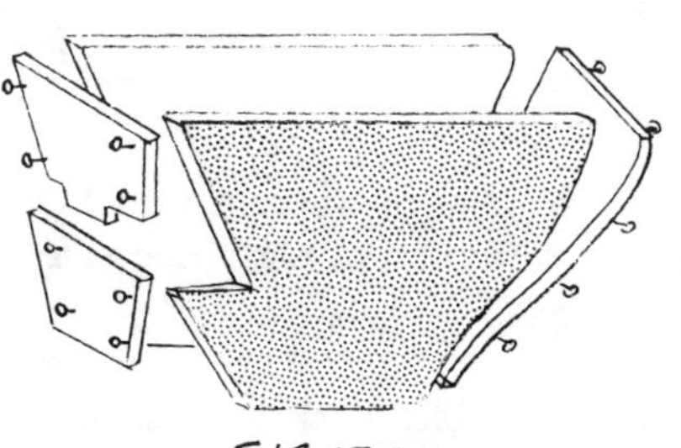

FIGURE 12

**Tools and Materials**

Wood: 19mm (3/4") thick, 31cm (12 1/2") wide and 152.5cm (5') long
Galvanized iron: 0.75mm x 32cm x 40.5cm (1/32" x 12 1/2" x 16")
Large nails: 18
Cement and clean sand
Galvanized wire: 5mm (3/16") in diameter and 30.5cm (1') long, for interior mold handle
Bamboo pole or iron rod: 30.5cm (1') long, to position interior mold

## *Making the Mold*

If the materials for the mold are cut according to Figures 13 and 14, the bowl is easy to make.

- o Nail the metal sheet around the curved back of the mold (see Figure 12).

- o Attach the two front pieces with large nails through the loose-fitting holes. These holes make it easy to remove the front pieces. The extension at the bottom of piece No. 1 is important in making sure that the bowl will seal well below the water level.

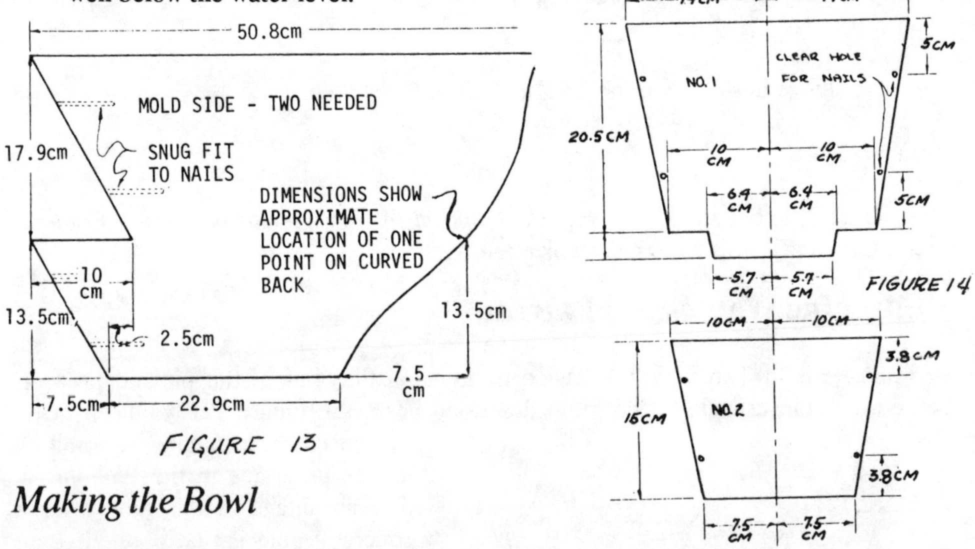


FIGURE 13

FIGURE 14

## *Making the Bowl*

Since the mold has no bottom, find a flat place to work where the mold can be propped against a wall. Fill the mold with a mixture of two parts fine sifted sand to one part cement.

Use only enough water to make the mixture workable. Pack it in so that there are no airpockets. Let it set for 15 to 20 minutes until the mixture is stiff. Next, with a ruler, measure a 38mm (1 1/2") wall around the top and outlet and dig out the inside with a tablespoon (see Figure 15).

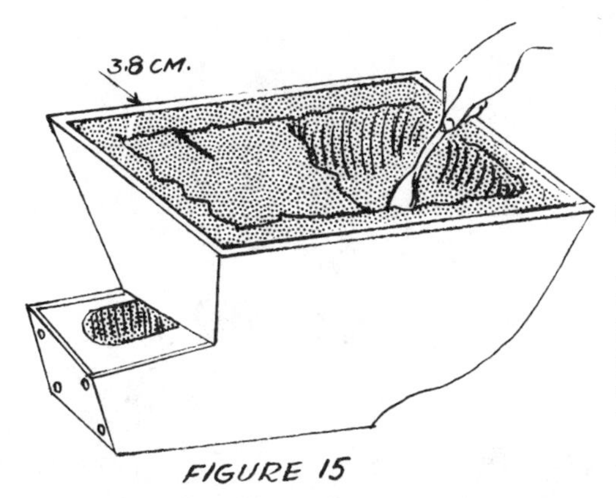


FIGURE 15

Keep a straw handy to gauge the thickness of the walls of the bowl while digging, because it is difficult to judge otherwise.

Dig out the large interior first, then the outlet. The finished interior of a bowl is shown in Figure 16.

Be sure you can insert three fingers vertically, 5cm (2"), through the hole leading to the outlet. Be careful to release front piece No. 1 by inserting the spoon around the edges (see Figure 16).

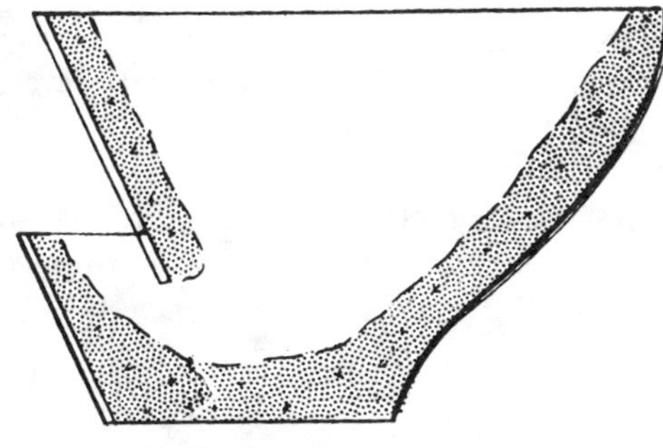

FIGURE 16

After the interior has been dug out, the walls will have slumped down about an inch. Use the cement taken from the interior to build the walls back up; then smooth all exposed surfaces with the back of the spoon as in Figure 18. To be sanitary, the bowl must be as smooth as possible so that germs cannot build up in crevices.

For a finishing coat, one of two methods may be used: (1) immediately after smoothing, sprinkle dry cement over the still wet surfaces and smooth again with the spoon (Figure 18); or (2) let the bowl set for half an hour and apply a mixture of pure cement and water–a coconut husk brush is good enough. Either method gives good results.

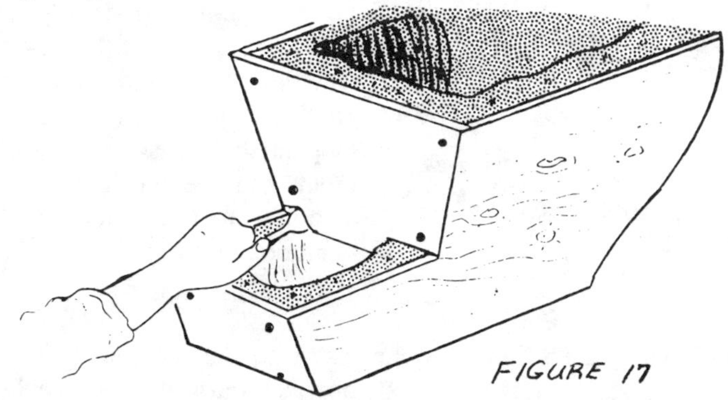

FIGURE 17

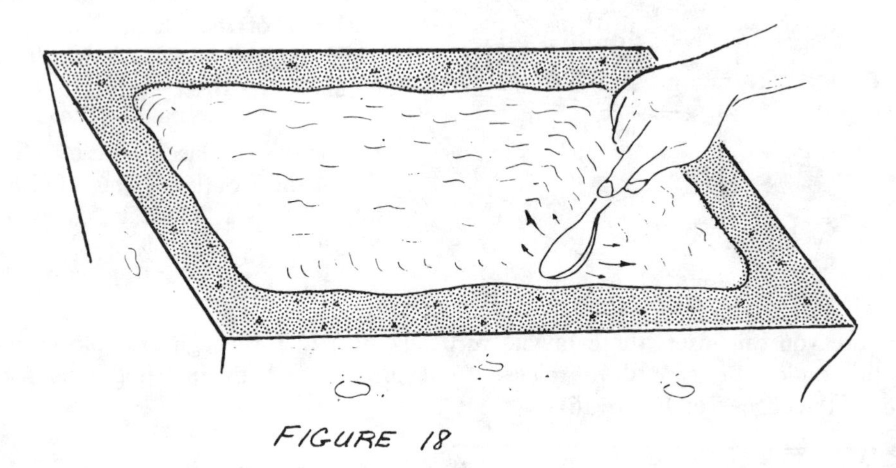
FIGURE 18

For a luxury product, use white or red cement for the finishing coat; several coats are necessary.

The finished bowl should be left in the mold to dry 48 hours. It can be removed after 24 hours only if extreme care is taken. Pull out the front nails and remove pieces No. 1 and No. 2; pull the sides and back away from the bowl.

## *Making an Interior Mold*

Because digging by hand is tedious and because it must be done very carefully to make the walls consistently thick, it is better and faster to use an interior mold.

After the first bowl has hardened thoroughly, fill the outlet with dry sand so that the cement cannot flow into it. This would make it impossible to remove the interior mold when it hardens (see Figure 19). Line the large interior with paper and fill it with cement–a 4 to 1 sand-cement ratio is good enough. Insert a heavy wire loop in the top so that the interior mold can be positioned on the exterior mold with an iron bar or bamboo pole.

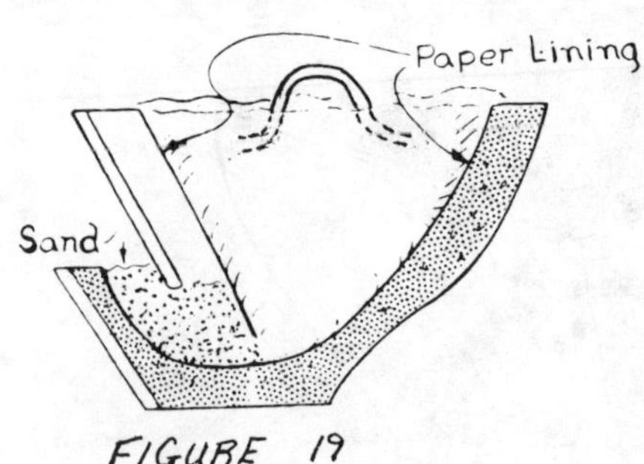

FIGURE 19

When an interior mold is used, it is only necessary to dig out the outlet. It is a good idea to have several interior molds, but not necessary to have one for each exterior mold. The interior mold should be removed after 15 to 20 minutes so that the bowl can be smoothed and finished. Then it can be used to make the next bowl.

## *Using the Interior Mold*

To use the interior mold, fill the wooden mold about 12.5cm (5") from the bottom and insert the interior mold in the correct position (see Figure 20). Push the cement around the mold with a stick and pack it well to get rid of air spaces.

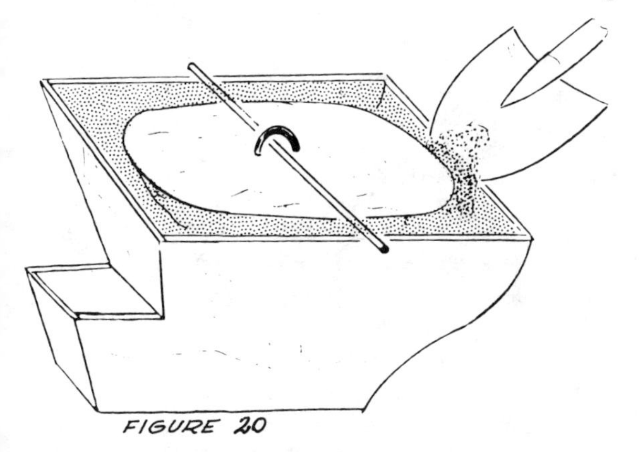

FIGURE 20

After the molds are removed, the finished bowl should be left to dry until it is rock hard–a week is usually safe–before delivery.

A sand-cement ratio of 2 1/2 to 1 has been used successfully with the bowls. A ratio wider than this may make them too expensive. There are many ways to strengthen cement; experiments may bring a cheaper solution. One possibility is to add short coconut husk or abaca fibers.

## *Installing the Toilet*

For use in private homes, dig a pit about 1.5m (5') deep and 1m (3') square. The deeper the pit and the smaller the width the better, since a small slab is cheaper (see "Pit Privy" Section). It can even be dug under the house–especially in cities–because the toilet gives off very little odor, unless this position endangers the household water supply. The pit may be lined or unlined, depending upon the soil. Hard clay soil need not be lined. But, if the house is near the sea or on sandy soil, the pit should be lined with, for example, bamboo poles or hollow blocks as shown in Figure 21.

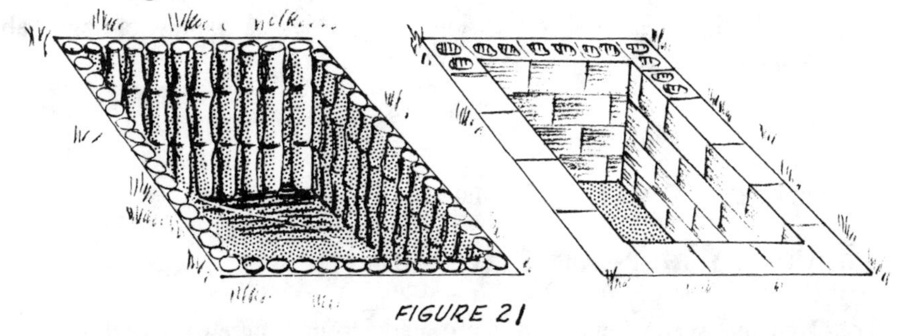

FIGURE 21

Place boards around the outside of the pit 15cm (6") from the edge of the pit to form the perimeter of the slab (see Figure 22). Place large pieces of bamboo split in half across the pit as a base for the slab. Place the bowl between two of the bamboo pieces with a piece of wood under the front and back; nail these to the bamboo. After the bowl is positioned in this way, pour water into it to be sure it will seal off the outlet. The top of the bowl should be 7.5cm (3") above the bamboo base.

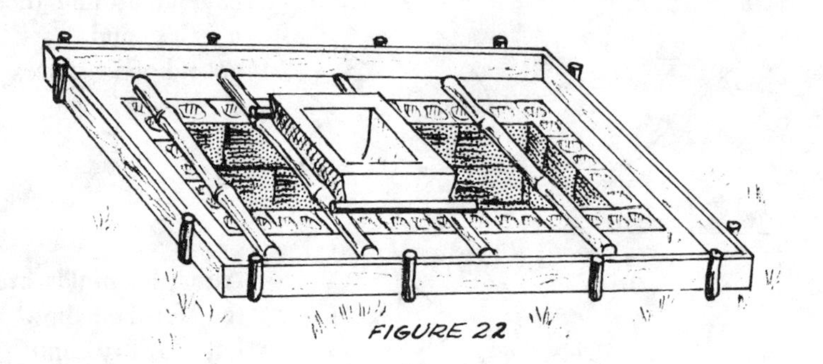

FIGURE 22

Now put bamboo slats across the pit at right angles to the large pieces of bamboo, completely covering the pit. Cover this with several thicknesses of newspaper. Pour cement around the bowl until the slab is about 5cm (4") thick. A mixture of two parts gravel, two of sand, and one of cement is good. The slab can be reinforced by placing bamboo slats between two layers of cement. Make sure that the outer edge of the slab is higher than the bowl and slants towards the center, so that the toilet can be easily cleaned. Apply a finishing coat of pure cement to the slab. Many people prefer to add foot rests and urine-guard–there is room for imagination.

It is extremely important to have an ample water supply at hand. About 1 liter (1 quart) of water is needed to flush the toilet, and people will be discouraged from using the latrine properly if they have to go some distance for water. It is a good idea to have an oil drum or a small concrete tank nearby to supply water for the latrine.

Do not use the latrine for at least 3 days–a week is best–after it is installed.

A pit with the suggested dimensions should last a family of eight about five years. One person uses about 28 liters (1 cubic foot) a year.

**Source:**

Gordon Zaloom, Peace Corps Volunteer.

## Thailand Water-Seal Privy Slab

The Thailand Water-Seal Privy Slab, made from concrete, is useful for large-scale privy programs. The slab, which includes a bowl and trap, is used to cover an ordinary pit privy.

FIGURE 1
SKETCH OF FINISHED PRIVY.

Master molds for the bowl and trap are used to make secondary molds from which the bowl and trap are actually made. The master molds can be made from the plans in the entry that follows. The master molds can sometimes be purchased from local health officers.

The finished slab is quite strong because its three parts are cast at the same time (see Figure 10). The method described here can be applied to other water-seal slab designs.

The water-seal trap is curved back under the bowl as shown in Figure 2a. This makes flushing more difficult, but prevents erosion of the back of the pit on loose soil. The same general method could be used to make a forward flushing trap (see Figure 2b).

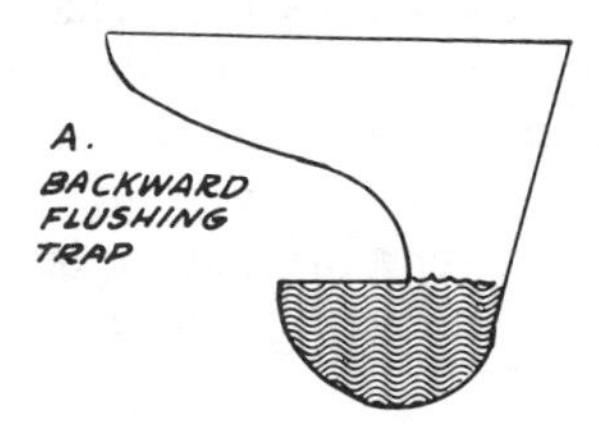


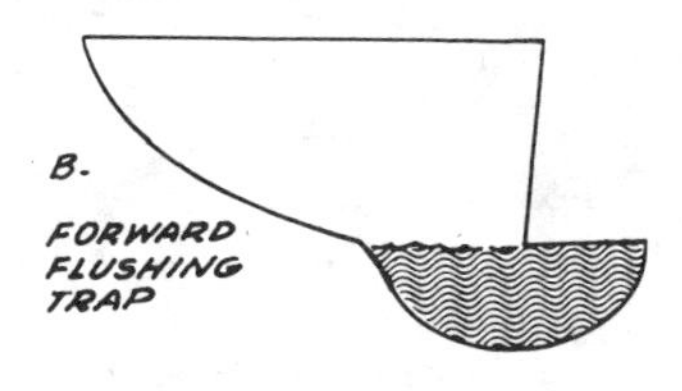


FIGURE 2. WATER SEAL TRAPS

The basic method for making these water-seal slabs is to cast the slab, bowl, and water-seal trap using three forms:

1. A wooden form for shaping the slab (see Figure 6).
2. A concrete bowl core for shaping the inside of the bowl (see Figure 3).
3. A concrete core for shaping the inside of the water-seal trap (see Figure 9).

**Tools and Materials**

Master molds
Materials for making concrete
Wool for platform forms
Reinforcing rod and wire
Clay
Crankcase oil
Beeswax and kerosene (optional)
Steel bars: 19mm x 19mm x 7.5cm (3/4" x 3/4" x 5")

The forms used when making a slab must stay in place until the concrete is strong enough, usually 24 hours. For this reason, many sets of forms are necessary if a reasonable number of slabs are to be cast every day. Here is where the three master molds are needed: one of them to cast the bowl core, and the other two to cast the trap core (see Figures 14 and 18).

## *Casting the Bowl Core*

Oil the inside of the master bowl mold and insert a 19mm x 19mm x 7.5cm (3/4" x 3/4" x 3") steel bar into the bottom.

Add a fairly loose mixture of cement and water, called neat cement, to a depth of about 15cm (6"). Then fill to brim with a 1:1 cement-sand mixture. The 1:1 should be firm, not runny, and should be laid into the loose neat cement without stirring to insure a smooth finish on the bowl core.

After the bowl core has become firm enough, scoop a depression into the surface to install two steel hooks made from the reinforcing rod. They should be about 22.5cm (9") apart, and should not protrude above the surface of the concrete (see Figure 3).

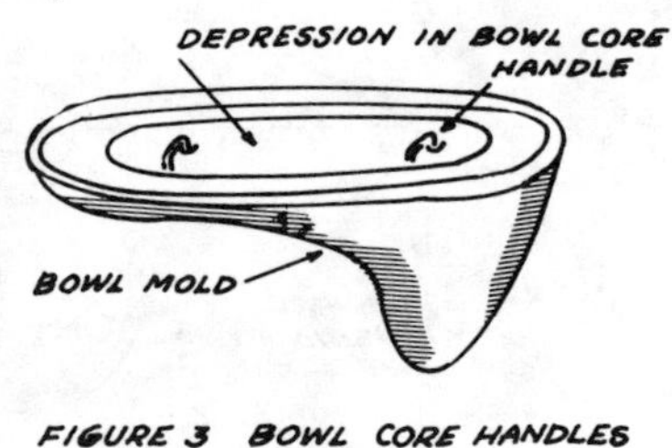

FIGURE 3 BOWL CORE HANDLES

Let the concrete set at least 24 hours before removing the bowl core from the master molds. The bowl core can be used to make another master mold and the master mold can be used to make more cores.

## *Casting the Trap Core*

Add about 2.5cm (1") of 1:1 cement-sand mix to the oiled trap master mold and put in some wire for reinforcing. Then fill it with 1:1 almost to the brim (see Figure 4).

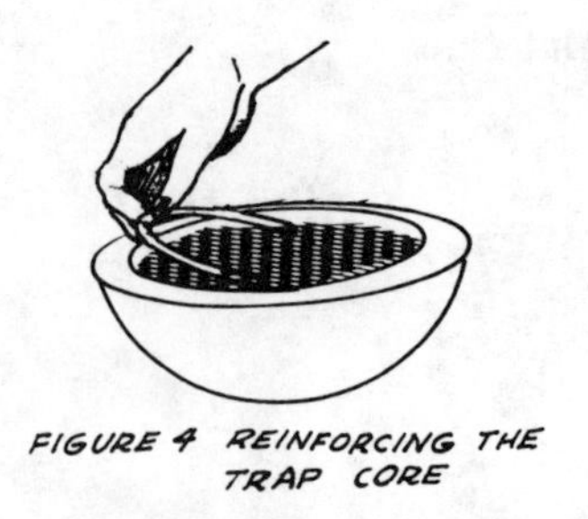
FIGURE 4 REINFORCING THE TRAP CORE

Put the oiled insert mold into place and scrape off excess (see Figure 5).

After 45 minutes, remove the insert and put a square sheet metal pipe 19mm (3/4") high into the cubical indentation left by the insert. The pipe is made by wrapping sheet metal around a 19mm x 19mm (3/4"

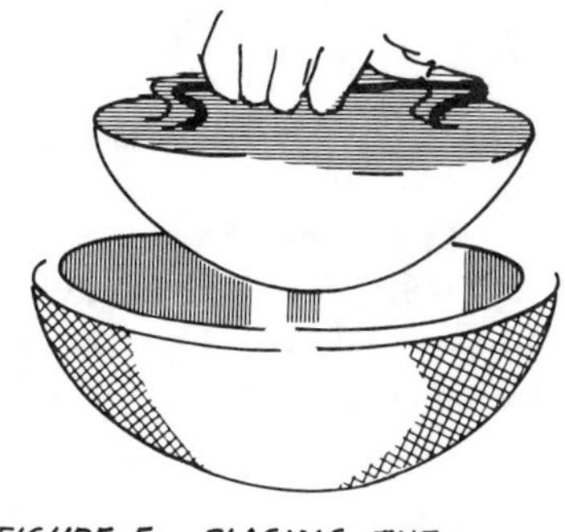

FIGURE 5 PLACING THE INSERT MOLD

x 3/4") steel bar. Let the concrete dry in the mold for 24 hours.

Remove the finished trap core by tapping the master mold gently with a wooden block.

## Making the Wooden Slab Form

Make a wooden platform 90cm x 90cm (35 1/2" x 35 1/2") out of 2.5cm (1") thick planks. This is the base of the form. The finished slab will measure 80cm x 80cm (31 1/2" x 31 1/2"). See Figure 6.

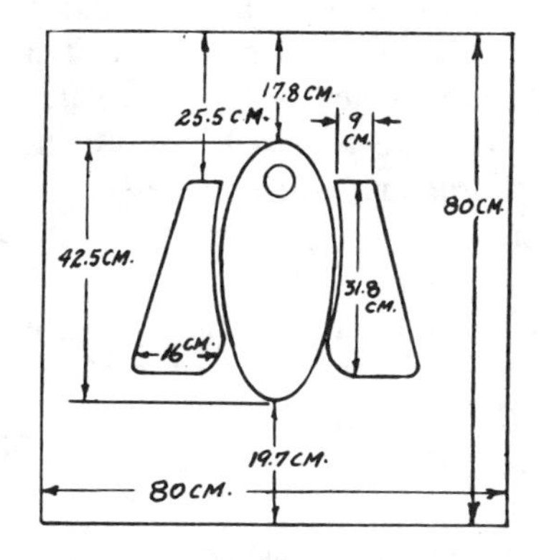


FIGURE 6 PRIVY SLAB OUTLINE

Cut out of the platform a hole 10cm x 33cm (4" x 13") for the hooks of the bowl core to extend into. The back of the hole should be 28cm (11") from the back of the platform. To determine the location of this hole, draw the outline of the bottom of the bowl on the platform, with the back of the bowl outline 23cm (9") from the back of the platform. (This is 17.8cm from the edge of the slab, as shown in Figure 6.) The back of the hole should be 28cm (11") from the back of the platform.

Using 38mm x 38mm (1 1/2" x 1 1/2") wood, make a frame with inside dimensions of 80cm x 80cm (31 1/2" x 31 1/2") (see Figure 7).

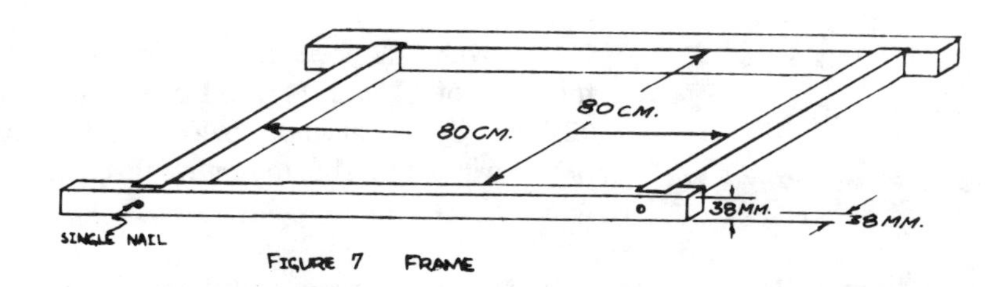


FIGURE 7 FRAME

Gouge out the footrest with a wood chisel. The inside of the foot-rests should be about 12.5mm (1/2") from the outline of the bowl.

## *Casting the Slab*

With these three forms finished, you are ready to cast the first waterseal slab.

If desired, coat the bowl core and the trap core with a layer of wax about 3mm (1/8") thick. Prepare the wax by dissolving 1kg (2.2 pounds) of melted beeswax in 0.5 liter (1 pint) of kerosene. Apply the wax with a paintbrush. The wax coating will last 5 to 6 castings. Wax makes removing the cores much easier, but it is not absolutely necessary. Let it dry before oiling.

Place the bowl core on the wooden slab form and fill all cracks with clay (see Figure 8). Oil the bowl, platform, and frame.

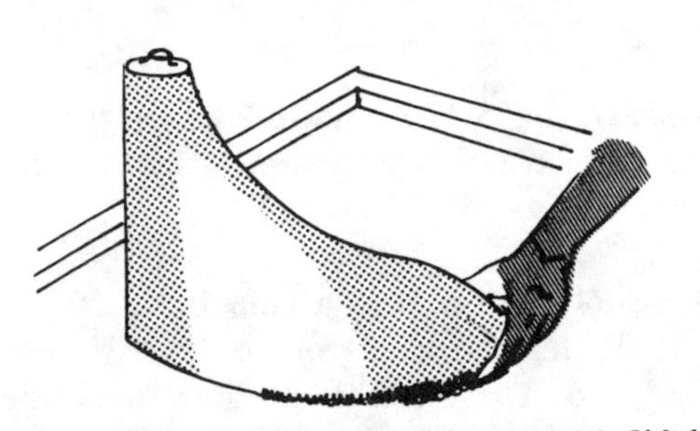

FIGURE 8. SEALING CRACKS WITH CLAY.

Apply a 6mm (1/3") thick coat of pasty cement and water mixture to the bowl core and platform. (Many people prefer to spend a little more for an attractive polished slab. To do this, use a mix of 5 cement: 5 color: 1 granite chips instead of a mixture of cement and water. After the forms are removed, polish with a carborundum stone and plenty of water.)

Cover the bowl core with a 1:2 cement-sand mixture to a total thickness of 12.5mm (1/2"). Make a smooth lip on the cement 10mm (3/8") from the top of the bowl core as in Figure 9. This lip is your water seal. Use fairly dry cement; let it set for 15 minutes before cutting the lip.

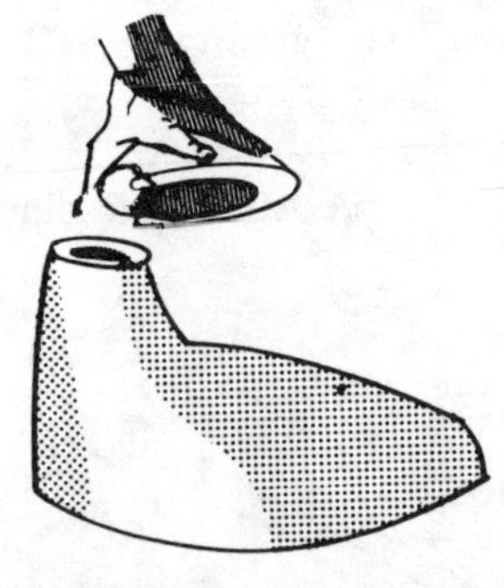

FIGURE 9. MOUNTING THE TRAP CORE.

Place the trap core on the bowl core and seal the crack with clay. Also add a little clay on each side of the form (near the thumb in Figure 9) to prevent cement from getting to the front lip.

Cover with 1:2 cement-sand mixture to a thickness of 12.5mm (1/2"). Do not exceed the 12.5mm (1/2") thickness below the trap core or you will not be able to remove this core.

Fill the slab form with a mixture of 1 cement: 3 clean gravel or crushed rock almost to the top. In preparing the concrete, first mix cement and sand, then add gravel and water. Use water conservatively. The looser the mixture, the weaker the concrete will be.

Press in 4 pieces of 6mm (1/4") steel reinforcing rod (see Figure 10). Fill to top of frame and smooth. Allow at least 24 hours for setting. Remove the frame by tapping lightly with hammer.

Turn the slab form over on a wooden stand and use simple levers to remove the bowl core. You must remove the bowl core before the trap core (see Figure 11). Tap the trap core gently and slip it out. Add a little water and check to see if your seal is 10mm (3/8").

Keep the slab damp and covered for a minimum of three days and preferably a week to gain strength.

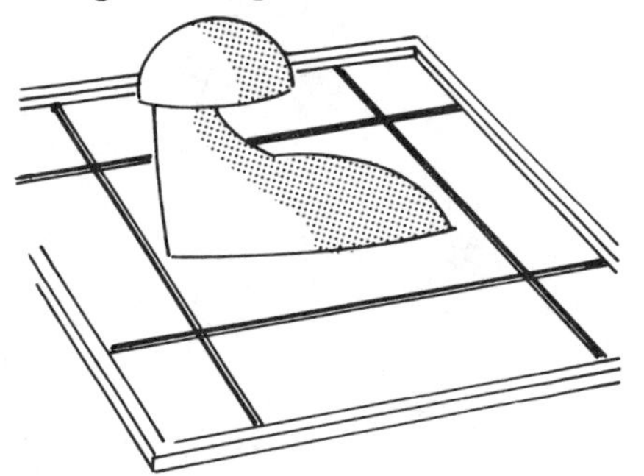

FIGURE 10. PLACING REINFORCING ROD.

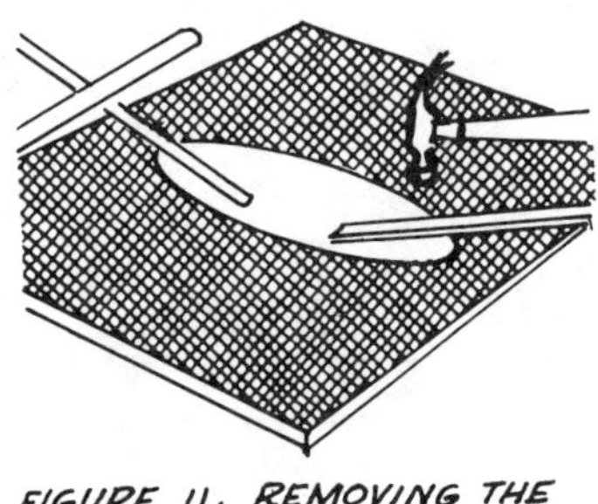

FIGURE 11. REMOVING THE BOWL CORE.

## *Master Molds for the Thailand Water-Seal Privy Slab*

This entry describes how to make the three master molds from which cores can be cast. The cores in turn are used for casting Thailand Water-Seal Privy Slabs.

**Tools and Materials**

Cardboard
Materials for making concrete
Steel rod, 19mm (3/4") square
Sheet metal (tin-can metal is satisfactory)
Reinforcing wire
Clay
Oil (used crankcase oil is satisfactory)
Paint brush

It may be necessary to make master molds rather than to purchase them. Study the entry "Thailand Water-Seal Privy Slab" before starting to make these master molds:

- o The Master Bowl Molds,
- o The Master Trap Molds, and
- o The Trap Mold Insert.

## Making the Master Bowl Mold

Enlarge the templates of the bowl outlines on Figure 12 (increase all dimensions by one third). Cut out profiles from your larger templates.

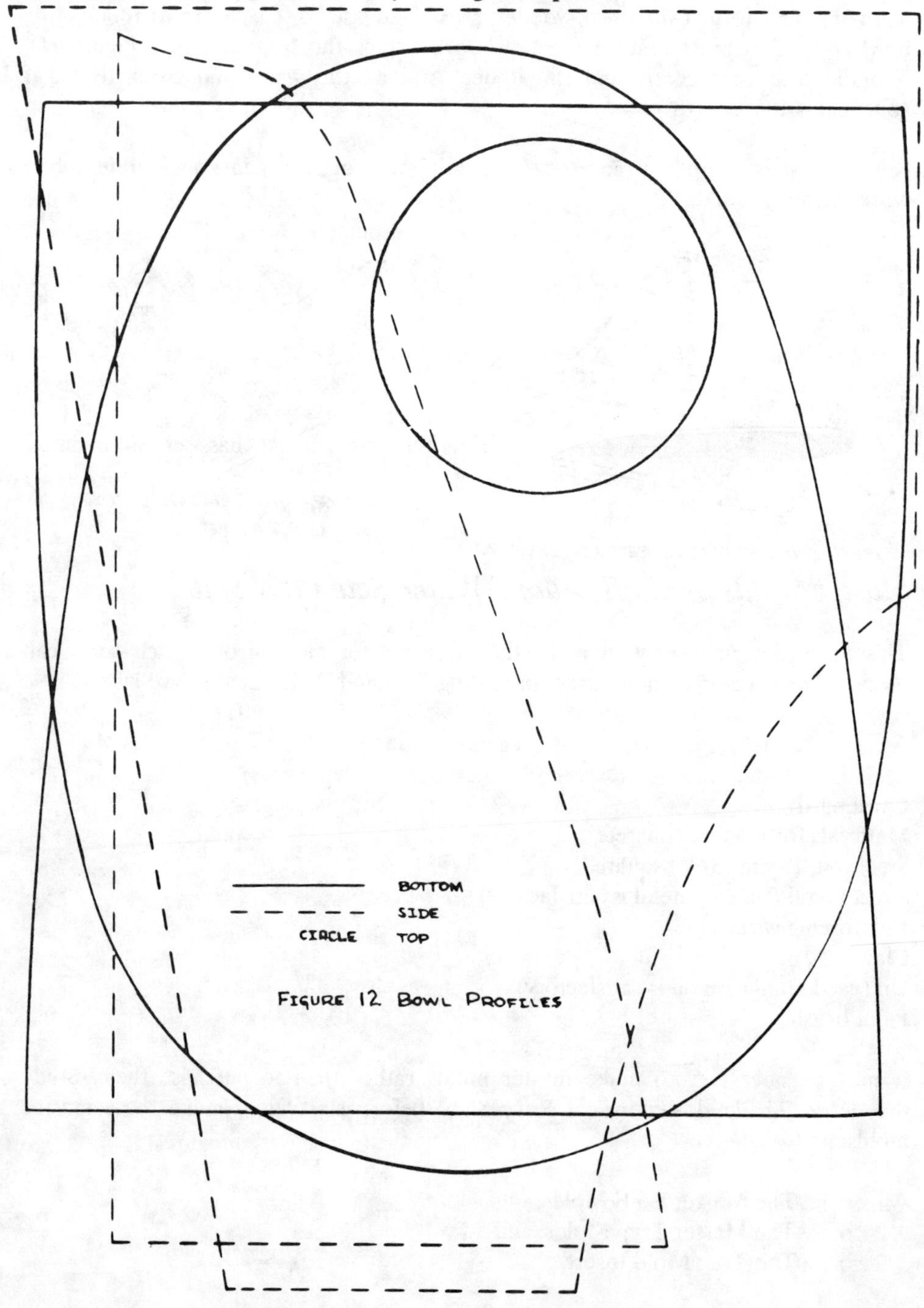


FIGURE 12 BOWL PROFILES

Shape a mound of clay using the cardboard profiles as a guide (Figure 13). Form a little square pipe, 19mm (3/4") long, of sheet metal on the 19mm (3/4") square steel rod. Make several of these as they will be used later when casting the cores. Fill the square pipe with clay and press it into the top of the clay mound a little bit. This will be used later to "key" the cores together.

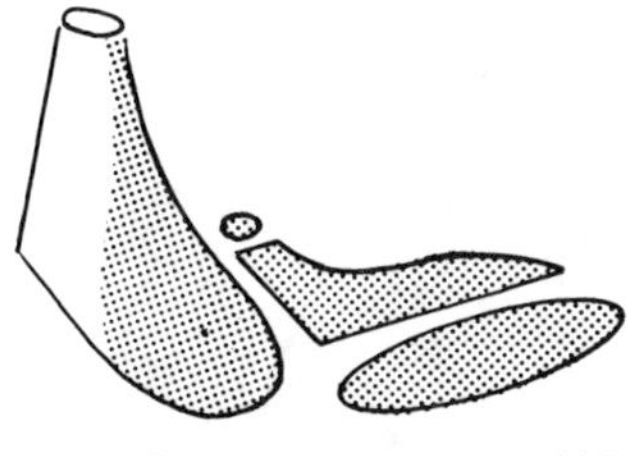

FIGURE 13 CLAY MOUND

Use a paint brush to paint the clay mound with oil; old crankcase oil is fine.

Cover the clay mound with a stiff mixture of cement and water to a thickness of 12.5mm (1/2"). If the clay mound was properly prepared, the inside finish of the bowl mold will need no further smoothing.

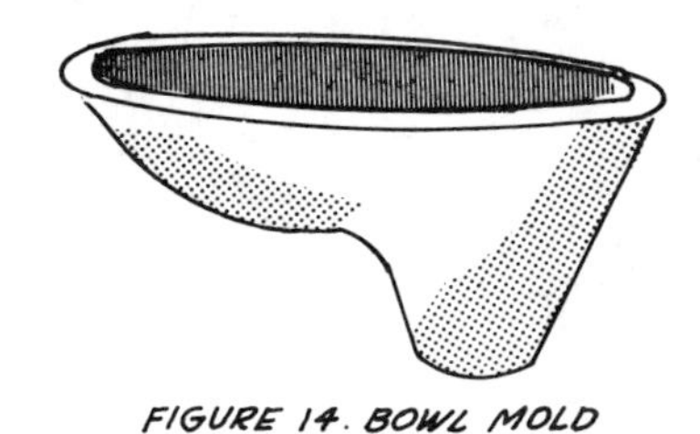

FIGURE 14. BOWL MOLD

After this cement has set 30 minutes, build up the thickness to 38mm (1 1/2") with 1:1 cement-sand mix. Let this set 24 hours and carefully lift the finished master bowl mold from the clay mound. The finished bowl mold is shown in Figure 14.

## Making the Master Trap Mold

Make cardboard profiles of the trap from Figure 17 as you did above with the bowl. Shape the outside of the trap from clay and let it harden overnight.

Shape the under side by hand with a trowel using Figure 15 and the insert profile from Figure 17 as guides. Mark the locations for a 19mm (3/4") square metal pipe by holding the clay trap over the clay mound used to shape the bowl mold, and letting the square sheet metal cube mark the trap.

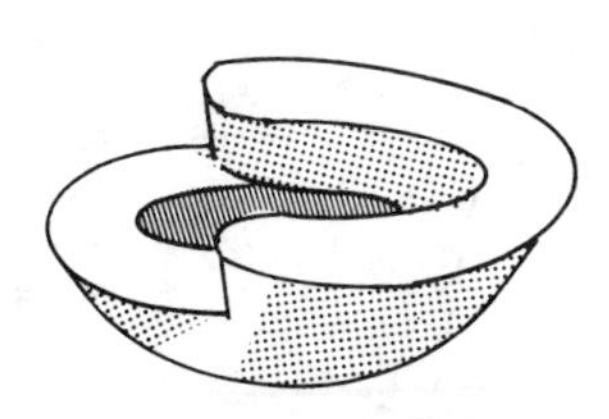

FIGURE 15. CLAY TRAP

Insert the sheet metal pipe into the clay trap and scoop out the clay from inside (see Figure 15).

Check the clay trap on the bowl mound again to be sure it lines up properly.

Oil the clay trap.

FIGURE 17. TRAP PROFILES

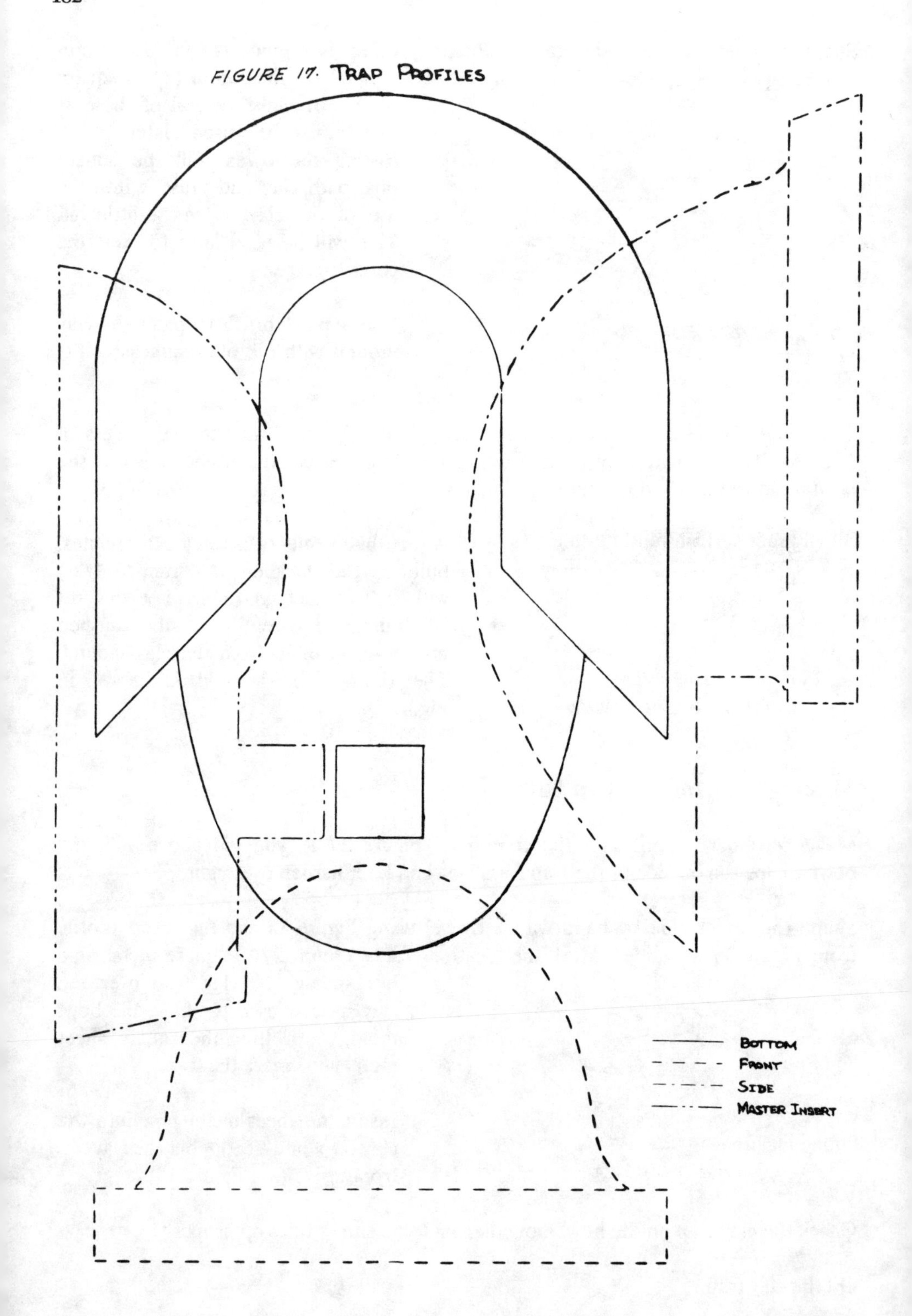

Put a heel-shaped piece of clay under the clay trap and trim the sides. This will prevent the cement from running under the mold (see Figure 16).

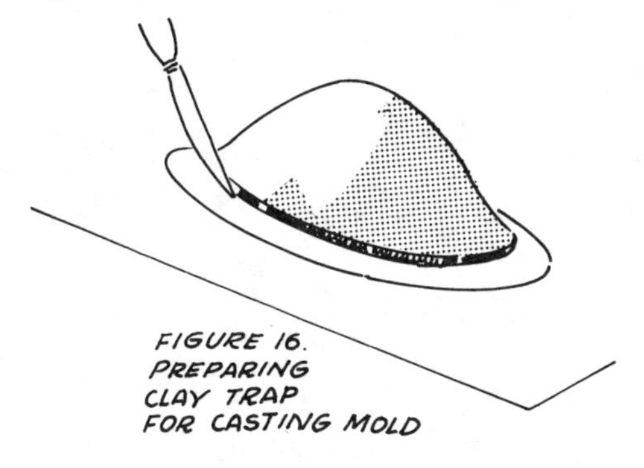

FIGURE 16. PREPARING CLAY TRAP FOR CASTING MOLD

Cover with cement and water to 19mm (3/4"), add steel reinforcing wire, and cover with 19mm (3/4") more of 1:1 cement-sand mixture.

Flatten the top and insert wire handles. Let it set at least 24 hours. This completes the master trap mold.

## Making the Trap Mold Insert

Turn the master trap mold over carefully and remove the heel-shaped clay plug.

Oil all inner surfaces and fill to the brim with 1:1 cement-sand mix.

Insert a small wire handle and let the concrete set for at least 24 hours before separating the finished molds.

Figure 18 shows the completed master trap mold and insert.

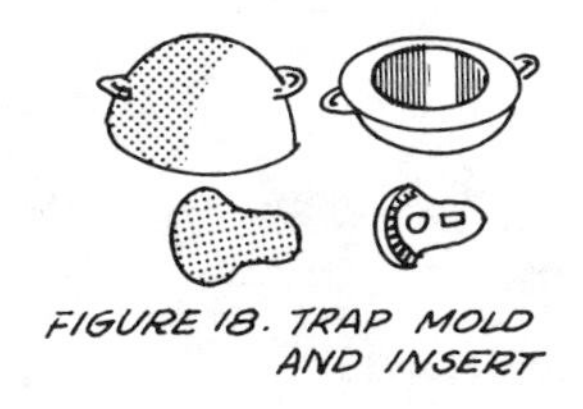

FIGURE 18. TRAP MOLD AND INSERT

**Source**:

Karlin, Barry. *Thailand's Water-Seal Privy Program.* Korat, Thailand: Ministry of Public Health.

# Bilharziasis

Bilharziasis (also called schistosomiasis) is one of the most widespread human diseases caused by parasites. This entry explains in general terms what is necessary for personal protection from bilharzia and for ridding an area of the disease. Further information from the references given is needed. Cooperation with government or other programs is essential.

An estimated 150 to 250 million people suffer from the disease. It is found in much of Africa, the Tigris and Euphrates valleys, parts of Israel, northern Syria, Arabia, Iran, Iraq, parts of Puerto Rico, Venezuela, Dutch Guiana, Brazil, Lesser Antilles, Dominica, Taiwan and parts of China, the Philippines, Japan, and a few villages in southern Thailand.

## THE PARASITES

A basic understanding of the life cycle of the parasites, called schistosomes, and the characteristics of each phase is the first step in preventing the disease (see Figure 1).

The disease has been found, besides in humans, in baboons, monkeys, rodents, water buffalo, horses, cattle, pigs, cats, and dogs. When water is contaminated by urine or feces from a victim of the disease, the eggs contained in these hatch out larvae that penetrate certain types of fresh-water snails. In the snail host, the larvae develop into **cercariae**, which work their way out of the snail and become free-swimming; this is the form that infects people. It can survive in water for a few days under favorable conditions.

The disease is contracted by contact with water containing cercariae. Typical ways are bathing, drinking, washing teeth, washing pots and clothes, walking through water, irrigating, and cultivating crops. Once the parasite has contacted a host, five minutes may be enough for it to penetrate the skin.

It is important to note that bilharziasis cannot be passed from human to human; it depends on the snail intermediary. A victim must live in or have visited an area where the parasite is found.

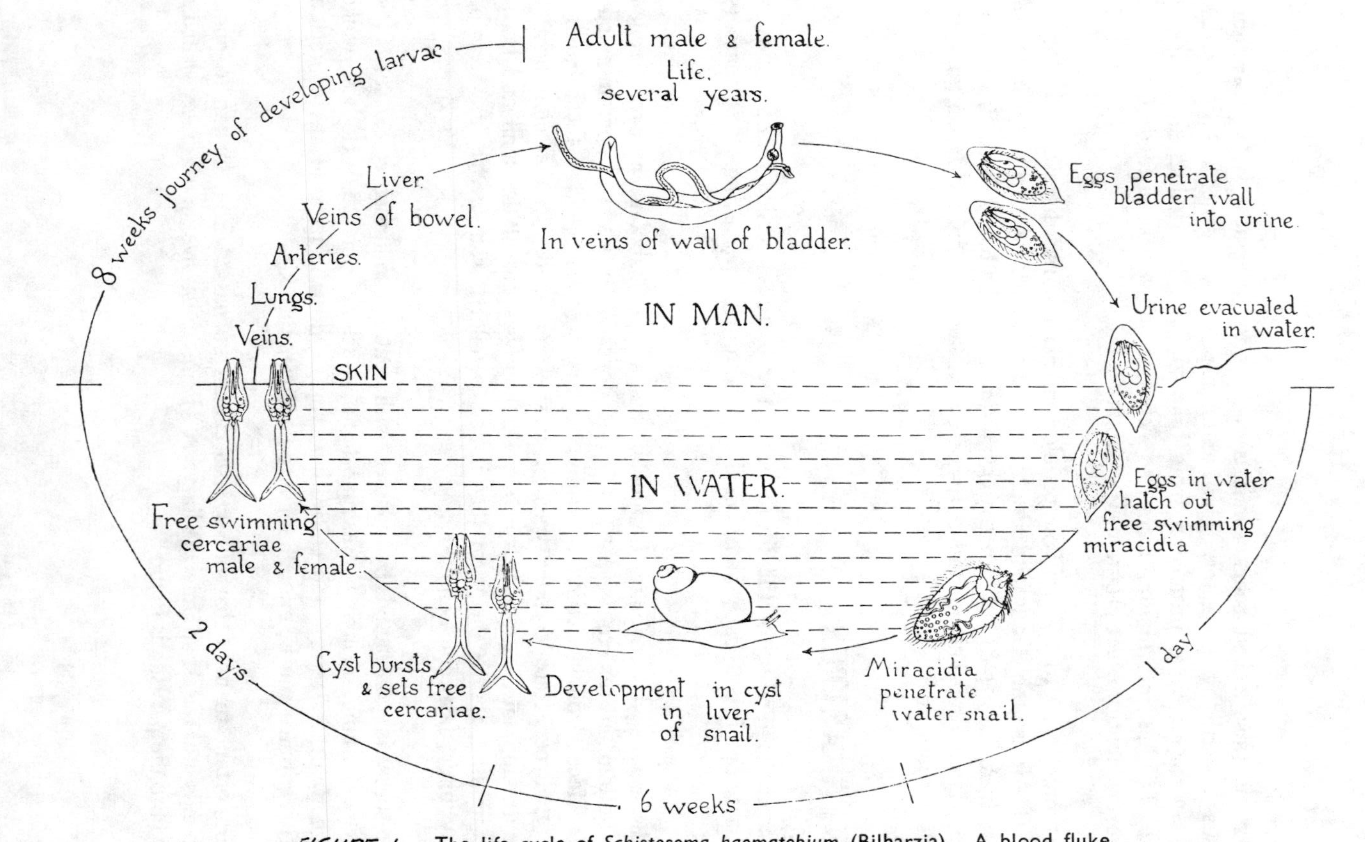


*FIGURE 1* —The life cycle of *Schistosoma haematobium* (Bilharzia). A blood fluke.

## SYMPTOMS AND DIAGNOSIS

At the spot where the parasite penetrates the host, a red itching eruption lasting several days usually develops. After the host is infected, symptoms relate particularly to the large bowel, the lower urinary tract, liver, spleen, lungs, and the central nervous system. The most characteristic symptoms are bladder and colon irritation, ulceration, and bleeding. Three to 12 weeks after infection, a victim will likely develop fever, malaise, abdominal pain, cough, itchy skin, sweating, chills, nausea, vomiting, and sometimes mental and neurological symptoms. Later developments may include frequent painful urination with blood in the urine, dysentery with blood and pus in the stool, loss of weight, anemia, and enlargement of the liver and spleen. Numerous complications are possible.

Typically the acute phase subsides and host and parasite live together over a period of years, sometimes as long as 30, with the host suffering a variety of symptoms of intermittent and variable types. Bladder and bowel troubles are the most characteristic symptoms in this period.

The variety of vague and general symptoms is considerable and may not be very specific. Examination of urine and/or feces is very important; special concentration techniques may be necessary to reveal the eggs. Tissue tests and skin tests can be used by medically-trained personnel to identify the disease.

## TREATMENT

The disease can be treated with drugs, but only well-trained persons should undertake to treat a victim. Supportive treatment, which includes good diet, nursing care, rest, and treatment of other ailments and infections, is important.

## PREVENTION

The disease can be prevented by:

- Using uncontaminated water–a properly built sealed well or an improved sealed spring is safe. (See section on "Water Resources.")

- However, it is important to remember that **all** water used must be safe. Never bathe in or touch water you wouldn't drink. Avoid suspected water. If it is necessary to use questionable water, boil it, or treat it with iodine or chlorine. If you must enter suspected waters, wear rubber gloves and wading boots, and put repellent on your skin; insect repellent (either diethyl toluamide or dimethyl phthalate), benzyl benzoate, cedar wood oil, or tetmosol give effective protection for about eight hours if applied to the skin before contact with the water. In case of accidental contact, rub your skin immediately with a dry cloth. Once cercariae have penetrated the skin, no preventive measures are possible.

- o Chlorination–Chlorine kills cercariae slowly, but properly chlorinated water systems are almost always free of the larvae. Use 2 halazone tablets in a liter (quart) of clear water; 4 tablets if the water is cloudy. In a water system, use 1 part per million chlorine. Iodine is even more lethal to cercariae. See section on "Chlorination of Polluted Water."

- o Filtering–Cercariae are just big enough to be seen with the unaided eye, and can be filtered from the water. However, dependence on filtration is questionable, since improperly made or operated filters will not only allow cercariae to pass, but may even provide a place for the host snail to live. In short, filtering is a poor technique.

- o Storage–Storing water at temperatures over 21C (70F) completely isolated from snail hosts for four days will allow the cercariae to die; at cooler temperatures they may live as long as six days. This is seldom a practical approach.

Eliminating the snail intermediate host is at present the most effective single method of controlling bilharziasis. The following methods are recommended:

- o Use a sealed, covered well or properly developed spring for a water supply. Make sure it is covered; this prevents access of organic matter that snails eat, cuts out light that would allow plants to grow for snail food, and prevents infected people from bathing in or contaminating the water.

- o If surface water must be used, put long-lasting (copper) screens on the intake; draw lake water far from vegetated shorelines, and preferably 2.4m (8') deep; take stream water from a fast moving spot.

- o Be sure filters and reservoir tanks are kept covered and dark and keep them clean.

- o Since snails prefer the stagnant water of canals, irrigation ditches, and dams, control has been possible where the water level in ditches has been varied, where it has been turned off completely for periods, and where canals have been lined with cement or pipes have been used. Although the latter is initially expensive, it pays dividends not only in better health, but also in less water evaporation.

- o Poison the snails with copper sulfate, copper chromate, or other copper salts. Use a dose of 15-30 parts per million by weight of copper and try to hold the copper-treated water over the snails for 24 hours. All or most of the aquatic vegetation should be stripped from the stream bed or pool before treatment. Results for other than small controlled pools have been poor. Before attempting to treat streams, lakes, or other natural waters, study the reference material and seek experienced help.

# RIDDING AN AREA OF BILHARZIASIS

Education is a major step in a continuing campaign against bilharziasis. Basic steps involved in improving your local waters so they will not spread the disease are as follows:

- Inform yourself. Study this article, locate reference material cited below, consult any available health officials.

- Learn to identify dangerous snails; for Africa, Professor Mozley's book is very helpful. To find the percentage of snails harboring schistosomes, collect a large sample of suspects (use rubber gloves, repellant, and snail scoop), put individually in test tubes or glass jars of water. Those shedding cercariae are readily detected, as the cercariae (0.5mm long and easily visible to the naked eye) are released in clouds. This test reveals only the snails harboring mature cercariae. Observe precautions at all times when collecting and handling snails!

- Find dangerous snails locally, collect (again using rubber gloves, repellent, and snail scoop) and kill them. Mail empty shells to an expert to confirm your identification. Visit the expert if possible. Find out about government or other programs and participate in these.

- Make a personal survey on foot (wearing boots) of local waters, using maps and keeping exact records to locate all dangerous snails. Local people can often help here. Aerial photographs are also helpful.

- Survey types and intensity of bilharzia present in populace. Differences may help localize infection points. Keep special records for three- to six-year-olds, who are the most recently infected; these records will show most accurately the incidence of new infections.

- Educate the public as much as possible, and get them to participate in the program. Better sanitation facilities, medical care, and improved nutrition are critical, but improved sanitary facilities are worthless if nobody uses them. Encourage people to live in villages away from infected waters, and to construct culverts or bridges at places where paths cross streams. The number of such crossings should be reduced. Any improvement should cater to local customs or offer an attractive alternative.

- o Personally supervise, participate in, and measure the effectiveness of poisoning the snails.

- o Take continuing steps to destroy the natural breeding places of snails, particularly at sites where humans and snails congregate. For example, the place where a stream crosses a road is a focal point: people stop to drink and bathe; they cook and wash out pots, providing food for snails. The culvert and embankments slow and impound the water, making ideal breeding conditions. Finally, a favorite sheltered place to defecate is under a bridge. Filling in places where water stands, changing drainage patterns, and eliminating snail food sources are possible techniques.

- o Maintain a continuing surveillance of focal spots and repeat poisoning periodically when necessary.

**Sources:**

Mozley, Alan. *The Snail Hosts of Bilharzia in Africa: Their Occurrence and Destruction.* London: H. K. Lewis & Co. Ltd.

*Schistosomiasis*, Bulletin No. 6. London: The Ross Institute, The London School of Hygiene and Tropical Medicine.

**Acknowledgements:**

Mason V. Hargett, M.D., Hamilton, Montana
Dr. Guy Esposito
Dr. Thomas W. M. Cameron, Montreal, Canada

**Other References:**

Craig, C. F. and Faust. *Clinical Parasitology.* Philadelphia: Lea and Fibeger, 1964.

Hinman, E.H. *World Eradication of Infectious Diseases.* Springfieldm Illinois: Charles C. Thomas, 1966.

Markell, Edward K. and M. Voge. *Medical Parasitology.* Philadelphia: W.B. Saunders Co., 1965.

*The Merck Manual of Diagnosis & Therapy.* Rahway, New Jersey: Merck.

Manson, Patrick. *Tropical Diseases.* Baltimore: William & Wilkins Co., 1966.

In addition, up-to-date information can be obtained from the World Health Organization, Geneva, Switzerland.

# Malaria Control

A second major sanitation-related disease is malaria. A serious resurgence of malaria is taking place in many countries. Between 300 and 400 million people suffer from malaria, and five million die from it annually. The disease is caused by the malaria parasite, *Plasmodium falciparum* (and three other *Plasmodium* species), which are transmitted by anopheline mosquitoes from an infected person to a healthy person. Tropical and subtropical regions of the world suffer the most from malaria.

## COMMUNITY PREVENTIVE MEASURES

Mosquitoes generally stay within about one mile (1.6 km) of where they hatch. The cycle from egg laying to hatching as mosquitoes usually takes about eight days. These facts make it easier for local mosquito eradication or control programs to be effective. But over time, persons infected with malaria can visit the local area or mosquitoes carrying the malaria parasite can be brought in with vegetable baskets, water containers, etc. Therefore, to be effective, anti-mosquito programs must be ongoing, and any spraying should be done on a regular basis. Other community based anti-malaria activities include:

- Eliminate or reduce the amount of stagnant water near the community by digging drainage ditches. The malaria mosquitoes must have water for their egg, larval, and pupal stages of development. Even small accumulations of water, as in wheel ruts or hoofprints of cattle may increase mosquito breeding if the water remains a week or more.

- Plan for the elimination of standing water in new water and flood control projects.

- "Supercharge" unlined irrigation ditches about every 6 days. To do this, raise the water level of the irrigation ditch three inches (8 cm) or more for a period of about an hour. This will cause mosquitoe larvae to float upward on the vegetation that lines the ditch. Do this in the morning on a sunny day. Then quickly drop the water level about five inches (13 cm.) or more and leave it at this level for several hours. The mosquito larva will be hung up on the dry vegetation and will die.

- Develop a voluntary reporting system for persons in the community who develop fevers, so that health care can be provided to them, and so that trends in the occurrence of malaria will be evident.

Mosquito-eating fish can reduce the number of mosquitoes in rice fields. This is not practical where rice cultivation includes alternate flooding and drying.

Regular use of mosquito-proof bed nets by all or most community inhabitants has been shown to reduce malaria rates. Programs with community participation in local production and repair of bed nets deserve field trials.

# PERSONAL PREVENTIVE MEASURES

To reduce the probability of malaria:

1. Inspect your living and sleeping quarters and install or repair screens in doors and windows.

2. Spray the walls, floors, and ceilings of your residence with insecticides.

3. Sleep under a mosquito-proof bed net.

4. Use mosquito repellents when you walk in the woods or other likely mosquito areas.

To reduce the risk of malaria, you should begin taking chloroquine two weeks prior to departing for regions of the world where malaria is found. Up to date information on the status of malaria and drug resistance can be obtained from references (1) and (2) below.

# TREATMENT

No vaccine is currently available against malaria. Breakthroughs have been made, but pharmaceutical availability is still many years away. The most effective drug against malaria is chloroquine, but in some areas of the world, the parasite is beginning to show some resistance to the drug. An alternative drug that is much more expensive is sold under the label "Fansidar." This drug is effective, but can cause serious allergic reactions in some people. Local health care providers should be consulted as to what drug to use.

The search for a vaccine against malaria is complicated by the fact that while *Plasmodium falciparium* is responsible for most malaria deaths, there are other plasmodium species, and each species may react differently to the drugs used to treat it.

In addition, the parasite goes through a series of stages of growth as it passes from the mosquito into the human bloodstream, back to the mosquito, and then back into a human host. Each stage requires its own separate defense.

For example, at one stage of the parasite's life it is called a gametocyte, a tiny body that will produce gametes or mature sexual reproduction cells. The gametocytes must pass into an anopheles mosquito to develop.

The mosquito bites a person whose blood contains the gametocytes. The gametocytes develop in the body of the mosquito and eventually produce sporozoites, tiny bodies that will grow into adult plasmodia. The infected mosquito then passes the sporozoites to another human host and the cycle begins again.

A vaccine against the sporozoite would keep the second person from getting the disease from the mosquito. It would not, however, defend against, say, contaminated blood used in a transfusion, nor one of the other infectious stages of the parasite's life.

The challenge to scientists is to develop vaccines that would be effective in three different ways. One would work against the sporozoite, preventing it from developing in its human host. Another would work against the gametocyte to prevent its growth in the body of the mosquito. Both of these vaccines could effectively block the transmission of the disease.

They would not, however, protect the person who was infected as a result of a blood transfusion. Such a person could become ill with malaria and would then be a source of infection to mosquitoes and ultimately to other people. Thus scientists are also working on a third type of vaccine, which would protect against this type of transmission.

In the meantime, the best protection for people living in malaria areas is to interrupt the cycle by getting rid of the mosquitoes or by trying to keep from being bitten. Malaria control is a community problem, not just a challenge to science. Use the measures described above to eliminate mosquito breeding areas around your home, farm, and community. Remember to protect yourself and your family from the mosquitoes by using window screens and mosquito-proof bed nets. Use mosquito repellents, and spray with appropriate insecticides where needed.

**Sources:**

Dr. Donald Pletsch, VITA Volunteer, Gainsville, Florida

Dr. Alan Greenberg, Center for Disease Control, Atlanta, Georgia

"Taking the Bite Out of Malaria," *VITA News,* January 1986, pp. 4-5.

**References:**

1. Tropical Disease Office, Pan American Health Organization (PAHO/WHO), 525 23rd Street, N.W., Washington, D.C. 20037 USA

2. Malaria Division, U.S. Public Health Service Center for Infectious Diseases, Chamblee, Georgia 30333 USA

3. "Malaria: Meeting the Global Challenge," USAID Science & Technology in Development Series. Boston, Massachusetts: Oelgschlager, Gunn & Main Inc., 1985

4. *Viajar con Salud*, Division of Public Information, World Health Organization, Geneva, Switzerland.

5. Manual on Environmental Management for Mosquito Control, World Health Organization, 1211 Geneva, Switzerland.

# Oral Rehydration Therapy

Every parent knows that diarrhea is one of the commonest ailments of childhood. It affects hundreds of millions of children around the world an average of three times a year. And especially in areas where water and sanitation are poor, it can be a problem for adults also.

But children are most vulnerable to the problems caused by diarrhea, especially children who are poorly nourished and in poor health to start with. UNICEF and the World Health Organization estimate that more than three million children in developing countries die each year from serious bouts of diarrhea–**the most important single cause of death and malnutrition among young children.**

## DEHYDRATION–A LIFE-THREATENING CONDITION

Most of the children who die from diarrhea die because their bodies have become dehydrated. That is, they have lost more fluid than they have taken in. As body fluids are lost, essential salts, minerals, and other nutrients are also lost and the body is no longer able to function properly. Severe dehydration may cause rapid weak pulse; fever; fast, deep breathing; or convulsions. Untreated, it is fatal.

The diarrhea that causes the dehydration can and should be treated before the problem becomes so serious. The idea is to give the child (or adult) as much fluid as possible and to restore the balance of salts and other nutrients. The treatment is called oral rehydration therapy (ORT). It works almost as fast as an intravenous (IV) feeding and is safer, simpler, and cheaper. Any mother can treat her child at home for just a few cents, versus the high cost of an IV or other medications. WHO estimates that use of ORT saved over 200,000 lives in 1984.

Use of ORT is so effective that as of January 1988 some 90 countries around the world had national programs to promote its use and it is becoming the treatment of choice in many hospitals in industrialized countries. Many organizations have programs to teach medical workers as well as parents about the treatment and to train them in its use.

## TREATING OR PREVENTING DEHYDRATION

A mixture–called rehydration salts–of salt, sugar, sodium, potassium (and perhaps other nutrients), and water is fed to the child frequently throughout the day and night. The salt-sugar mix is usually available in packets or tablets to be mixed with clean water. In some places, the bottled mixture may also be available. If the salt-sugar mixture is not available, you can make your own rehydration drink at home (see box).

Mix up the drink at the first signs of diarrhea. Give the person sips of the drink every few minutes, day and night, that they are awake–even if they don't feel like drinking it and even if they vomit. An adult should drink three or more liters a day and a small child should have at least one liter a day or one glass for each watery stool.

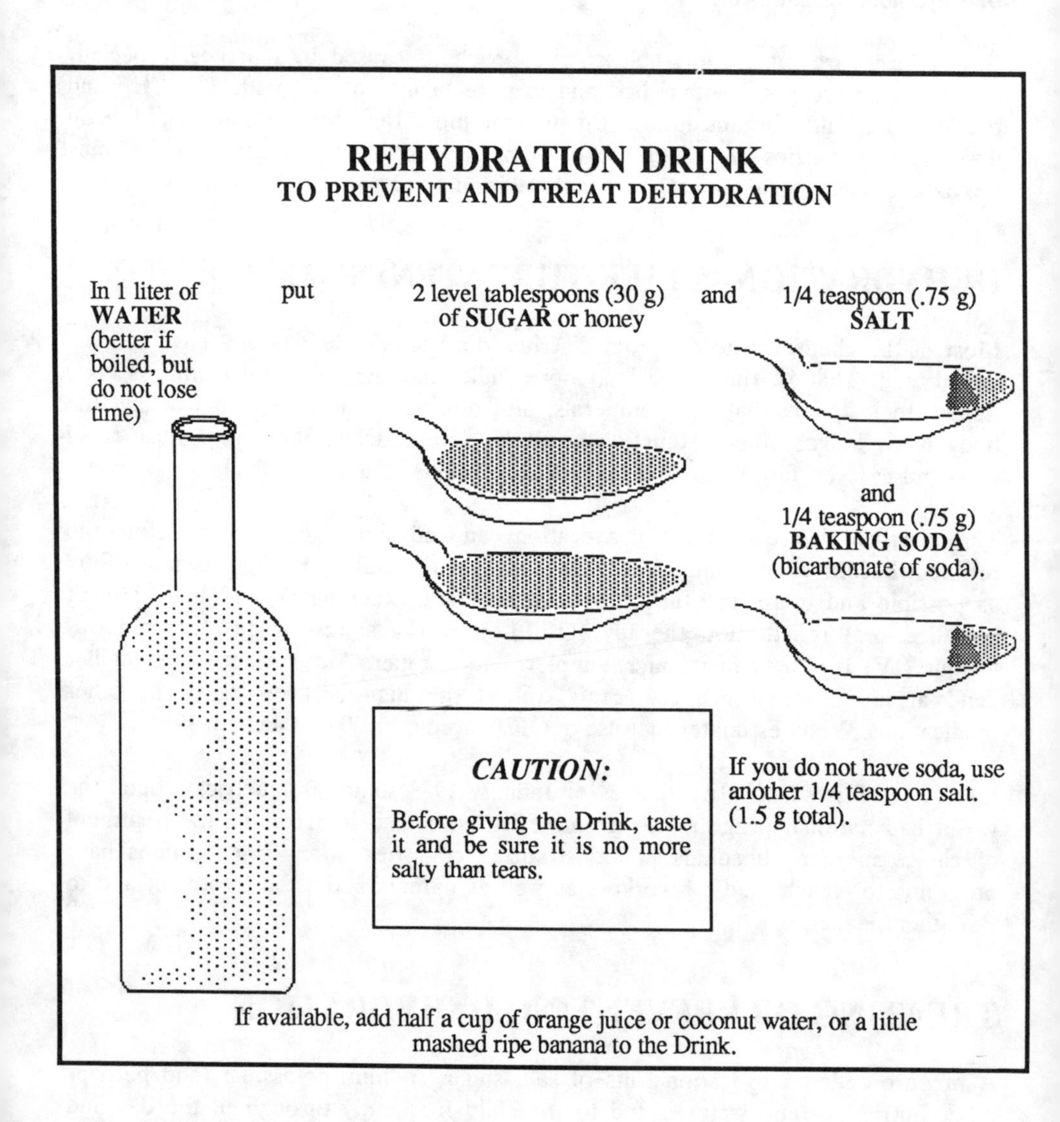

Diarrhea is often caused by malnutrition, but if it goes on long enough the diarrhea itself contributes to the malnutrition. Be sure that the person who has diarrhea eats good, easily digestible food along with the rehydration drink. This is especially important for children, but anyone who is thin and weak should get plenty of protein and energy foods all the time that they have diarrhea. If they are too sick to eat much, they should take broth, porridge, rice water, and/or cooked and mashed beans or fruit, in addition to the rehydration drink. Babies should continue to be fed breast milk. As soon as they can, the sick persons should begin eating well again.

(It should be noted that doctors often have different ideas about how to treat people with diarrhea, especially regarding the types and quantities of food the sick person should eat. Many doctors feel that people with diarrhea should not eat anything but thin soups or cereals. Other doctors say that the sick person should be allowed to eat almost any good healthful food they feel like eating. You should be prepared to follow the advice of your doctor or health worker.)

Unless the diarrhea is caused by some other disease, such as amebic dysentery, the person should respond quickly to the treatment. If the diarrhea gets worse, or if there are other disease symptoms such as fever, and the person seems to be dehydrating, get help from a doctor or health worker immediately. Remember that children are affected more quickly than adults, and dehydration is very dangerous for babies.

Look for these signs of dehydration:

- dry, tearless, sunken eyes
- sudden weight loss
- dry skin, mouth, and tongue
- sudden weight loss
- sunken "soft spot" on a baby's head
- little or no urine, and what there is is dark yellow

Dehydration also causes the skin to lose its elasticity: a pinch of skin does not fall back to normal, but stays up in a lump. Dehydration may also cause rapid, deep breathing; a fast but weak pulse; fever; and/or convulsions.

**Source:**

Werner, David. *Where There Is No Doctor.* Palo Alto, California: Hesperian Foundation, 1980. First published in Spanish as *Donde No Hay Doctor.* Now available in English, Spanish, French, Portuguese, and Swahili. Available through VITA in English, Spanish, and French.

*The Project for Appropriate Technology for Health,* Seattle, Washington USA.

Grant, James F. *State of the World's Children 1988.* New York: Oxford University Press, for UNICEF (United Nations Children's Fund), 1988.

# Agriculture

IAF/Marcelo Montecino

# Earth Moving Devices For Irrigation and Road-Building

Moving soil for irrigation and road-building is important to good farming. Careful preparation of land for irrigation and good water usage saves water, labor, and soil, and improves crop yields. Improved roads make communication easier between farmers, their suppliers, and their markets.

Although modern heavy equipment is often sought for such work, it is not necessary. Land can be prepared effectively with small equipment that can be made by farmers or small manufacturers and can be pulled by animals or farm tractors. Descriptions of yokes and harnesses are given in *Animal Traction,* by Peter R. Watson, published by Peace Corps and TransCentury Corporation (1981).

The following seven entries describe such small equipment:

- o Drag Grader
- o Fresno Scraper*
- o Barrel Fresno Scraper
- o Float with adjustable blade
- o Buck Scraper*
- o V-Drag*
- o Multiple Hitches

* The fresno scraper, buck scraper, and V-drag are designed for use with large horses.

## DRAG GRADER

This simple metal-edged wooden grader is designed for two medium-sized work horses or oxen. The grader can be scaled down for use with one horse or with smaller animals.

Road-building does not require giant tractors and earth movers. The grader described here was used for dirt and gravel roads in the midwestern United States in the 1920s. Similar graders were used in the original construction of U.S. Highway No. 1 from Maine to Florida.

**Tools and Materials**

Lumber: 7.5cm x 30.5cm (3" x 12")
  2 pieces: 243cm (8') long
  1 piece: 152cm (5') long
  2 pieces: 30.5cm (1') long
Lumber: 7.5cm x 15cm (3" x 6")
  1 piece: 37 cm (4 1/2") long
4 Metal edges: 6mm to 12.5mm (1/4" to 1/2") thick, 10cm (4") wide, 243cm (8') long
17 Lag screws: 16mm (5/8") in diameter, 18cm (7") long
2 Eye bolts, 7.5cm (3") diameter, and large lock washers
Heavy chain: 3.7m (12')
32 Flathead steel wood screws, 7.5cm (3") long. (Carriage bolts with lock washers would strengthen the grader.)

Construction details for the grader are shown in Figure 1. The metal edge overhangs the surfaces of the 243cm (8') beam by 2.5cm (1"). Each edge is attached with eight large wood screws or carriage bolts. Lock washers should be used throughout to keep strains and tensions from loosening the nuts. The metal edges are attached to both top and bottom so the grader can be turned over to reverse the direction in which the soil is cast.

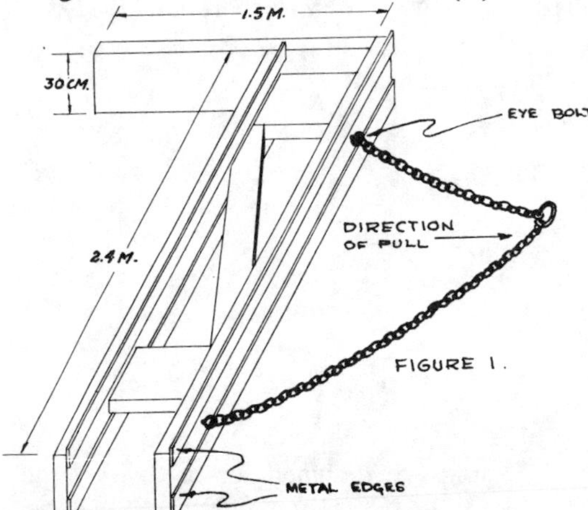

FIGURE 1.

If the grader is to be used for cleaning ditches, the angle between the 152cm (5') and 243cm (8') beams should be 30 degrees.

The drawing position of the grader is adjusted by changing the hitching point on the chain. The hitch link should be such that when the small end is put over a link it will not slide. Reverse the hitch ring to slide it along the chain.

If welding equipment is available, the same design can be used for making steel road graders, with cutting edges hard-surfaced to make them last longer.

**Source:**

Richard Hunger, John McCarthy and John Rediger, VITA Volunteers, Peoria, Illinois.

Vernon E. Moore, VITA Volunteer, Washington, D.C.

# FRESNO SCRAPER

This scraper is used for moving larger amounts of earth from higher spots to low areas. It can be made at low cost by farmers or small manufacturers, if materials and a well-equipped blacksmith shop are available. The scraper can do the work of larger, more expensive equipment.

Implements that slide soil on soil are inefficient. They require a large amount of power to move a small amount of soil. The fresno scraper can move soil more easily because it slides on its metal bottom. It is a large metal scoop that can be built in a number of sizes, depending on the number of animals that can be used to pull it. Good results will be obtained by using the size described here with two oxen or two to four horses. Construction details are given in Figures 2 to 5.

To use the scraper, first plow the high spots that you want to remove. This will make it easier to load the fresno and save a great deal of power.

The fresno is made so that the power used to pull it will also help in loading and unloading. The rope in the handle is used for pulling the scoop into position for loading and to spread the soil evenly when unloading.

Always be careful when operating the fresno. Do not have any part of your body directly above the handle. Always keep a firm grip on the handle, while loading or getting ready to unload. A sudden jerk or unseen rough spot may cause the handle bar to fly up and strike you.

FIGURE 1

To load the fresno, simply lift the handle until the front of the bit goes into the ground at a depth the animals can pull. Do not try to make too deep a cut or the fresno will be pulled over or the animals pulled to a stop. You will soon learn how to hold the handle for the proper cut and smooth loading.

When the fresno is full, push down on the handle and it will go forward without touching the ground until you are ready to unload it. Lift the handle when you are ready to unload and the pull of the animals will turn it into the dumping or spreading position. The stop bar across the top of the fresno can be moved to change the depth of the spreading of soil. Move it forward for a shallow depth or back for a deeper spread.

After it has been emptied and returned to the point of loading give the rope a hard pull and the fresno will fall back into position for loading.

Usual hitches for animals pulling the fresno are:

- o two horses
- o two oxen
- o three horses

Two lines are used. Each outside horse is tied back to the hame or collar of the center horse. The center horse is then guided by the inside strap from the lines.

**Tools and Materials**

2 steel plates for sides:
6mm x 40cm x 60cm
(1/4" x 15 3/4" x 23 5/8")

1 steel plate for blade:
6mm x 35cm x 1.24 meters
(1/4" x 13 3/4" x 48 7/8")

1 steel plate for backplate:
6mm x 52cm x 1.24 meters
(1/4" X 20 1/2" X 48 7/8")

4 steel plates for stiffener plates:
6mm x 10cm x 18cm
(1/4" x 4" x 7 1/8")

1 steel plate for stiffener plate:
6mm x 10cm x 28cm
(1/4" x 4" x 11")

2 draft bar rod stock:
20mm x 1.2 meters
(25/32" x 47 1/4")

2 draft bar rod stock:
20mm x 95cm
(25/32" x 37 3/8")

2 draft bar loop stock:
20mm x 45cm
(25/32" X 17 3/4")

2 draft bar loop stock:
20mm x 38cm
(25/32" x 15")

2 eye bolts with nuts and washers:
20mm x 25cm
(25/32" x 9 7/8")

75 flathead rivets:
15mm x 3cm (19/32" x 1 1/8")

12 flathead rivets:
20mm x 3cm (25/32" x 1 1/8")

2 angle irons for runner:
6mm x 45mm x 45mm x 1.57 meters
(1/4" x 1 3/4" x 1 3/4" x 62 13/16")

2 steel plates for shoes:
6mm x 12.5cm x 66cm
(1/4" x 5" x 26")

2 strap irons for bar brace:
10mm x 4cm x 32cm
(3/8" x 1 9/16" x 12 5/8")

1 iron bar for handlebar
15mm x 5cm x 1.6 meters
(9/16" x 2" x 63")

1 rope:
13mm x 2 meters
(1/2" x 78 3/4")

2 side plates, draft clamp stock:
20mm x 21cm
(25/32" x 8 1/4")

2 machine bolts with nut and washer:
13mm x 4cm (1/2" x 1 9/16")

2 strap iron clevis stock:
10mm x 4cm x 60cm
(3/8" x 1 9/16" x 23 5/8")

2 machine bolts with nuts & washers:
13mm x 10cm
(1/2" X 4")

1 oak draft bar:
6cm x 15cm x 1.52m
(2 3/8" x 6" x 59 7/8")

4 machine bolts with nuts & washers:
13mm x 6cm
(1/2" x 2 3/8")

8 machine bolts with nuts & washers:
13mm x 4cm
(1/2" x 1 9/16")

1 oak stop bar:
4cm x 8cm x 1.45 meters
(1 9/16" x 3 1/8" x 57 1/8")

2 stop bar stock:
threaded one end

2 stop bar ball nuts & washer:
13mm (1/2")

**Source:**

Forsberg, Carl M., Metzger, James D. and Steele, John C. *Construction and Use of Small Equipment for Farm Irrigation.* USOM/Turkey, in cooperation with the Turkish Ministry of Agriculture.

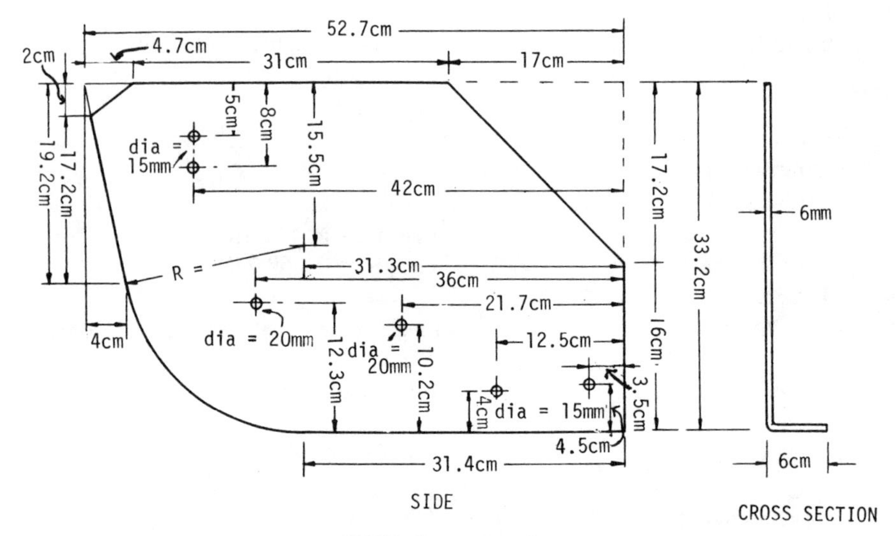


FIGURE 2 - END PLATE

FRESNO

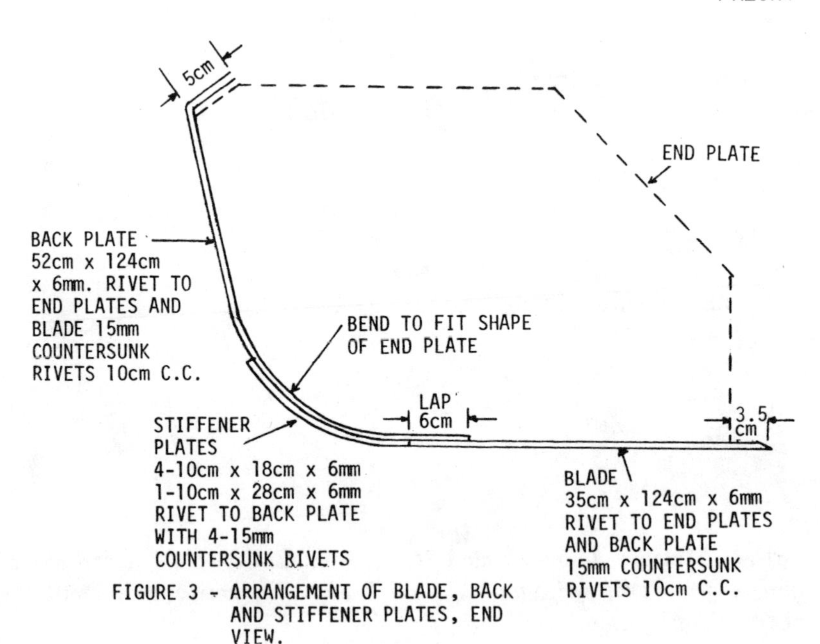


FIGURE 3 - ARRANGEMENT OF BLADE, BACK AND STIFFENER PLATES, END VIEW.

SHOES

12.5cm x 6mm x 66cm plate
Rivet to Runners, 6-20mm countersunk rivets

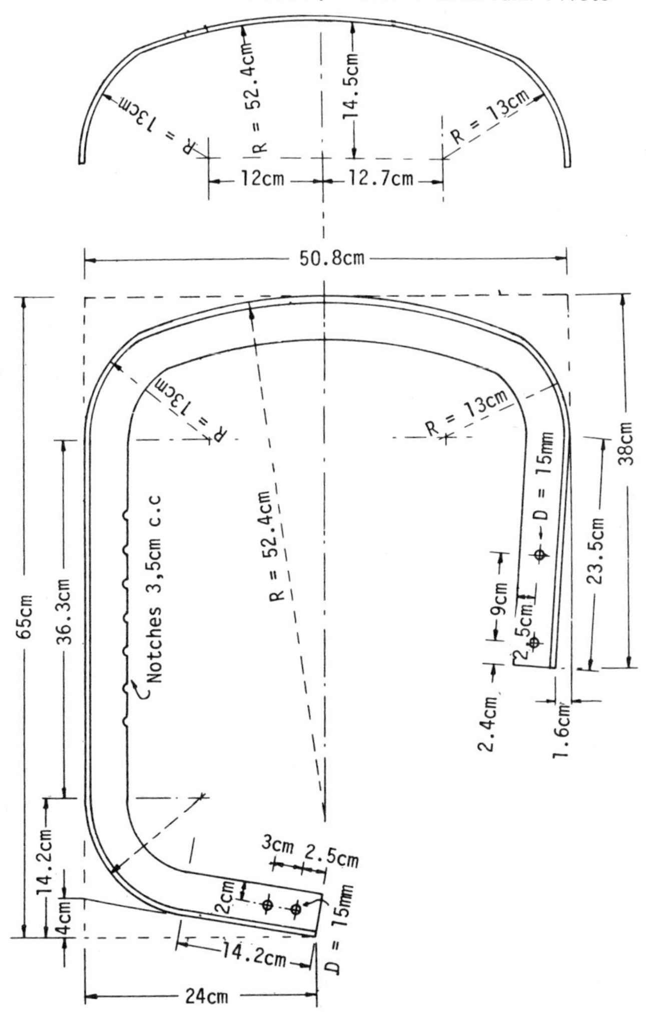


FIGURE 4 FRESNO RUNNER AND SHOES

<u>Runner</u>

4.5cm x 4.5cm x 6mm Angle
157cm long (overall)

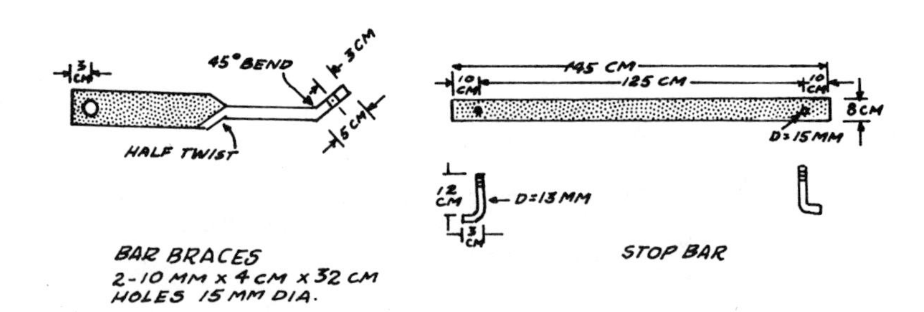


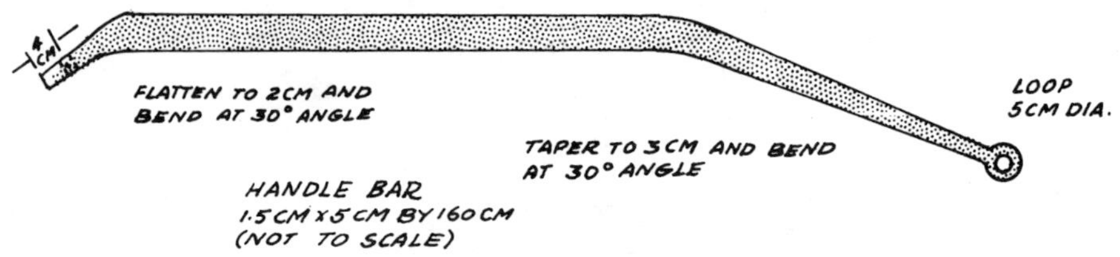


FIGURE 5 FRESNO BAR BRACES, STOP BAR AND HANDLE BAR.

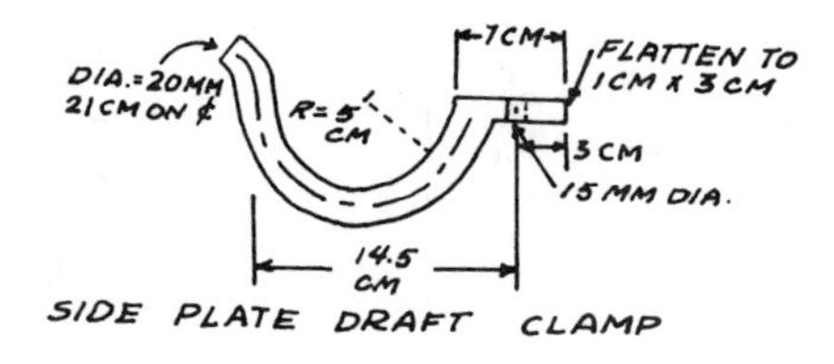


SIDE PLATE DRAFT CLAMP

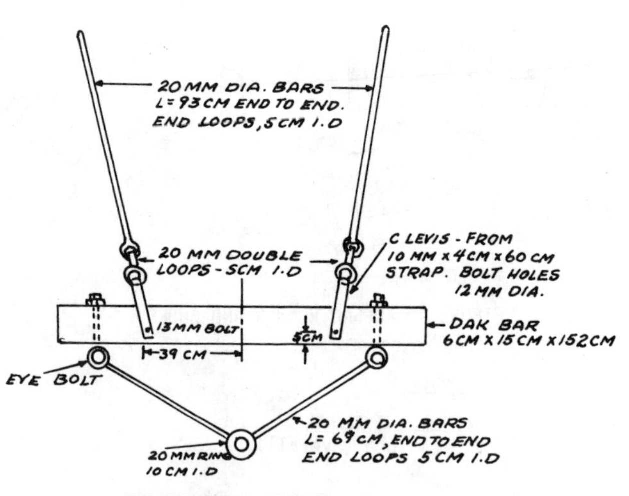


DRAFT BAR ASSEMBLY
FIGURE 6

# BARREL FRESNO SCRAPER

The barrel fresno scraper (Figure 1) is a lighter, simpler version of the fresno scraper described in the preceding entry. It is a low-cost implement to move soil efficiently. It can be pulled by a team of bullocks and is operated by one person.

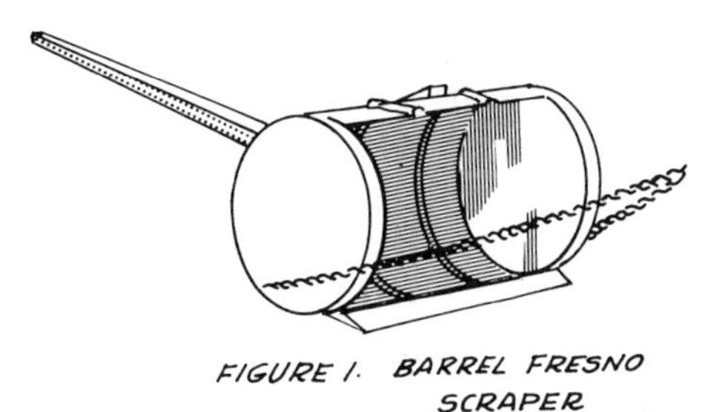
FIGURE 1. BARREL FRESNO SCRAPER

The scraper, which is well-adapted to production by a village blacksmith, is made from an old barrel and scrap metal. The scraper can be adapted for heavy duty use.

The barrel fresno scraper presented here was built and tested in Afghanistan. It was found that it could move approximately twice as much soil as the shovel board normally used by the Afghan farmer. The scraper worked better when the high spots were plowed with a mold-board plow, which breaks up the soil, making it easier to pick up. Using the local wooden plow was satisfactory but it left the soil cloddy.

It is estimated that it could be used for 8 to 10 years under normal farm use on. Under other conditions, particularly where the soil is sandy or where the scraper is used for road or terrace construction, its life would be much shorter.

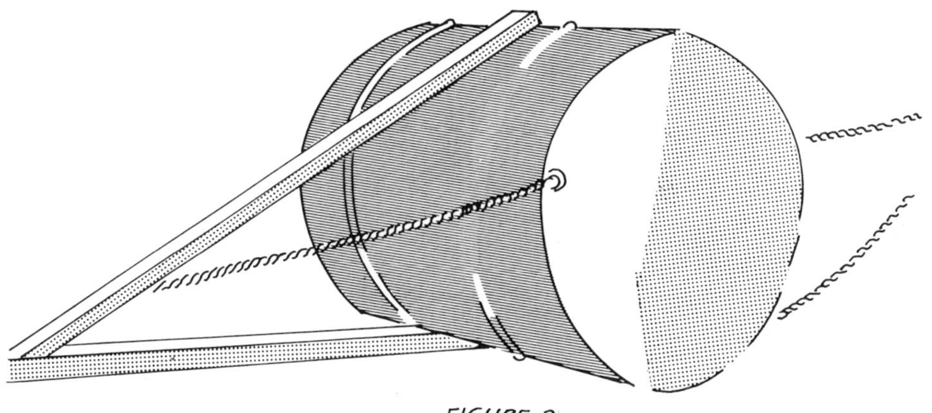
FIGURE 2.

**Tools and Materials**

Heavy hammer
Chisel - for cutting barrel
Punch - for making holes in barrel
Saw - for cutting wood
Drill - for boring holes in wood
Pliers

Welder or access to the services of a welder
Barrel, 208-liter (55-U.S. gallon) as new and strong as possible. Rust weakens the metal and a rusty barrel should not be used.
Blade, metal, 1 piece, 5 to 8mm (3/16" to 5/16") thick, 88cm (34 5/8") long. Have a blacksmith taper the blade until it has this shape when viewed from the end. The blade should be sharp. Old truck springs make good blades.
Blade holder, metal, 2 pieces, 5 to 8mm (3/16" to 5/16") thick
Handle, wood, 1 piece, if soft wood 4 by 8cm (1 9/16" x 3 1/8") or pole 8cm (3 1/8") in diameter at large end, 3m (9'10")
Handle brace, wood, 1 piece, 3cm by 8cm by 150cm (1 3/16" x 3 1/8" x 59")
Block wood, 1 piece, 3cm by 8cm by 12cm (1 3/16" x 3 1/8" x 4 3/4")
Bolt, 1 piece, 1cm diameter by 10cm (3/8" x 4")
Nails, 5 pieces, 9cm (3 1/2") long
Wire, heavy - at least 3mm (1/8") thick, 12m (39') long
Chain, 4m (13'), made from 7mm (9/32") rod, with hook at each end. See Figure 1.
Rope, 12mm (1/2") diameter, 3m (9'10") long

## Construction

Cut the barrel, starting next to the welded seam, as shown in Figure 3 below (also see Figure 1). The cut is exactly half way around the barrel.

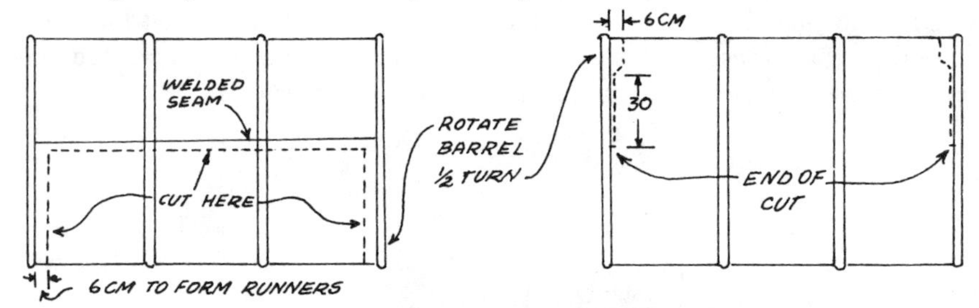


FIGURE 3 CUTTING THE BARREL

Pull the cut-out section forward and flatten it with a hammer (Figure 4). Fold the cut-out section back 17 to 20 cm (6 3/4" to 7 7/8") from the end of the cut, depending on the width of the blade, to form a double bottom (see Figure 5).

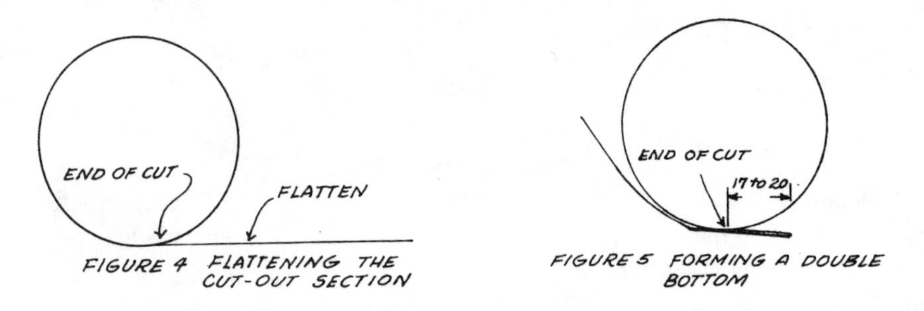


FIGURE 4 FLATTENING THE CUT-OUT SECTION

FIGURE 5 FORMING A DOUBLE BOTTOM

The blade can be installed by welding or by riveting.

To install the blade by welding (see Figure 1):

- Butt the blade (see Tools and Materials) against the barrel fold and spot weld it. Five spots of welding 3cm (1 3/16") long, evenly spaced, are enough.

- The lower tip of the blade holder (see Tools and Materials) should be even with the end of the cut).

- Weld the blade holder at the outside of the barrel to the heavy rim.

- Weld the blade to the bottom of the blade holder.

To install the blade by riveting:

- No blade holder is required.

- The metal for the blade should be 5 to 8mm (3/16" to 5/16") thick, 8 to 12cm (3 1/8" to 4 3/4") wide and 164cm (64 1/2") long. Taper and sharpen the blade before bending.

- Bend the blade up at right angles 40cm (15 3/4") from each end. This will leave the main part of the blade 86cm (33 7/8") long to fit inside the barrel.

- Insert the blade.

- Drill holes and rivet as shown in Figure 6.

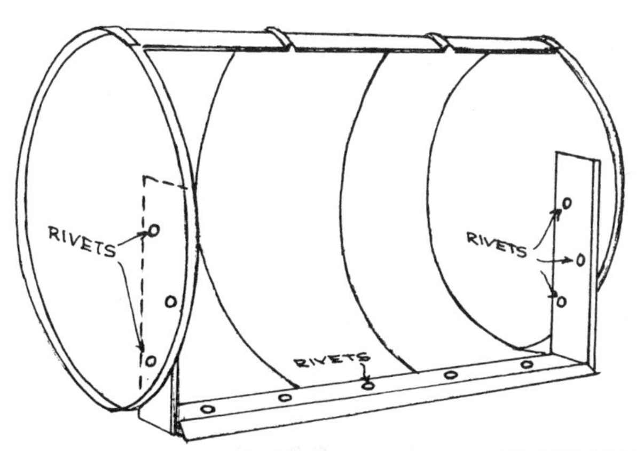


FIGURE 6 ATTACHING THE BLADE WITHOUT WELDING

- The folded part of the barrel bottom should extend 3cm (1 3/16") under the blade and be riveted to the bottom of the blade.

Install the handle and handle brace (Figure 7):

- Position the barrel so the edge of the blade is exactly 4cm above the floor.

- Taper the end of the handle and place it in the position shown in the sketch, making sure it is in the center of the barrel.

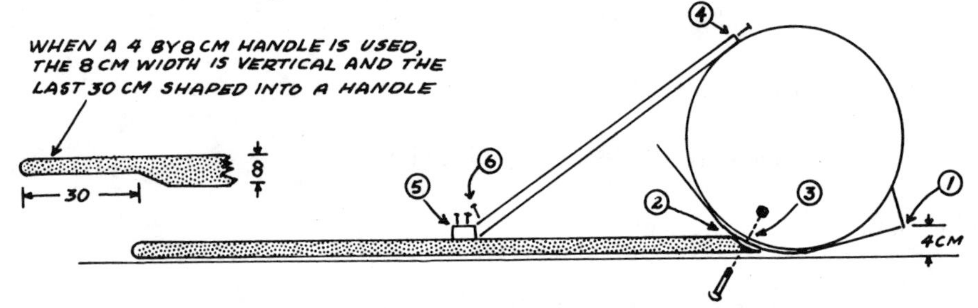


FIGURE 7 INSTALLING THE HANDLE AND HANDLE BRACE

- Punch a hole through the bottom of the barrel, drill a hole through the end of the handle, and bolt the handle to the barrel.

- Bend 2 1/2cm of the edge of the barrel metal up as shown in Figure 1. Punch 2 small holes in the metal and drive 2 nails through the holes into the end of the wooden brace.

- Making sure the blade is still 4cm from the floor, nail the wooden block against the end of the wooden brace.

- Drive a nail through the end of the brace into the handle.

Install the brace wires and rope (Figure 8):

- Punch holes through the side and end of the barrel halfway between the bolt and brace end.

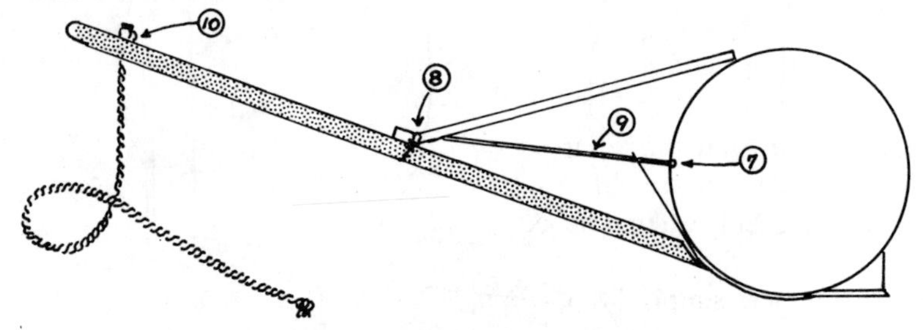


FIGURE 8 INSTALLING BRACE WIRES AND ROPE

- Fix 4 strands of wire through the holes and around the brace and handle.

- Twist the wires with a small stick to tighten the wire braces, making sure the handle is at right angles to the barrel.

- Drill a 1 1/2cm hole, 20cm from the end of the handle. Thread the end of the rope through the hole and make a knot at each end.

Make holes for installing chain (Figure 9).

Install chain (see Figures 1 and 2).

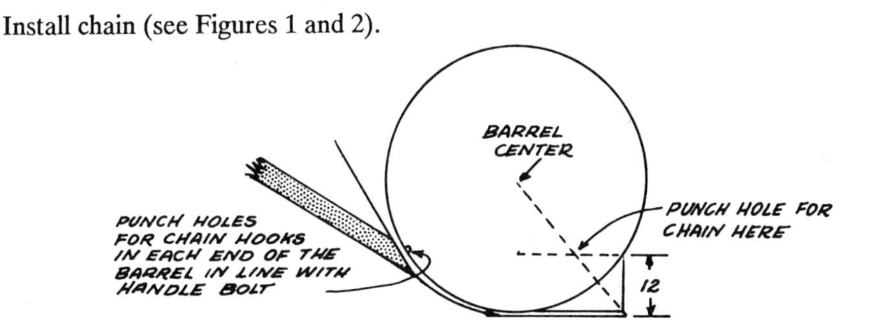


FIGURE 9 PREPARING TO INSTALL CHAIN

# Operation

When operating the barrel fresno scraper, always be careful not to have any part of your body directly above the handle.

Keep a firm grip on the handle while loading, during operation, and getting ready to unload. An unseen rough spot may cause a sudden jerk, which will make the handle fly up and strike you.

Before using the scraper, plow the high spots that you want to remove. This will make it easier to load the soil.

The power used to pull the scraper will also help in loading and unloading. Use the rope in the handle to pull the scraper into position for loading and for spreading the soil evenly when unloading.

To load the scraper, simply lift the handle to let the blade go into the soil. Do not make too deep a cut: this would either pull the scraper over or pull the animals to a stop.

You will learn by experience how to hold the handle for a proper cut and smooth handling.

When the scraper is full, push down on the handle to let the loaded scraper slide forward without picking up more soil to where you want to unload it.

To unload, lift the handle. The pull of the animals will move the scraper into dumping position. To spread the soil evenly, hold the rope tight. To dump the soil in a pile, let the rope go.

## Repairing the Barrel Fresno Scraper

To repair the scraper when the bottom starts to wear through, cut off the unworn part of the cut-out section of the barrel and weld or rivet it over the old bottom. When the rims of the barrel, which serve as runners, start to show wear, weld or rivet old truck springs or similar heavy strap iron over their entire length.

## Adapting for Heavy Duty

To adapt the barrel fresno scraper for heavy duty, the two wearing points, the bottom and the runners, must be reinforced. To reinforce the scraper bottom, cover it with a heavy iron plate 4 to 6mm (5/32" to 1/4") thick from the rear of the blade to the bolt that holds the handle. Weld or rivet the plate in place.Reinforce the runners by welding or riveting old truck springs or other heavy strap iron as described in the paragraph on repairing the scraper.

**Source:**

Dale Fritz, VITA Volunteer, Schenectady, New York

# FLOAT WITH ADJUSTABLE BLADE

The float is very useful for leveling a field before planting a crop. It can be made by a small manufacturer or a carpenter-blacksmith with locally-available materials (Figure 1).

FIGURE 1 A FLOAT FOR LEVELING A FIELD BEFORE A CROP IS PLANTED

All earth moving operations, where any quantity of soil is moved, leave the land surface in an uneven condition. The float is the best piece of equipment for obtaining a smooth, even surface. It is difficult to do a perfect job of leveling the first season after earth has been moved. The areas from which the soil has been removed are usually hard and the areas to which it has been moved are soft so that uneven settling results. Also, general tillage and plowing operations sometimes roughen the land surface. Using the float over the entire field each season before planting the crop will help answer these problems. Best results may be obtained by floating the field in both directions (at 90 degrees), going back and forth. The last floating should be in the direction of the irrigation flow.

When borders are built in a field for border irrigation it is usually best to use the float over the entire area between the borders before seeding.

The float can be built in various widths according to the available power. It is necessary, however, that the float be at least 5 meters (16") long to ensure a good job of leveling the earth. The adjustable blade is optional but it is often desirable if a buck scraper is not available. Common hitches for the float are the same as used with the fresno.

**Tools and Materials**

Wood:
- 2 runners, 5cm x 30cm x 5.5 meters (2" x 12" x 18')
- 3 blades, 5cm x 30cm x 1.8 meters (2" x 12" x 70 7/8")
- 2 cross braces, 5cm x 20cm x 1.9 meters (2" x 7 7/8" x 74 13/16")
- 2 diagonal braces, 5cm x 15cm x 3.75 meters (2" x 5 7/8" x 12'4")
- 2 diagonal braces, 5cm x 15cm x 3 meters (2" x 5 7/8" x 9'9")
- 4 side blocks, 5cm x 30cm x 45cm (2" x 12" x 17 3/4")
- 1 lever, 5cm x 10cm x 1.5 meters (2" x 4" x 59")

2 strap iron runner plates, 7mm x 50mm x 6 meters (9/32" x 2" x 19'7")
3 angle iron cutting edges, 7mm x 50mm x 1.8 meters (9/32" x 2" x 70 7/8")
2 steel tie rods (threaded both ends), 7mm x 2 meters (9/32" x 78 3/4")
4 nuts, 7mm (9/32")
8 washers, 7mm (9/32")
1 pipe axle, 5cm (2") diameter x 2 meters (78 3/4")
2 steel plates, 10mm x 20cm x 20cm (3/8" x 7 7/8" x 7 7/8")
3 u-bolts, with nuts and washers, 13mm x 20cm (1/2" x 7 7/8")
2 hitch stocks, 7mm x 50mm x 70mm (9/32" x 2" x 27 9/16")
50 flat head screws, 4cm (1 9/16") (No. 14)
15 flat head stove bolts, with nuts and washers, 6mm x 8cm (1/4" x 3 5/32")
4 carriage bolts, with nuts and washers, 13mm x 13cm (1/2" x 5 1/8")
1.5kg nails, 13cm (40d) (5 1/8")
1.5kg nails, 10cm (20d) (4")
1 rope, 10mm x 4 meters (3/8" x 13')
1 chain or cable hitch, 5 meters (16'5")

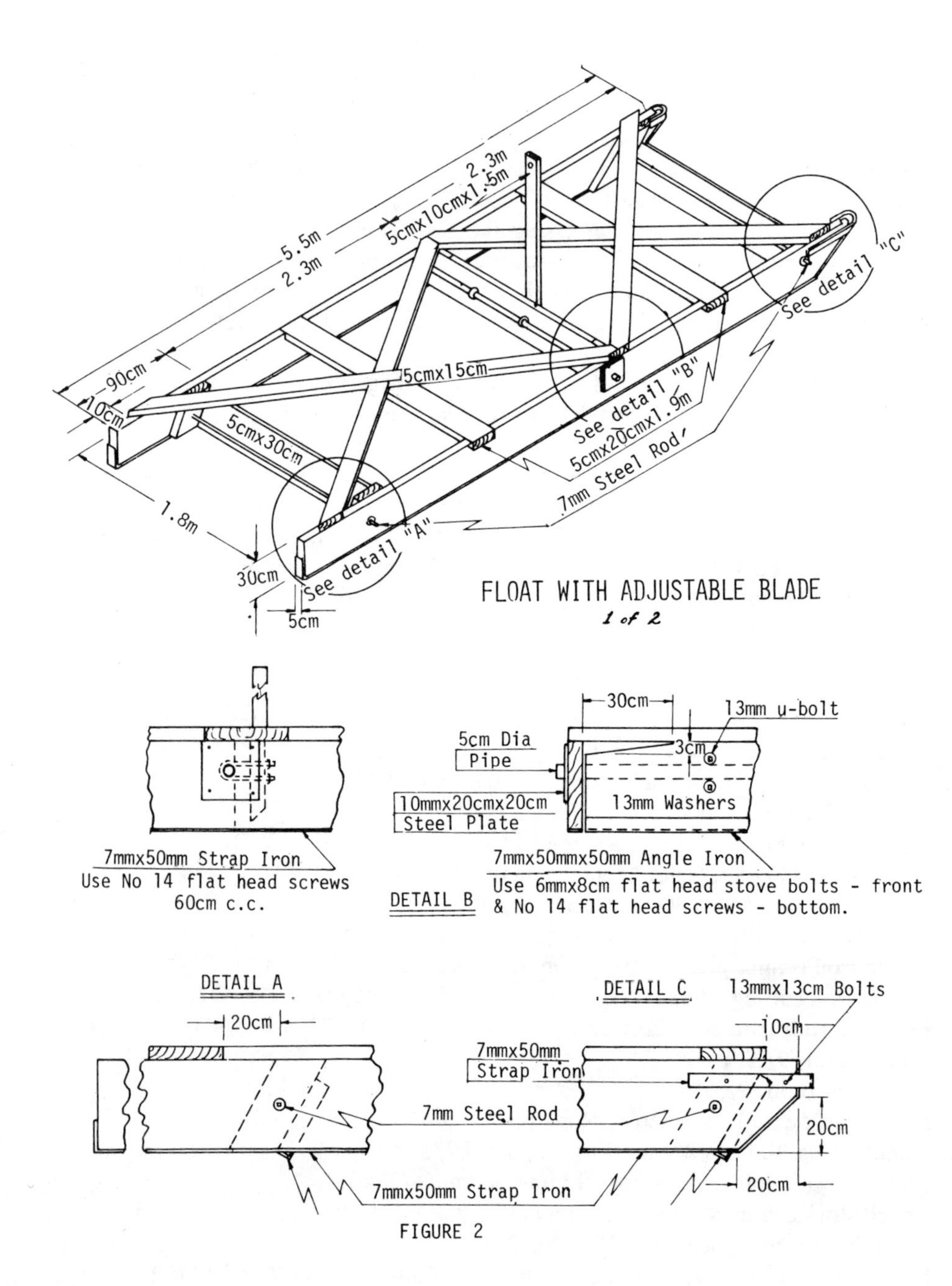


FIGURE 2

**Source:**

Forsberg, Carl M., Metzger, James D., and Steele, John C. *Construction and Use of Small Equipment for Farm Irrigation.* USOM/Turkey in cooperation with Turkish Ministry of Agriculture.

# BUCK SCRAPER

This buck scraper, which has been designed for use with large horses or oxen, may be used for leveling small humps of earth where the haul distance is short. It can be made by a small manufacturer or by a carpenter-blacksmith if equipment and materials are available.

FIGURE 1
BUCK SCRAPER FOR WET AND DRY LEVELING

After using the fresno to move large amounts of soil from high spots to low spots, the surface of the cut and fill areas will usually be rough. The buck scraper is useful for smoothing out the uneven spots caused by the fresno. It can be used for filling ditches or for smoothing border irrigation systems. After the border levee has been made, it is very important to smooth the area between and close to the levees. The buck scraper can be used very effectively for this purpose by shortening the hitch on one side and allowing the blade of the scraper to run at an angle, thus shoving the soil into the the rough areas around the newly-built levee.

Earth moving may be aided by loosening the soil to be moved by plowing before using the buck scraper.

The buck scraper is loaded by pushing down on the handle as the equipment moves forward. The handle must be held down while the soil is being transported. The scraper is unloaded by lifting up on the handle. A shallow spread is made by lifting the handle slightly and a deeper spread by pushing the handle farther forward.

The most common hitches with the buck scraper are:

- 2 oxen
- 2 horses
- 3 horses

The buck scraper may be made in different sizes according to the available power.

**Tools and Materials**

1 buck board, 5cm x 30cm x 183cm (2" x 12" x 6')
1 trailer board, 5cm x 30cm x 122cm (2" x 12" x 4')
1 iron pipe handle, 3cm x 2cm (1 3/16" x 3/4")
1 strap iron cutting edge, 6mm x 10cm x 183cm (1/4" x 4" x 6')
4 strap iron hinges, 6mm x 4cm x 30cm (1/4" x 1 9/16" x 12")
2 strap iron, 6mm x 4cm x 30cm (1/4" x 1 9/16" x 12")
2 strap iron pipe clamps, 6mm x 4cm x 15cm (1/4" x 1 9/16" x 6")
2 band strap iron pipe clamps, 6mm x 4cm x 20cm (1/4" x 1 9/16" x 7 7/8")
1 bolt for hinge, 16mm x 46cm (5/8" x 18 1/8")
2 eye bolts, 16mm x 9cm (5/8" x 3 1/2")
2 carriage bolts, 13mm x 13cm (1/2" x 5 1/8")
4 carriage bolts, 13mm x 10cm (1/2" x 4")
22 carriage bolts, 13mm x 8cm (1/2" x 3 1/8")
2 washers, 16mm (5/8")
28 washers, 13mm (1/2")
1 chain or cable hitch, 5mm (3/16")

**Source:**

Forsberg, Carl M., Metager, James D., and Steele, John C. *Construction and Use of Small Equipment for Farm Irrigation.* USOM/Turkey in cooperation with Turkish Ministry of Agriculture.

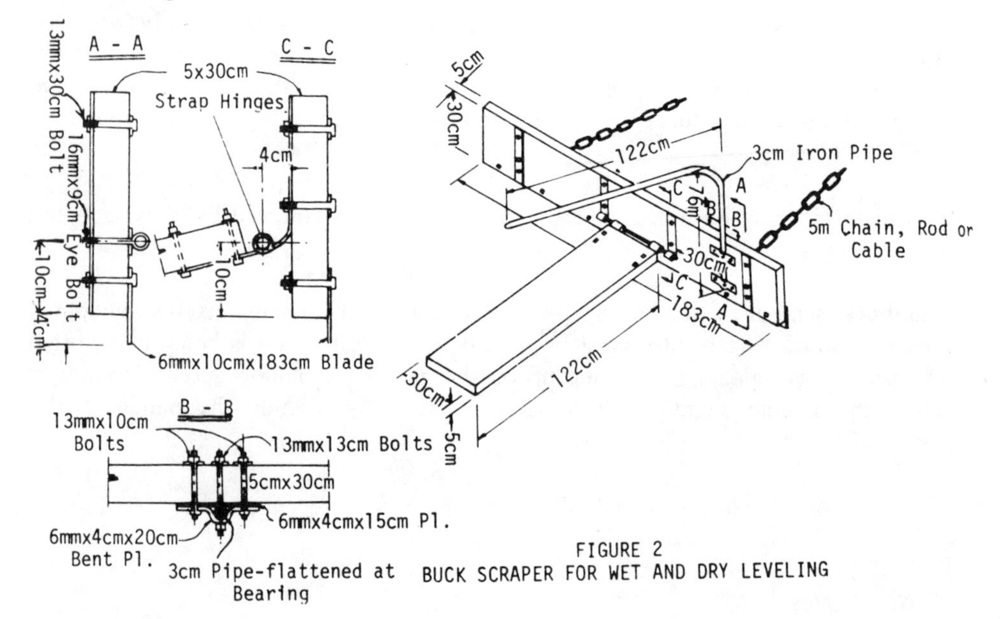


FIGURE 2
BUCK SCRAPER FOR WET AND DRY LEVELING

# V-DRAG

The V-Drag (Figure 1) is used for making ditches for irrigation and drainage of fields and roads. It can also be used to make levees (banks) or borders for border irrigation. The V-Drag can be made locally by carpenter-blacksmiths or small manufacturers if materials are available. (See Figure 2 for list of materials and construction details.)

FIGURE 1

## Plowing Ditches

After the desired ditch line has been established by means of a level or transit, the plow may be used to make a furrow where the line has been staked. Plow along the line one way, then turn and plow back again in the same furrow but throwing the soil the other way.

When the furrow has been plowed, use the V-Drag to move the soil out of the furrow. By making a complete round (down and back) the soil can be thrown out on both sides of the ditch. By alternately plowing and using the V-Drag to throw the soil out, any desired depth of ditch can be obtained.

The method of hitching the animals is important. If two horses are used, it is necessary to hitch them far enough apart so that both can walk outside the ditch. If two oxen are used it is important that the yoke be long enough to permit each animal to walk on the outside of the ditch.

If the soil is hard and more power is required, three horses can be used and one horse can walk in the ditch and one on each side.

The depth of cut made by the V-Drag can be adjusted to the available power. Shortening the hitch will reduce the depth of cut as will shifting your weight to the back of the drag.

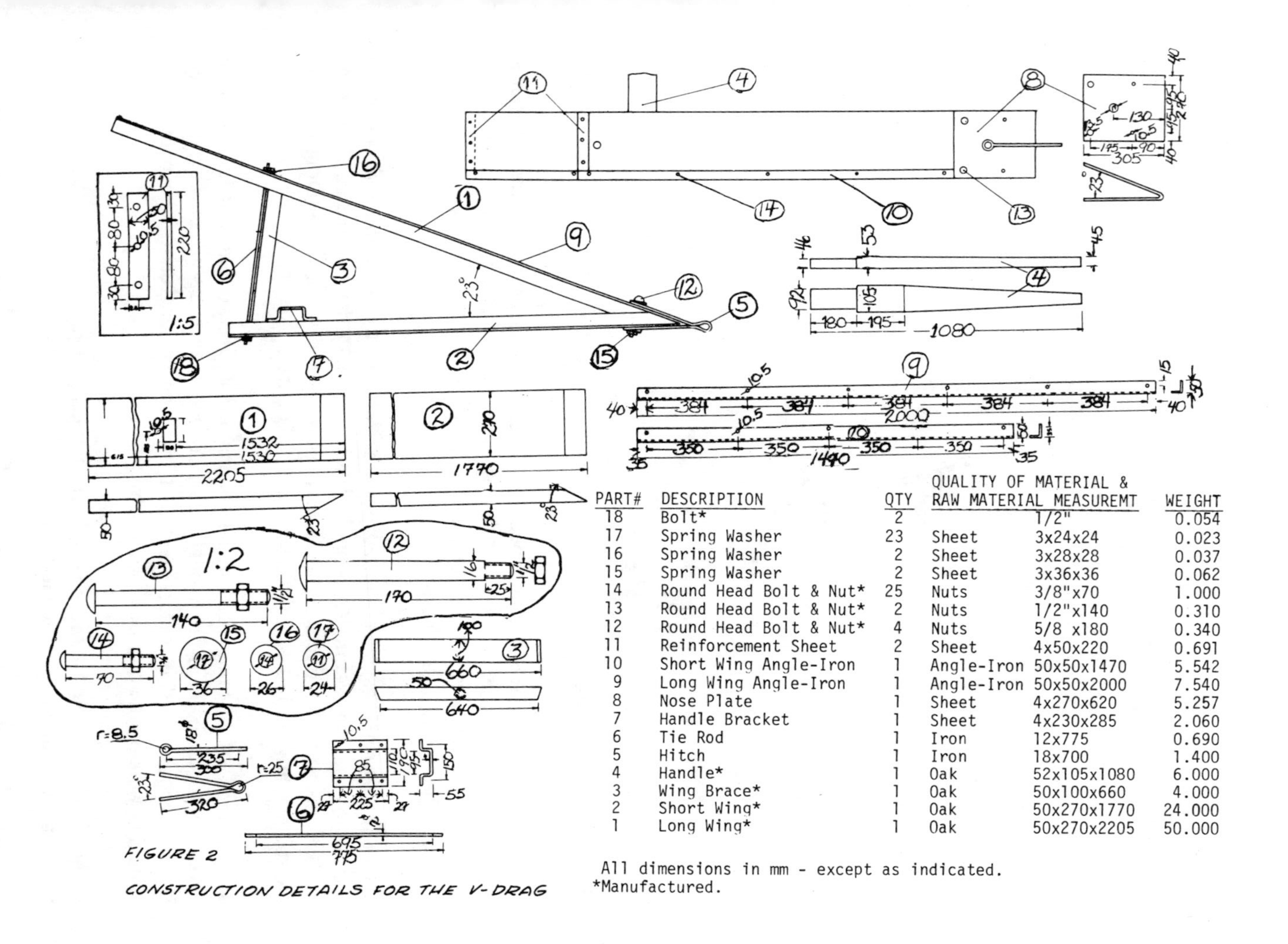

| PART# | DESCRIPTION | QTY | QUALITY OF MATERIAL & RAW MATERIAL MEASUREMT | | WEIGHT |
|---|---|---|---|---|---|
| 18 | Bolt* | 2 | | 1/2" | 0.054 |
| 17 | Spring Washer | 23 | Sheet | 3x24x24 | 0.023 |
| 16 | Spring Washer | 2 | Sheet | 3x28x28 | 0.037 |
| 15 | Spring Washer | 2 | Sheet | 3x36x36 | 0.062 |
| 14 | Round Head Bolt & Nut* | 25 | Nuts | 3/8"x70 | 1.000 |
| 13 | Round Head Bolt & Nut* | 2 | Nuts | 1/2"x140 | 0.310 |
| 12 | Round Head Bolt & Nut* | 4 | Nuts | 5/8 x180 | 0.340 |
| 11 | Reinforcement Sheet | 2 | Sheet | 4x50x220 | 0.691 |
| 10 | Short Wing Angle-Iron | 1 | Angle-Iron | 50x50x1470 | 5.542 |
| 9 | Long Wing Angle-Iron | 1 | Angle-Iron | 50x50x2000 | 7.540 |
| 8 | Nose Plate | 1 | Sheet | 4x270x620 | 5.257 |
| 7 | Handle Bracket | 1 | Sheet | 4x230x285 | 2.060 |
| 6 | Tie Rod | 1 | Iron | 12x775 | 0.690 |
| 5 | Hitch | 1 | Iron | 18x700 | 1.400 |
| 4 | Handle* | 1 | Oak | 52x105x1080 | 6.000 |
| 3 | Wing Brace* | 1 | Oak | 50x100x660 | 4.000 |
| 2 | Short Wing* | 1 | Oak | 50x270x1770 | 24.000 |
| 1 | Long Wing* | 1 | Oak | 50x270x2205 | 50.000 |

All dimensions in mm - except as indicated.
*Manufactured.

FIGURE 2

CONSTRUCTION DETAILS FOR THE V-DRAG

Either lengthening the hitch or shifting your weight to the front will increase the depth of cut.

The handle of the V-Drag can be used to vary the width of the ditch. Pressing down will widen it while lifting up will narrow the width.

## Building Levees or Borders for Border Irrigation Systems

After the desired location has been selected for constructing a levee, or border, the plow may be used to plow down and back twice and throw the soil into the border line. The V-Drag can then be used to crowd the soil into a ridge.

When a border irrigation system is constructed in this manner it is necessary to smooth around the border with a buck scraper (see page 217). If the hitch on the scraper is shortened on one side it will roll the dirt into the border.

## The Hitch

The hitch on the V-Drag is the same for construction of a ditch or a border. Two horses, two oxen, or three horses are usually satisfactory.

**Source:**

Forsberg, Carl M., Metzger, James D., and Steele, John C. *Construction and Use of Small Equipment for Farm Irrigation.* USOM/Turkey in cooperation with Turkish Ministry of Agriculture.

# MULTIPLE HITCHES

Multiple hitches or load eveners are necessary when more than one animal is used for pulling equipment to adapt the proper power to the load and the job. Correctly-made hitches enable each animal to do its share of the work and exert an even pull on a piece of equipment.

Various combinations of hitches may be used, according to the job. The most common hitches are:

- o double trees, or 2-horse evener
- o 3-horse evener
- o 4-horse evener
- o 6-horse evener

Figure 1 shows a four-horse evener and a three-horse evener. This illustration is helpful in reading the construction details in Figure 3.

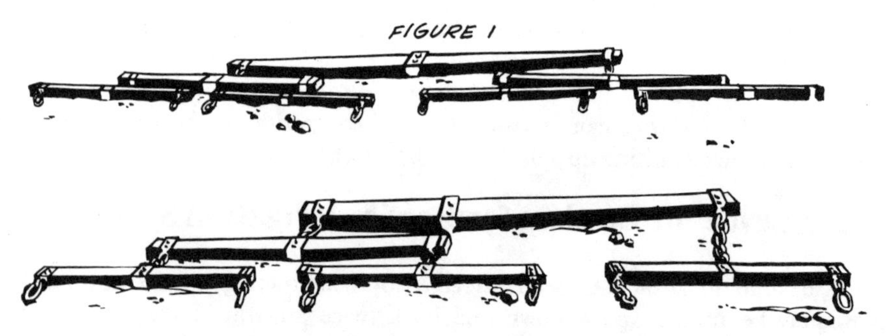


The major parts of the hitches can be adapted for use with oxen or bullocks. Figure 2 shows simpler eveners that can be used with horses, oxen, or bullocks.

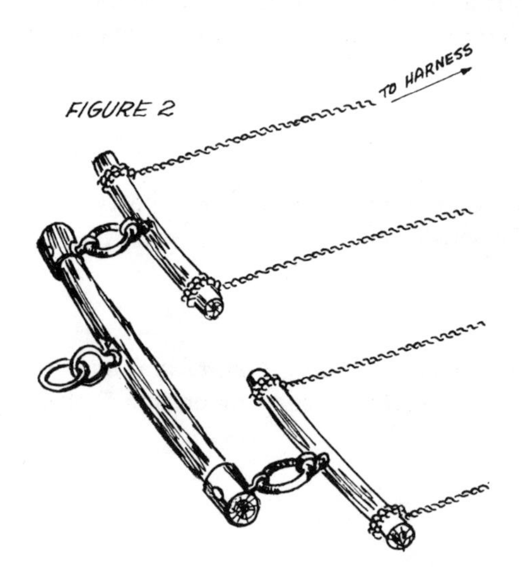


**Sources:**

Hoffen, H.J. *Farm Implements for Arid and Tropical Regions.* Rome: Food and Agriculture Organization of the United Nations, 1960.

Forsberg, Carl M., Metzger, James D., and Steele, John C. *Construction and Use of Small Equipment for Farm Irrigation.* USOM/Turkey, in cooperation with the Turkish Ministry of Agriculture.

Watson, Peter R. *Animal Traction.* Washington, D.C.: Peace Corps and TransCentury Corporation, 1981.

## Tools and Materials

**2-horse evener:**

1 oak plank, 4cm x 10cm x 1 meter (1 9/16" x 4" x 39 3/8")
2 oak bars, 4cm x 6cm x 77cm (1 9/16" x 2 3/8" x 30 5/16")
4 strap irons, 10mm x 4cm x 22.5cm (3/8" x 1 19/16" x 8 7/8")
4 machine bolts with nuts and washers, 13mm x 8cm (1/2" x 3 1/8")
2 carriage bolts with nuts and washers, 10mm x 12cm (3/8" x 4 3/4")

**3-horse evener:**

1 oak plank, 4cm x 12cm x 1.52 meters (1 9/16" x 4 3/4" x 59 7/8")
1 oak bar, 4cm x 6cm x 77cm (1 9/16" x 2 3/8" x 30 15/16")
2 strap iron, 10mm x 4cm x 46.5cm (3/8" x 1 9/16" x 18 5/16")

2 strap iron, 10mm x 4cm x 34cm (3/8" x 1 9/16" x 13 3/8")
4 machine bolts with nuts and washers, 13mm x 8cm (1/2" x 3 1/8")
2 carriage bolts with nuts and washers, 10mm x 14cm (3/8" x 5 1/2")
Plus material for *one* 2-horse evener

**4-horse evener:**

1 oak plank, 4cm x 16cm x 1.96 meters (1 9/16" x 6 5/16" x 78")
4 strap iron, 10mm x 4cm x 40cm (3/8" x 1 9/16" x 15 3/4"
4 machine bolts with nuts and washers, 13mm x 8cm (1/2" x 3 1/8")
2 carriage bolts with nuts and washers, 10mm x 18cm (3/8" x 7 1/16")
Plus materials for *two* 2-horse eveners

**6-horse evener:**

1 oak plank, 6cm x 20cm x 2.84 meters (2 3/8" x 7 7/8" x 9' 3 3/4")
4 strap iron, 10mm x 5cm x .45cm (3/8" x 12.5cm x 5/32")
2 machine bolts with nuts and washers, 20mm x 8cm (3/4" x 3 1/8")
2 machine bolts with nuts and washers, 20mm x 10cm (3/4" x 4")
2 carriage bolts with nuts and washers, 10mm x 22cm (3/8" x 8 5/8")
Plus materials for *two* 3-horse eveners

**Clevis** (U-shaped piece by which draft animal is connected to hitch):

1 clevis stock, 20mm x 70cm (3/4" x 27 1/2")
1 machine bolt with nut and washers, 20mm x 12cm (3/4" x 4 3/4")
(one clevis is needed for each horse)

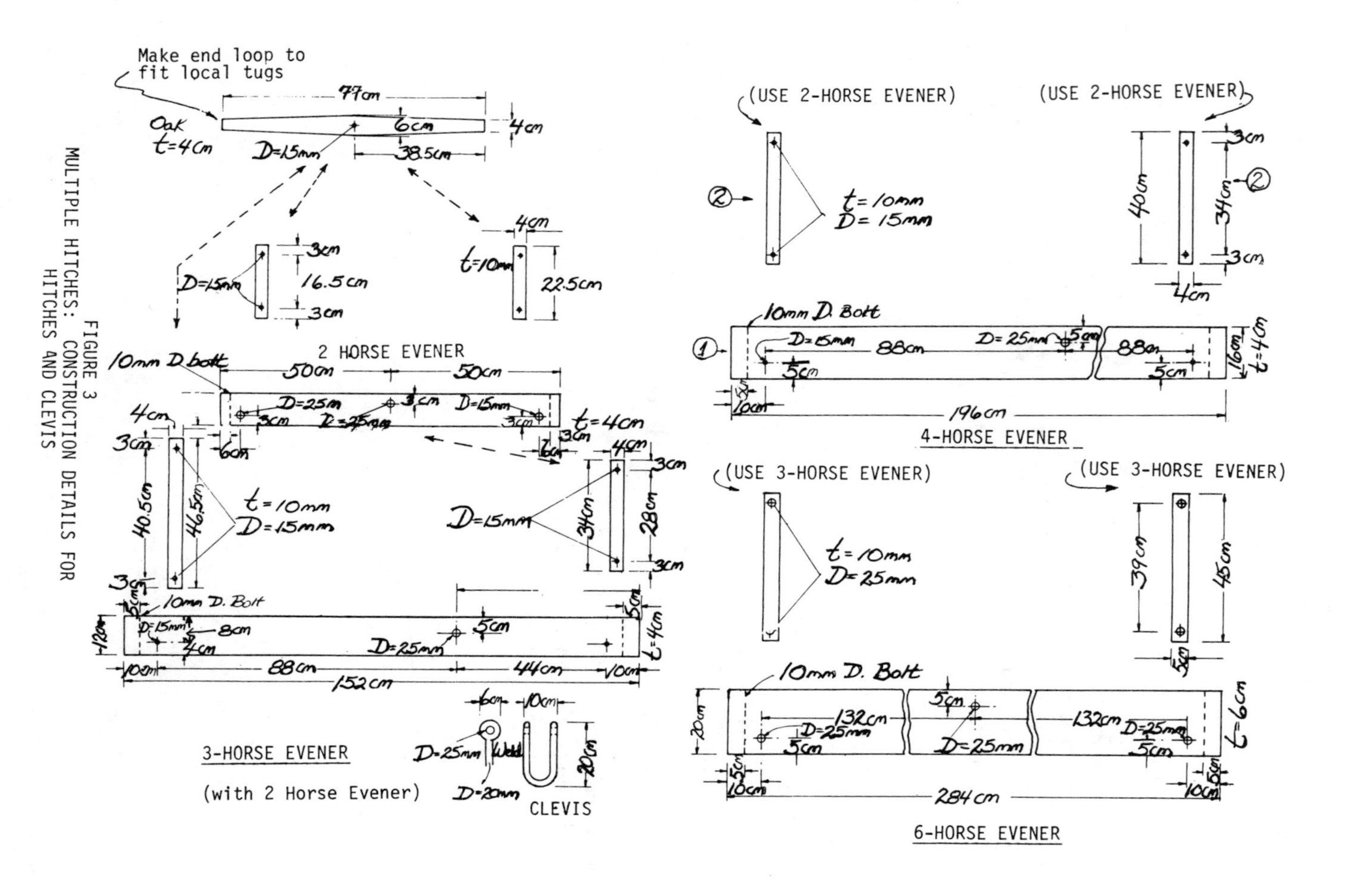


FIGURE 3
MULTIPLE HITCHES: CONSTRUCTION DETAILS FOR HITCHES AND CLEVIS

# Irrigation

## SIPHON TUBES

The galvanized metal siphon tube described here can be used for irrigation (see Figure 1). It can be easily made and repaired by tinsmiths. A siphon can also be made from a piece of rubber hose or by bending a piece of plastic tubing. Construction details are given in Figure 2.

The purpose of this siphon tube is to carry water out of a ditch without cutting a hole in the ditch bank. In many soils a small hole cut in the ditch bank soon becomes a large hole because of erosion. Imported plastic siphons are often expensive, easily broken and usually impossible for local people to repair.

There are several good ways to start a siphon tube. The simplest way is to put the tube in the ditch until it fills with water. Holding one hand over the end of the tube, so that air cannot get in, lift the tube out and place it as shown in Figure 1. Be sure the other end of the tube does not come out of the water while placing the tube. When the tube is in place, remove your hand and the water will begin to flow. The end of the tube outside the ditch must be lower than the level of the water in the ditch.

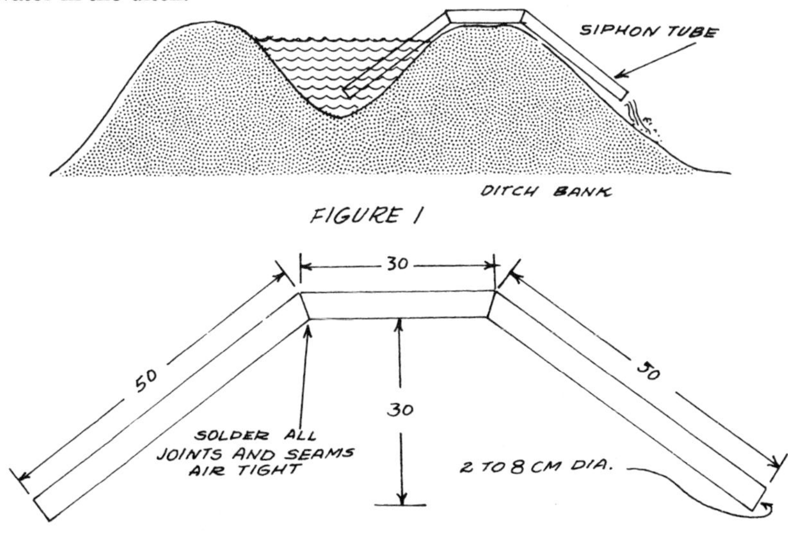


FIGURE 2

**Source:**

Dale Fritz, VITA Volunteer, Schenectady, New York

# USING TILE FOR IRRIGATION AND DRAINAGE

An irrigation or drainage system made with the concrete tiles described here can help to keep a garden in production during both wet and dry seasons. It will make good use of irrigation water and, during the wet season, will drain off surplus water.

The entries that follow explain how to make a concrete-tile machine and how to use the machine.

In regions of heavy rainfall, the tile drainage can be combined with good surface drainage by making raised beds in gardens, shoveling out 30cm (1') wide pathways that will be 15cm (6") lower than the beds. Put the beds over the tile lines and make them 1 meter (3') wide. Use the pathways also as drainage ways and connect them with a good outlet to lower ground.

This system of under-ground irrigation (and drainage) can serve under fruit trees or gardens. It can also be used around the foundations of buildings where drainage is a problem.

Concrete irrigation tiles, whether for irrigation or drainage or both are laid 30cm (12") deep in lines 1.2m (4') apart (the latter measurement depending on the texture of the soil: more distance between lines for clay soils and less for sandy soils). The garden should be almost level, with good surface drainage. Upright "elbows" at the ends of the lines give access to the tile at either end (see Figure 1). A garden hose can deliver the water from its source to the upright ends of

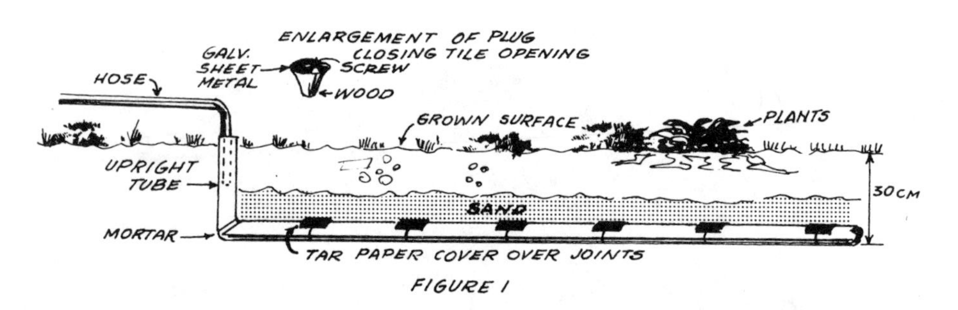


FIGURE 1

the tile lines. While tile lines must be level, they do not have to be straight; they can follow a contour line or double back to make a more convenient system of installation with four or more lines connected to make one unit (Figure 2).

In dry seasons, the tiles supply water to the plant roots. In wet seasons, the water escapes through the sand and gravel around the tile and follows the concrete tube formed by the tiles to a drainage outlet (see Figure 2). While passing downward through the soil to the tile, the water draws air into the soil and supplies oxygen to the helpful bacteria and to the plant roots.

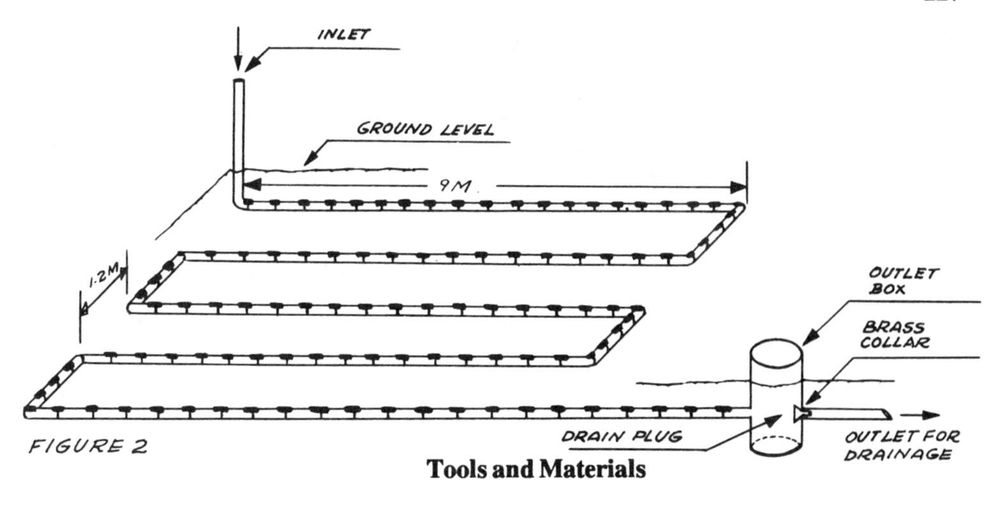


FIGURE 2

**Tools and Materials**

Concrete tile
Cement for mortar, concrete
Sand for mortar and tile covering
Gravel or crushed stone for concrete
Wood for plugs
Optional - Brass outlet box collar
Shovels, concrete-mixing tools

To install the tiles:

- o Grade the garden plot to within 5cm to 7cm (2" to 3") of level and make trenches 30cm "12") deep, according to the design in Figure 2. This will give an even distribution of the water. Check the bottom of the tile ditches to be sure they are level. Only the drainage outlet will have a drop.

- o Lay the tile end to end in the bottom of the trench. Use an "elbow" (made of two tiles cut to 45-degree angle) to make a place for putting the hose at one end, and use other elbows to turn corners.

- o Put a piece of tar paper or used linoleum over each joint (Figure 3) to keep the dirt out of the line. A piece 5cm x 12.5cm (2" x 5") is large enough. .

TAR PAPER OR LINOLEUM IS PLACED OVER EACH TILE JOINT TO KEEP DIRT OUT OF THE LINE.

- o Cover the tile with sand to give the water an opportunity to soak out into the soil or (in the case of drainage), to seep into the tile. The bottom 12.5cm (5") of the trench are filled with sand or gravel (around the tile) and the top 17.5cm (7") are filled with soil.

- o Near the outlet, make an upright concrete box with two holes near the bottom to let drainage water run through and on out to an outlet. The box should be large enough so that one can reach into it to install a plug in the drain side of the box when the system is used for irrigation. A brass or aluminum collar installed in the concrete will make it easier to close this hole completely and thus avoid a loss of water.

- o Put covers over both ends to keep out small animals (see Figure 1).

- o Do not water more frequently than once or twice a week, so that plant roots will not enter the tile line to obstruct it.

- o Be careful not to damage the tile with tillage equipment.

- o For irrigation, the tile system is used with its drain plug securely closed (see Figure 2). Water is run into the line once or twice a week, by means of a hose, until the soil becomes moist. For drainage, simply pull the plug.

## Making a Concrete Tile Machine

This all-steel tile-making machine (Figure 1) can be made of scrap metal in any shop with welding equipment. The machine makes 80 to 100 tiles to a sack of cement. One worker can make about 300 tiles in an 8-hour day. Construction of the machine is a good welding project for students.

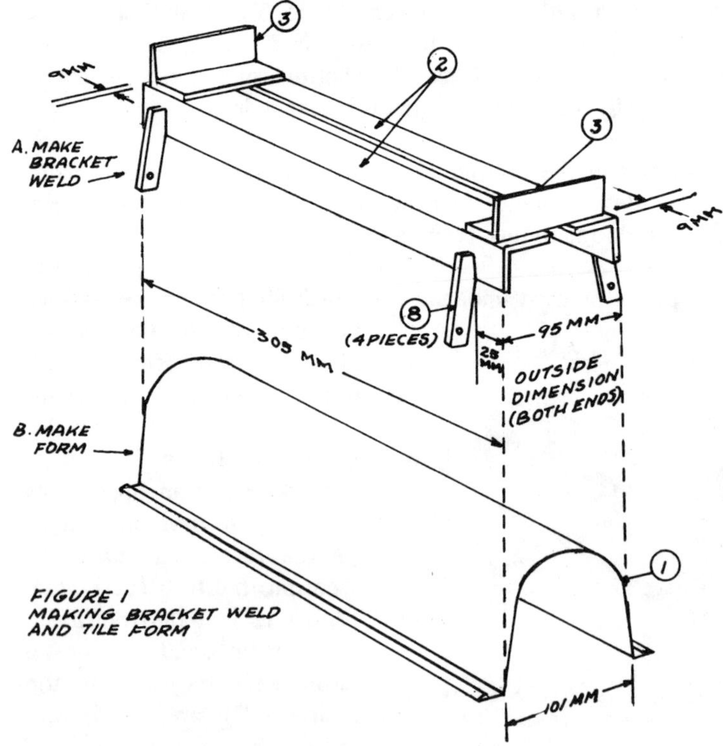


FIGURE 1
MAKING BRACKET WELD AND TILE FORM

A tile-making machine made from wood is illustrated in Figure 15. The tiles made with this machine are the same size as those made with the all-metal machine.

All the drawings of the form and its several parts in this entry show the form in its upside-down, or emptying position.

The machine can be made of used or new materials. To make the form, it is desirable to have both electric and acetylene welding equipment, although either will serve. The thicker parts are assembled by arc welding and the thinner parts have to be put through other parts before welding, as will be explained below. We shall refer to each individual part by its number, which appears on the sketches.

The assemblies made of parts No. 10, 11, and 12 (Figures 8 and 14) are simply a convenient means of taking hold of the levers to open the end doors. These levers are made of parts No. 5 and 13 as described below and shown in Figures 9, 10, and 11. They work against the two springs that hold the doors shut--the tension being made sufficient to hold the doors closed against the force of tamping.

The hole in the end door is shown as 3mm (1/8") larger than the diameter of the pipe that shapes the interior surface of the concrete tubes. This 3mm (1/8") is an allowance of clearance necessary to keep the sand particles from making the pipe difficult to remove after the mortar is tamped around it. Greater clearance would hurt the uniformity of the tile. The finished tile should have a uniform 13mm (1/2") wall and part No. 1 must be shaped and so related to the pipe that the thickness of the tile wall will be correct (see Figure 6).

Parts No. 7 are bronze welded to the sides of No. 1 (see Figure 6). These parts, like other parts that touch the hands, should be dressed to a smoothness sufficient to avoid injury to the operator. The outside of the form should be well painted but the inside cannot be painted, as paint would cause the mortar to stick to the inside. When the form is not in use, the inside should be kept oiled.

The pipe may need to be dressed lightly in the lathe to make it easier to remove from the form after the mortar is tamped around it. In turning, it is advisable to make the end opposite the handle end 0.5mm (1/64") smaller, as this will facilitate its removal in the emptying process. This lathe work should be done after the end of the pipe opposite the handle end has been welded shut with a disc of galvanized sheet metal. If this end is not closed, cement will enter the pipe and thus be spilled into the inside of the tile to become an obstruction there.

Part No. 19 is a wire of 3mm (3/32") diameter steel welding rod with the shape shown in Figure 2, but one of the eyes has to be formed after the part has been threaded through the hole in part No. 8 (see Figures 1 and 8).

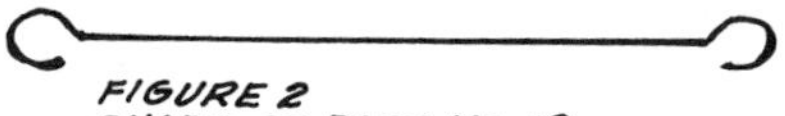

FIGURE 2
SHAPE OF PART NO. 19

The following paragraphs are listed by part numbers:

1. The inside walls of the form are made of 16-gauge galvanized iron. Part No.

1 as shown in Figure 1 is made from a sheet cut to a true rectangle, 26.6cm x 30.5cm (10 1/2" x 12"). This is bent to shape by putting a 6mm (1/4") fold on each of the 30.5cm (12") sides; bending 19mm (3/4") more of same sides to a right angle; and then shaping the sheet according to the curve shown in Figure 3. This lining is then fitted into the cradle made of parts No. 2 and 3. Parts No. 6 will be the end doors, which are also made of 16-gauge sheet iron. The inside of the form should not be painted, as this interferes with its operation.

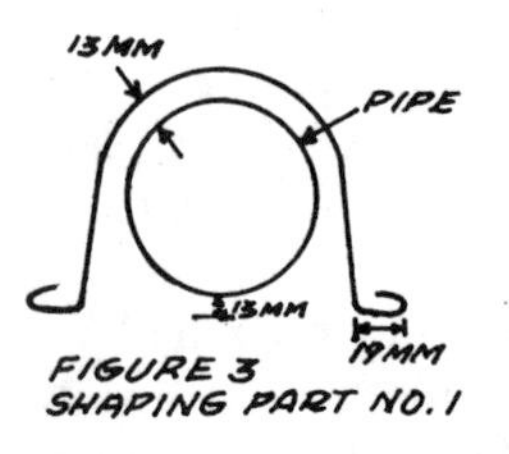

FIGURE 3
SHAPING PART NO. 1

2. For part No. 2, two pieces of angle iron, 38mm x 38mm x 3mm x 30.5cm (1 1/2" x 1 1/2" x 1/8" x 12") are needed.

3. Angle iron, 38mm x 38mm x 5mm (1 1/2" x 1 1/2" x 3/16"), 95mm (3 3/4") long. Two are needed. Parts No. 2 and 3 are welded together to form the cradle. Parts No. 8 are welded in place on parts No. 2 and corrections are made for shape before No. 1 is tack welded into the cradle thus formed. The design above gives some idea of final relationship to be kept between the sheet metal lining of the form and the metal pipe. Notice that the tile wall will be uniformly 13mm (1/2") thick (see Figures 4 and 8).

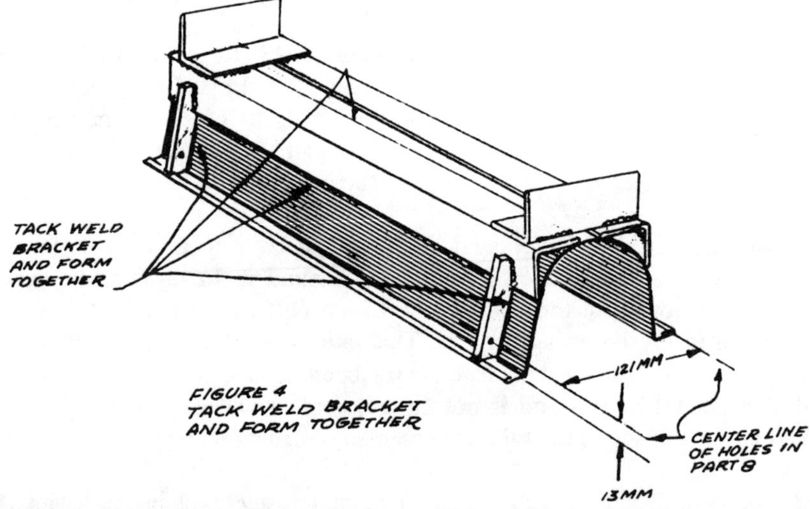

FIGURE 4
TACK WELD BRACKET
AND FORM TOGETHER

4. Mild steel rods, 10mm x 15.2cm (3/8" x 6") (see Figure 13). Two are needed. These are welded in place to make the form stand a little taller so the levers will not touch the work bench while the mortar is being tamped into the form. They also provide a wider base.

5. Mild steel rods, 10mm x 22.9cm (3/8" x 9") (see Figure 10). Four are needed. These are bent to form the levers and are welded into pairs by means of the

connecting piece, No. 13 (see Figure 9). Notice the tiny tabs welded to the handle end of the levers. These are to keep the hand hold from turning or sliding endwise from its proper position. By the "hand hold" we mean the assembly made of Parts No. 10, 11 and 12.

6. Galvanized sheet metal, 16-gauge, 14cm x 16.5cm (5 1/2" x 6 1/2"). Two are needed. These are the doors and the parts that hold the center pipe in its proper position. They should be cut and shaped after Part No. 1 has been tack-welded in its place (see Figure 5).

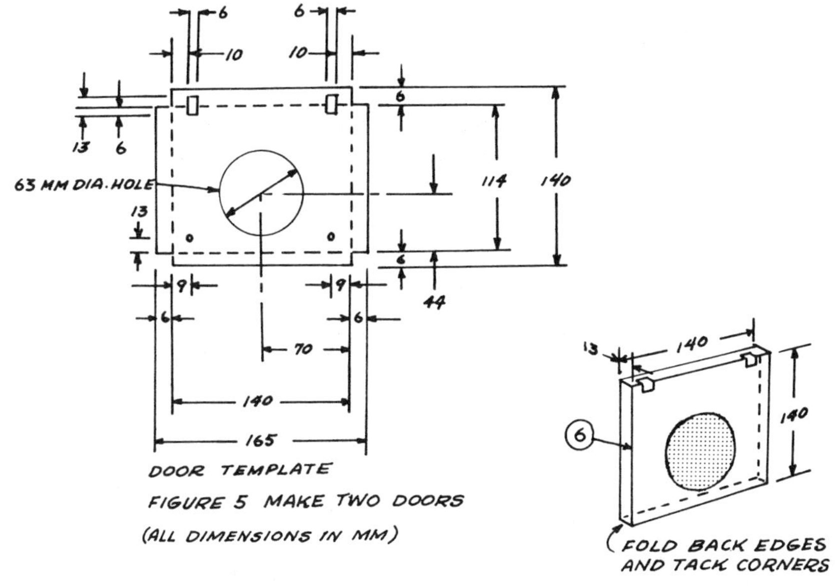


DOOR TEMPLATE
FIGURE 5 MAKE TWO DOORS
(ALL DIMENSIONS IN MM)

7. Galvanized sheet metal, 16-gauge, 38mm x 10.2cm (1 1/2" x 4"), bent to angle as shown in Figure 6. Two are needed. These are handles for lifting the form. They are dressed smooth and bronze welded to the sides of No. 1 after the doors are properly installed as explained under No. 15 below.

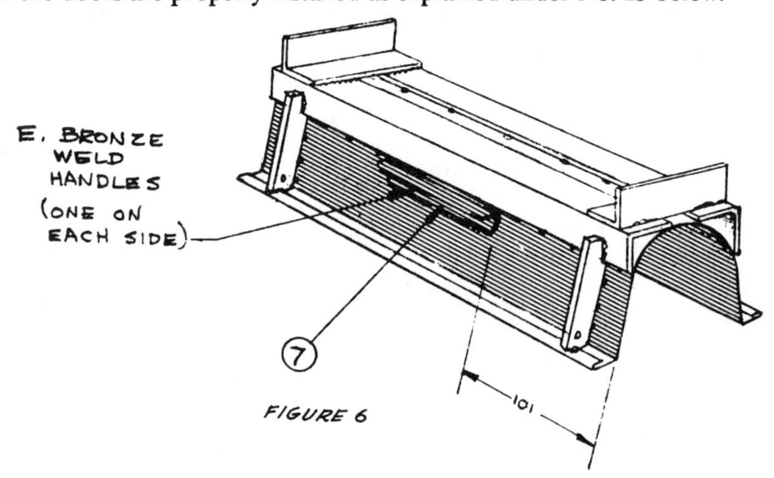


FIGURE 6

8. Mild steel bar, 19mm x 6mm x 7cm (3/4" x 1/4" x 2 3/4"). Four are needed (see Figure 1). They are welded to No. 2 to complete the cradle for the lining of the form. Then the lining, part No. 1 is welded to No. 8 at the fold in the edge of No. 1. Check to see that the space for the thickness of the tile wall remains 13mm (1/2").

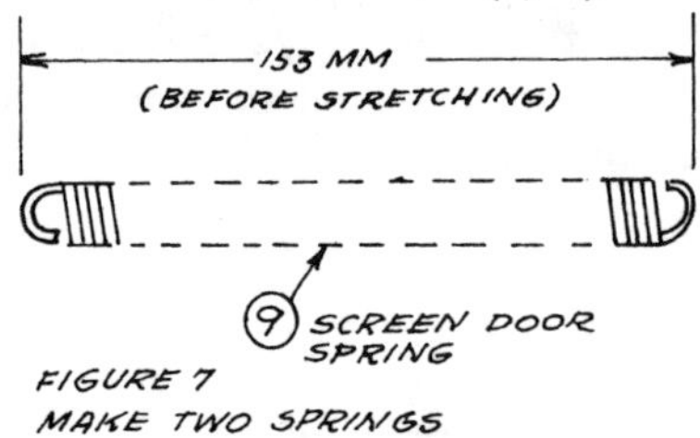

FIGURE 7
MAKE TWO SPRINGS

9. Screen door spring, cut into coils as shown, 14cm (5 1/2") long with the end loops bent out to form eyes. Two are needed (see Figure 7).

10. Channel iron, 31mm x 19mm x 8.2cm (1 1/4" x 3/4" x 3 1/4"). Two are needed. Countersink hole for screw head. Dress parts No. 10 and 11 smooth as they are handles.

11. Strap iron, 2.5cm x 3mm x 8.2cm (1" x 1/8" x 3 1/4"). Two are needed (see Figures 8 and 14). Drill and thread hole to match the screw hole in part No. 10. Make guide holes for the round tabs that are welded to the end of the levers, No. 5. The tabs on No. 5 is made by sawing off a 10mm (3/8") length of 10mm (3/8") diameter rod and bronze welding it to the end of the handle as shown.

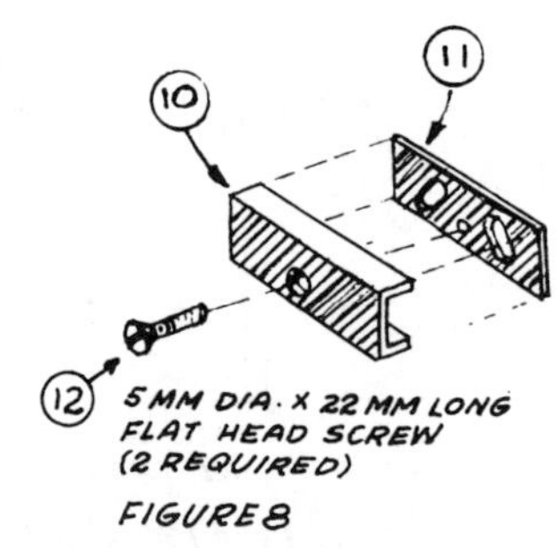

FIGURE 8

12. Machine screw, flat head, 6mm x 19mm (1/4" x 3/4"). Two are needed. This unites No. 10 and 11.

13. Mild steel rod, 9mm x 12.7cm (3/8" x 5"). Two are needed (Figure 9 and 11). Parts No. 5 are made in pairs by welding to the ends of part No. 13. Before welding, insert part 13 in the tube, No. 14, which will become the pivot (after No. 14 is welded to the inside angle of No. 3). Thus we have the levers that open the doors.

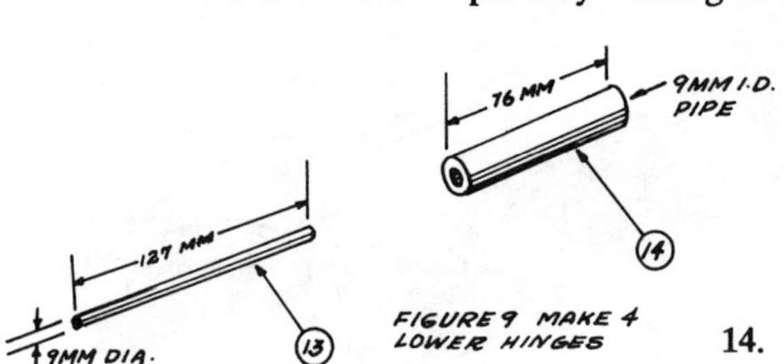

FIGURE 9 MAKE 4 LOWER HINGES

14. Pipe, 10mm (3/8"), 7.6cm (3") long; two are needed. They form the pivots for levers.

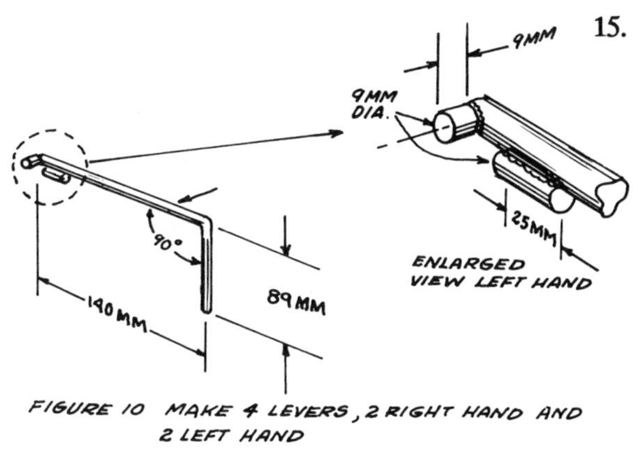

FIGURE 10 MAKE 4 LEVERS, 2 RIGHT HAND AND 2 LEFT HAND

15. Steel welding rod, 6mm x 10.8cm (1/4" x 4 1/4"). The ends are ground flat and smooth. Two are needed (see Figure 14). These are the hinge pins for the doors.

After the hinge holes, No. 16, are welded to part No. 3, parts No. 15 are put in place in the holes.

Then parts No. 6, the doors, are put in place, checked for exact position and bronze welded to the hinge pins, No. 15. This weld extends almost the entire distance between one pivot hole (part No. 16) and the other. The weld holds the door to the hinge pin and prevents the hinge pin from sliding out of place.

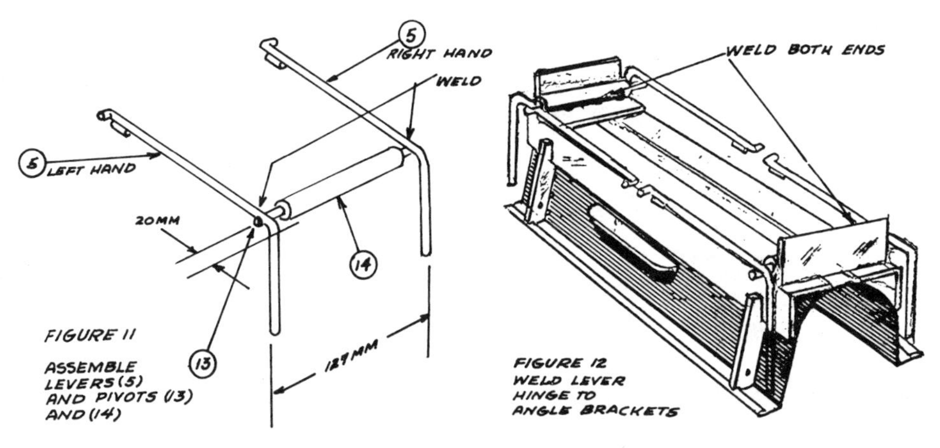

FIGURE 11 ASSEMBLE LEVERS (5) AND PIVOTS (13) AND (14)

FIGURE 12 WELD LEVER HINGE TO ANGLE BRACKETS

16. Steel bar, 19mm x 2.5cm x 6mm (3/4" x 1" x 1/4") (see Figure 13). Four are needed. Bore 6mm (1/4") hole for the hinge rod as shown. No. 15 pivots in these holes to make hinges for the doors. Parts No. 16 are welded to part No. 3 in such position as to be as far to the outside edge of the door as possible. It is best to make a trial positioning of the door and parts No. 15 and 16 by tack welding No. 16 lightly before welding it permanently. Then it is possible to make sure that the door is going to be in such place that the pipe will have its proper position.

17. Common nails, 6 penny, with strong heads (see Figure 14). Four are needed. Connect the nail to the spring by a wire through the hole in No. 8. Put the wire through the holes before forming the second end loop.

18. Piston, 5cm (2") galvanized pipe, 40.6cm (16") long. (The 5cm (2") measurement is the inside diameter of the pipe.) Weld one end shut by bronze welding a metal disc to the end. Then dress lightly in the lathe, making the closed end 0.5mm (1/64") smaller than the other. It will serve well without turning, but will be easier to operate it dressed.

19. Wire or welding rod, 2mm (3/32") to make the connection between parts No. 9 and 17 (see Figures 2 and 14).

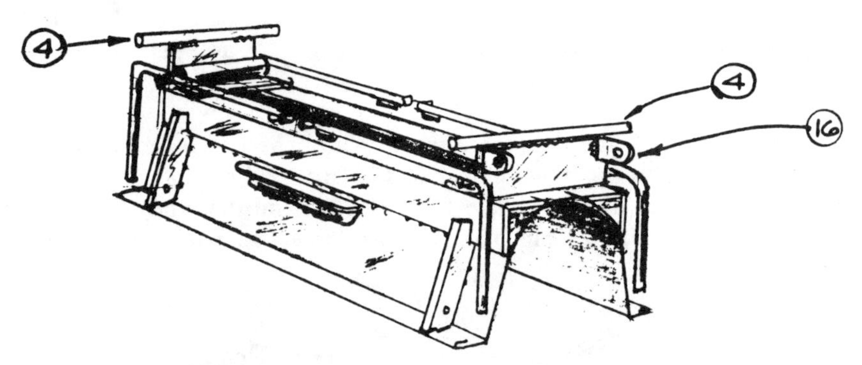


FIGURE 13 WELD ON HINGE PIECES (16) ON BOTH ENDS USING DOORS (6) AS GUIDES. ALSO WELD ON TWO RODS (4).

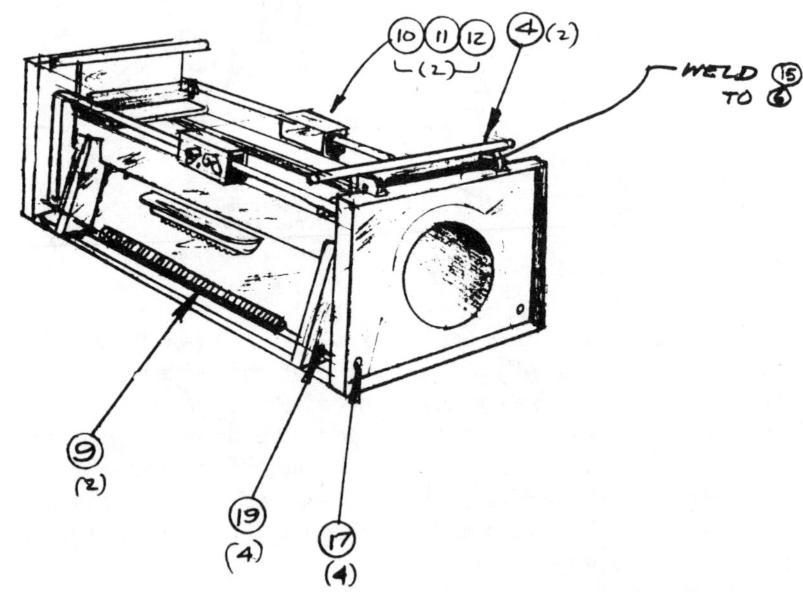


FIGURE 14. ATTACH DOORS WITH HINGEPIN (15), WELD IN PLACE, AND PUT SPRINGS (9), CONNECTORS (19), AND ATTACHMENTS (17) IN PLACE. UNIT IS NOW COMPLETE EXCEPT FOR PARTS 10, 11 AND 12, WHICH NOW CAN BE INSTALLED.

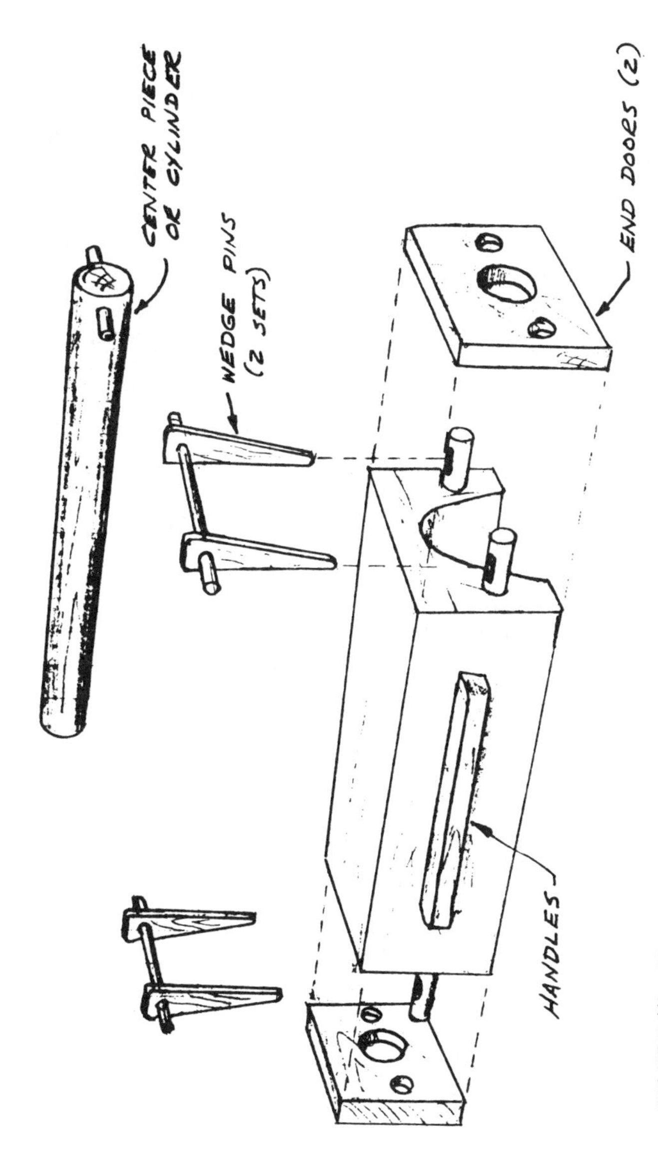


FIGURE 15 ALL-WOOD CEMENT-TILE MACHINE (DIMENSIONS OF TILE SAME AS ALL-METAL MACHINE). BLOCK IS STRONGER IF MADE FROM 3 PIECES JOINED LENGTHWISE.

# Making the Tile

It is possible for one worker to make two tiles per minute, although a good day's work would be 300 or more. The mortar remains in the form only a few seconds. The cement mixture is tamped into the form with a tamper. Then the form is immediately turned upside-down on a (slightly oiled) concrete floor and emptied, leaving the tile completed and ready to start its curing process. The same general method can be adapted for the wooden tile-making machine in Figure 15 of the preceding entry.

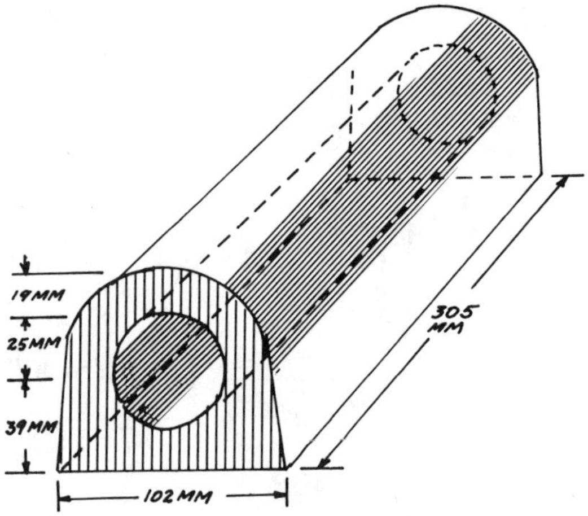

FIGURE 1 THE FINISHED TILE

### Tools and Materials

Fresh Portland Cement
Clean sand, screened through a 6mm (1/4") screen
Clean water
All-metal tile machine
Metal tamper
Plastering trowel
Work bench
Shop with concrete floor
One (11-liter) bucket
D-handled shovel (square point)
Large hoe for mixing cement
A strong dust pan without a handle.
Gloves

Make the tile by following these steps:

1. Screen the sand and spread out 28 liters (1 cubic foot) on the shop floor. Use a 28-liter (1 cubic foot) measuring box without a bottom.

2. Spread 7 liters (1/4 cubic foot) cement over the sand. Measure in the box, filling it 1/4 full.

3. Mix thoroughly with shovel and hoe. Turn over the pile four to six times.

4. Spread the pile out and scatter the mixing water over it. The amount of water should be no more than 2/3 the volume of cement, including any water in the damp sand. The mix should be as dry as possible and still be plastic.

5. Make the batch into tile before 45 minutes of time elapses. Cement loses its strength if put into the form too long after mixing.

6. Fill the form (without the pipe) 1/4 full and tamp the ends with two strokes with the (gloved) left hand. This gives the tile perfect ends.

7. Insert the pipe and fill the form with mortar, using one dip from a strong dust pan without a handle.

8. Tamp the sides of the tile. Make three strokes with the iron tamper.

9. Fill the form again, with another dip from the dust pan.

10. Turn the tamper over and pack the cement again. Give three strokes with the flat surface of the tamper.

11. Use the trowel to finish the tile. Strike off the surplus with one stroke and leave the surface trowelled level with a second stroke.

12. Carry the tile and form to a place where the floor has been lightly oiled. In carrying the form, do not touch the pipe.

13. Place the form carefully on its side on the floor and then tip it quickly to an upside down position. Hesitation in the middle of the tipping action may cause the mortar to fall out.

14. Pull out the pipe, turning it slightly first. Hold the form down with one hand. If the pipe is too hard to remove, it may have irregularities and need to be dressed lightly in the lathe.

15. Lay the pipe on top of the form. This gives the form a slight jar.

16. Gripping the sides of the form with both hands, push down on the levers, which open the hinged ends, and then lift the form off the tile. In lifting use leg action and hip action. Bending the elbows may knock an end off the tile.

17. Leave the tile in its place on the floor over night. Sprinkle very lightly with water if it begins to get dry. To dry at this stage would ruin it.

18. The next day the tile can be picked up by gripping it at its middle with the hand. Stack the tile at the side of the shop to clear the center floor space for another day of production. The first day, stack only two layers high, as the tile is not strong yet. The second day, they can be stacked as high as desired.

19. When tiles are one day old, it is a good time to make 45-degree ends on tile that have been injured in manufacture. About 5 percent (or more) of the tile made will need a 45-degree end for use in turning corners in the tile line.

20. **Keep the tile wet at least a week.** The strength is increased by each day that the tiles are kept wet.

If you need further instruction on the fundamental principles of good concrete construction, study the entries on concrete.

**Source:**

Brown, J. Oscar. *A Machine for Making Concrete Tile for Irrigation and Drainage.* O.T.S. Information Kit, Vol. 2, No. 2. Washington, D.C.: U.S. Department of Commerce, 1961.

# Seeds, Weeds, and Pests

## SEED CLEANER

This seed cleaner was developed in Afghanistan to remove round seeds of weeds from wheat grains. The round seeds could not be separated by a sieve because they were the same size as the wheat grains. The cleaner described here takes advantage of the round shape of the weed seeds to separate them from the wheat. The wheat grains, which roll down the chute slowly, collect at the base of the inclined platform ("x" in Figure 1); while the round seeds roll faster and fall off the side opposite the chute ("y" in Figure 1).

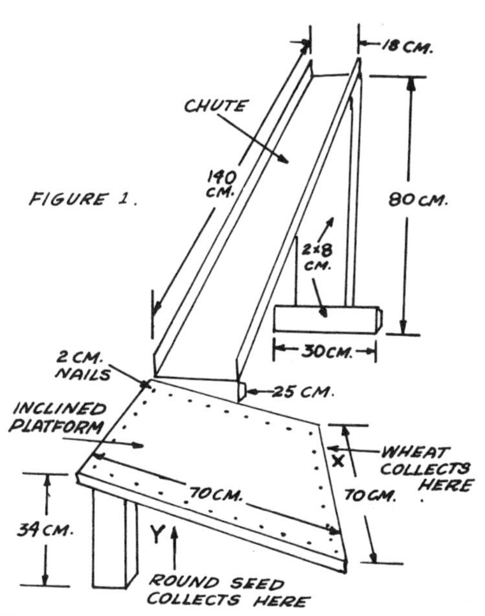


**Tools and Materials**

Hammer, saw
Nails or screws

**Inclined Platform:**

Galvanized iron sheet: 70cm x 70cm (2'3" x 2'3")
Wood: 2cm x 4cm x 68cm (4 pieces) (3/4" x 1 1/2" x 2'2 3/4")
Wood: 2cm x 4cm x 25cm (1 piece) (3/4" x 1 1/2" x 10")
Attached to platform to support chute
Wood: 2cm x 8m x 34cm (2 pieces) (3/4" x 3" x 1'3 1/2")
Legs for platform

**Chute:**

Galvanized iron sheet: 24cm x 140cm (9 1/2" x 4'7")
Wood: 2cm x 8cm x 80cm (1 piece) (3/4" x 3" x 2'7")
Wood: 2cm x 8cm x 80cm (1 piece) (3/4" x 3" x 12")

As shown in Figure 1, the chute is attached at the top of the 80cm (2'7") support by nails whose heads have been removed. This makes it easy to remove the chute when it is not being used. The chute's lower end sits on the 2cm x 4cm x 25cm (3/4" x 1 1/2" x 10") support attached to the platform.

The seed should first be cleaned with sieves to remove as much dirt and chaff as possible. To use the seed cleaner, drop the seed **very slowly** onto the top of the chute.

**Source:** Dale Fritz, VITA Volunteer, Schenectady, New York

# SEED CLEANING SIEVES

An important step for improving crop production is the effective cleaning of crop seeds. The sieves described here have been found effective in many countries.

**Tools and Materials**

Wood: 12 pieces: 2.5cm x 5cm x 46cm (1" x 2" x 18")
Wood strips: 12: 1cm x 2.5cm x 43.5cm (1/2" x 1" x 17")
Galvanized screen:
  6mm (1/4" mesh: 46cm (18") square
  5mm (3/16") mesh: 46cm (18") square
  3mm (1/8") mesh: 46cm (18") square

Hammer, saw, nails

The exact size of these sieves is not importat, but 3mm (1/8"), 5mm (3/16"), and 6mm (1/4") mesh make convenient sizes for cleaning wheat, barley, corn, and seeds of similar size. The sieves are also useful for grading certain seeds. Grading consists of removing the small, weak seeds, which will produce small weak plants or will not grow at all. Less seed can be planted per acre, if it is properly cleaned and graded, and still produce a good crop.

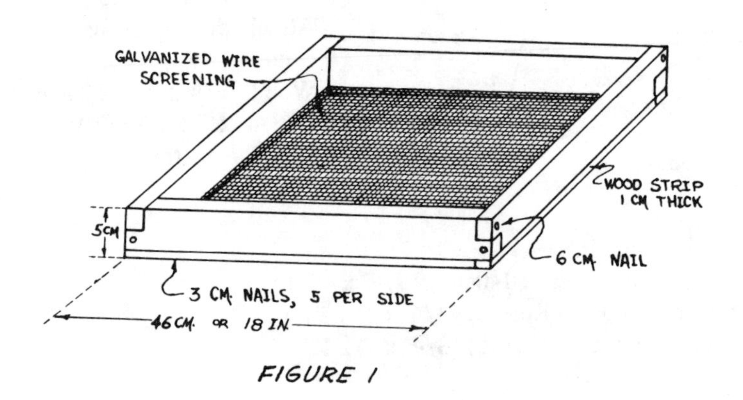


FIGURE 1

**Source:**

Dale Fritz, VITA Volunteer, Schenectdy, New York

# DRYING GRAIN WITH WOODEN BLOCKS

Small blocks of wood treated with calcium chloride, a low-cost chemical, can be used to dry grain to be used as seed. The blocks, which absorb moisture from the grain, can be used repeatedly by drying them in an oven after use. The blocks can absorb water up to one-fourth their weight.

In a test using balsa blocks, the moisture content of grain dropped from 17 percent to 12 percent in three days. The blocks were not dried at this point; in the next five days, moisture content did not change. The blocks were then dried in an oven and put back in with the grain. Three more days of drying brought the moisture content down to 10 percent, at which grain resists mold and insects.

### Tools and Materials

- Balsa or cedar: Cedar absorbs water and is durable. Balsa absorbs more water, but it breaks easily. Other wood can also be used.

- Calcium chloride ($CaCl_2$): Add enough to a liter of water to make the solution weigh 1/2kg (or to a quart of water to make the solution weigh 2.5 pounds).

- Waterproof chest that will keep out vapor, to dry and store the grain. A steel drum or sheet metal cabinet would be good. A wooden chest can be used if it is vapor-proof, as in Figures 1, 2, and 3.

- Coarse Screen: 2.5cm (1") mesh

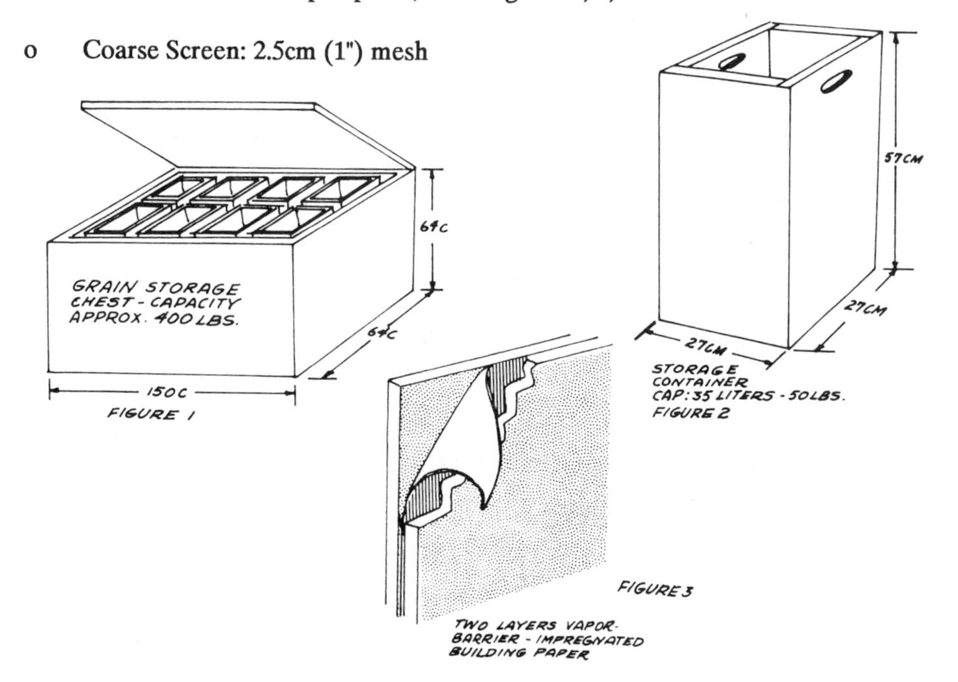

## Preparing the Blocks

- Cut the wooden blocks so that as much as possible of the surface is end grain. A good size is 3cm x 3cm x 0.75cm (1" x 1" x 1/2").

- Dry the blocks in a 90-100$^{o}$C (194-212$^{o}$F) oven or double boiler to remove all moisture (see Figures 4 and 5).

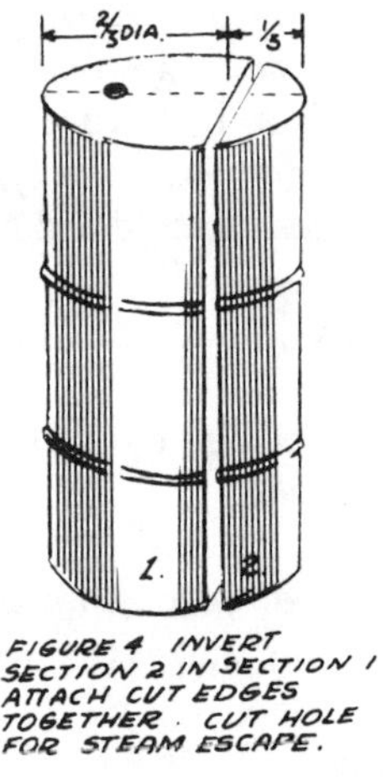


FIGURE 4 INVERT SECTION 2 IN SECTION 1 ATTACH CUT EDGES TOGETHER. CUT HOLE FOR STEAM ESCAPE.

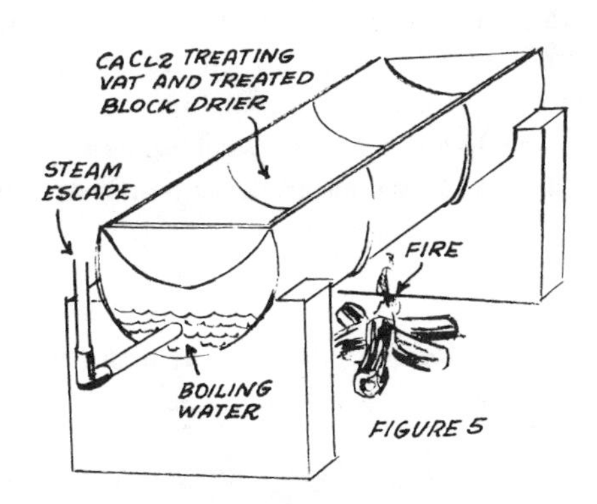


FIGURE 5

- Cook the blocks in the calcium chloride solution for four hours at a temperature just below the boiling point, 100$^{o}$C (212$^{o}$F).

- Let the solution cool; let the blocks soak in the solution for 24 hours.

- Dry the blocks again.

- When the blocks are dry, wipe off any calcium chloride on their surface before putting them in the grain.

## Using the Blocks

- Mix the blocks with grain in a container. The blocks should be spaced throughout the container so that the grain will dry evenly in the shortest time possible. The blocks should not take up more than 10 percent of the container's space. Small containers (see Figure 1) are helpful when there are several kinds of grain to dry. They also make it easier to remove and replace the blocks. These containers are placed in the waterproof chest.

- After three to five days, remove the blocks. They can be separated from the grain easily with a coarse screen. Dry the blocks again.

- Continue re-drying the blocks in an oven or double boiler and placing them back in the grain until the blocks no longer absorb moisture. To find out when this point is reached, weigh the blocks after three or four days in the grain: if they weigh the same as dry blocks, the grain is dry.

**Source:**

Ives, Norton C. *Grain Drying and Storage for Warm, Humid Climates.* Turrialba, Costa Rica: Inter-American Institute of Agricultural Sciences, 1951.

# BUCKET SPRAYER

The bucket sprayer described here has been designed primarily to meet the need for a sprayer that can be built in an area where production facilities are limited. This sprayer, which can be made by the local artisans, is intended only for water solutions of insecticides or fungicides.

Two people operate it; one sprays while the other pumps.

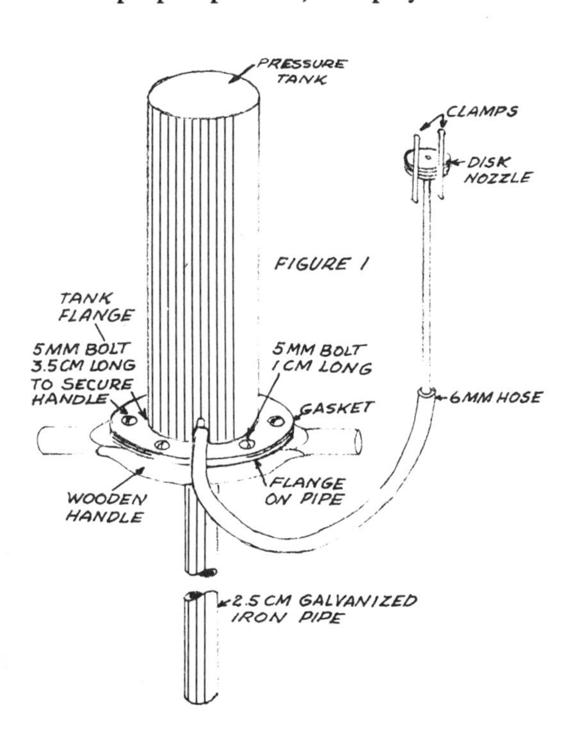


**Tools and Materials**

Galvanized iron: 30cm x 30cm (1' x 1') plus 10cm x 20cm (4" x 8")
Barrel metal: 10cm x 20cm (4" x 8")
6mm (1/4") hose (high pressure) 4m (13') long
6mm (1/4") pipe (truck brake line may be used) 50cm (19 5/8") long
Wood for handle: 2cm x 15cm x 30cm (3/4" x 6" x 12")
2.5cm (1") Galvanized iron pipe (thin-walled) 120cm (4') long
4mm (5/32") wire: 20cm (8")
Truck inner-tube material: 10cm x 20cm (4" x 8")
1mm (1/32") Galvanized wire, 30cm (12") long
4 - 5mm (3/16") bolts x 1cm (3/8")
2 - 5cm (3/16") bolts x 3.5cm (1 3/8")

The sprayer pump operates on the same principle as the Inertia Pump (see page 101). The top of the 2.5cm (1") iron pipe is plugged and a simple valve is located 8cm (3 1/8") from the top. The valve is a piece of truck inner-tube rubber wrapped around the pipe and held in place by wire. One corner of the rubber is over a hole in the pipe. Some careful adjustment is necessary when placing the rubber to make sure it works properly and does not leak.

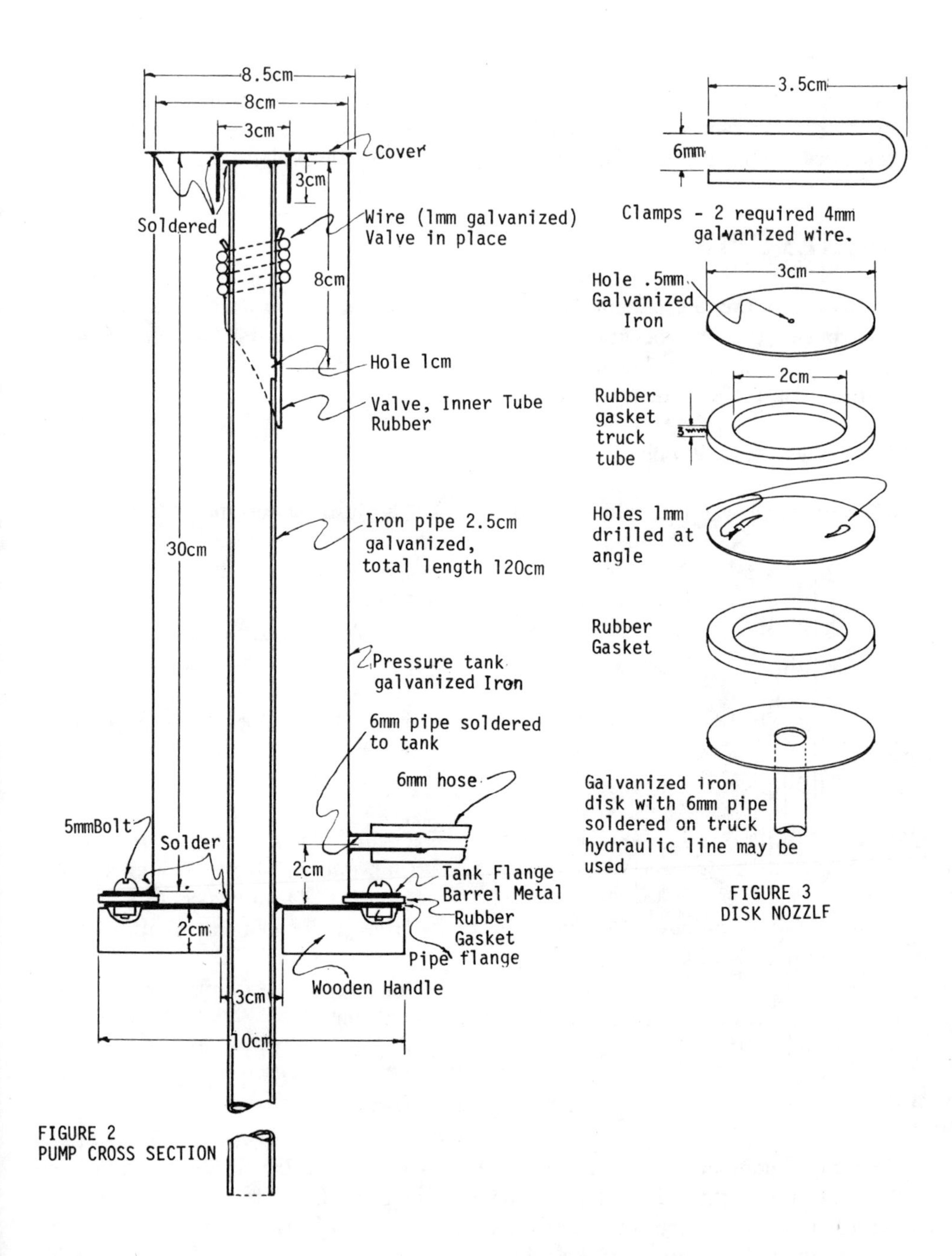


FIGURE 2
PUMP CROSS SECTION

FIGURE 3
DISK NOZZLF

The pressure tank encloses the valve assembly and, as the liquid is pumped into the tank, builds up pressure sufficient to operate the simple disk type spray nozzle. The tank is built so that it can be removed in order to service the valve.

The length of the hose can be determined by the maker of the sprayer but it should be about 4m (13') to allow the worker doing the spraying to cover quite a large area before having to move the bucket. Also, the length of the small pipe and the angle of the spray nozzle will be determined by the kind of crops being sprayed.

At times it will be necessary to "prime" the sprayer pump: if the valve rubber is too tight and the air cannot be forced through the valve, or if the rubber is stuck to the pipe. To prime the pump turn it upside-down and fill the pipe with water. Holding your thumb over the pipe, turn the pump over, lower it into the bucket of liquid and start pumping in the usual manner. If priming does not start the pump it will then be necessary to remove the pressure tank to inspect and repair the valve.

Only very clean water should be used to make the mixture for spraying. It should be strained through a cloth after mixing to remove any particles that might cause the nozzle to plug. If a very fine brass screen is available, it should be put in the nozzle to keep the dirt from plugging the holes.

**Source:**

Dale Fritz, VITA Volunteer, Schenectady, New York

# BACKPACK CROP DUSTER

The backpack duster described here, designed so that it can be easily made by tinsmiths, has been used by Afghan farmers to dust sulfur on their grapes to control powdery mildew. The duster is made from easily available materials. Its feed rate is adjustable (see Figure 1).

The springs needed for the duster can be made with the simple spring winder shown on p. 251.

**Tools and Materials**

Soldering equipment
Sheet-metal working tools
Carpentry tools

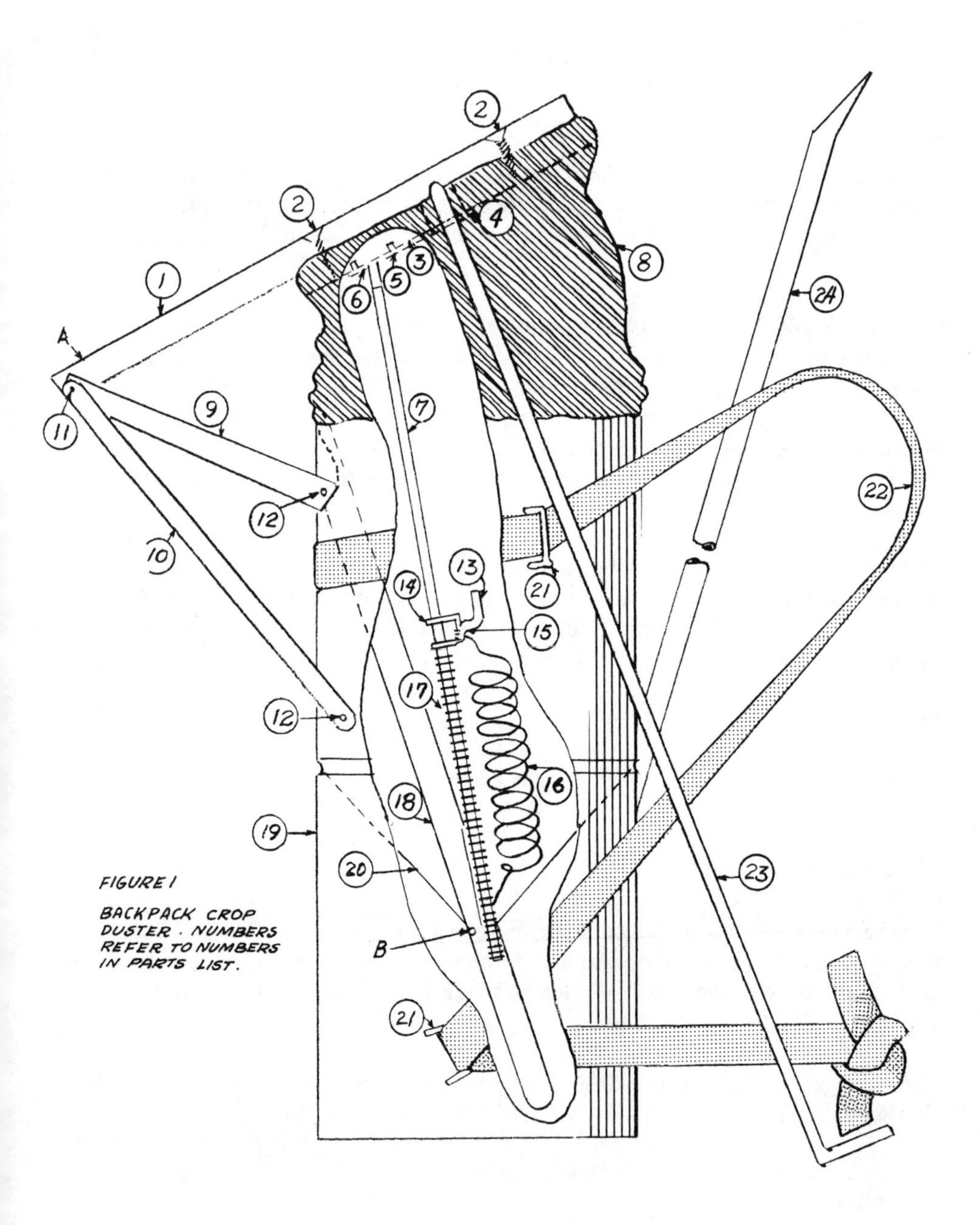


FIGURE 1

BACKPACK CROP DUSTER. NUMBERS REFER TO NUMBERS IN PARTS LIST.

| Part No. | Name | Material | Description |
|---|---|---|---|
| 1 | Bellows Support | Wood | 38cm x 7cm x 2cm (15" x 2 3/4" x 3/4"). |
| 2 | Screws | | 4cm (1 9/16") long. |
| 3 | Bellows Plug | Wood | 22cm (8 5/8") in diameter, 2.5cm (1") thick. |
| 4 | Valve | Rubber | 4cm x 5cm (1 9/16" x 2"). See Figure 2. |
| 5 | Screws | | 2cm (3/4") long. |
| 6 | Feeder Rod Anchor | Barrel Metal | See Figure 3. |
| 7 | Feeder Rod | 6mm (1/4") rod | See Figure 3. Total length 50cm (19 3/4"). |
| 8 | Bellows | Truck inner-tube rubber | 30cm (12") long on long side. Tube measures 29cm (11 3/8") from edge when laid flat. |
| 9 | Bellows Support | Barrel metal | 20cm (8") long. See Figure 4. |
| 10 | Brace | Galvanized tin | 33cm (13") long. See Figure 4. |
| 11 | Nails | | 3cm (1 3/16") long. |
| 12 | Rivets | | |
| 13 | Bolt | 6mm (1/4") rod | See Figure 5. |
| 14 | Clamp | Barrel metal | See Figure 5. |
| 15 | Nut | 6mm (1/4") nut | See Figure 5. |
| 16 | Agitator Spring | Tire bead wire | 3.5cm (1 3/4") diameter. See Figure 6. |
| 17 | Feeder Spring | Tire bead wire | 9mm (11/32") diameter. See Figure 3. |
| 18 | Pipe | Galvanized tin | 3.5cm (1 3/4") diameter, 71cm (28") long. See Figures 6 and 7. |

| 19 | Hopper | Galvanized tin | 22cm (8 5/8") diameter, 48cm (18 7/8") high. See Figure 7. |
|---|---|---|---|
| 20 | Floor | Galvanized tin | Make to fit. See Figure 7. |
| 21 | Strap Holder | Galvanized wire | 4mm (5/32") diameter. Soldered to hopper. |
| 22 | Strap | Webbing | 6cm (2 3/8") wide, 3m (9'10") long. Tied at waist. |
| 23 | Handle | 8mm (5/16") rod | Total length 1 meter (39 3/8"). |
| 24 | Pipe | Galvanized tin | 3.5cm (1 3/4") diameter, 140cm (55 1/4") long. See Figure 1, 6 and 8. |

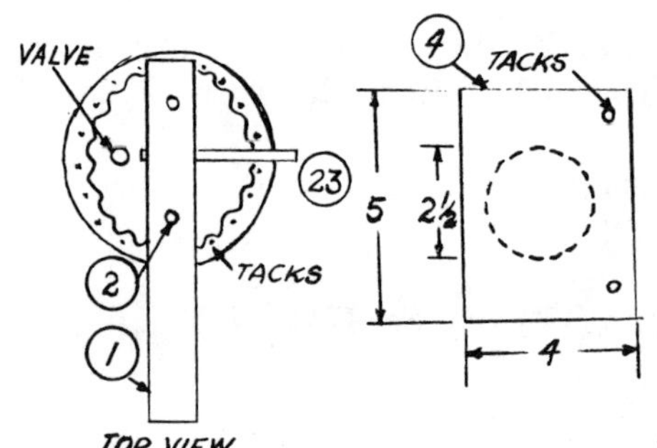

FIGURE 2 DETAIL OF PART 4, VALVE

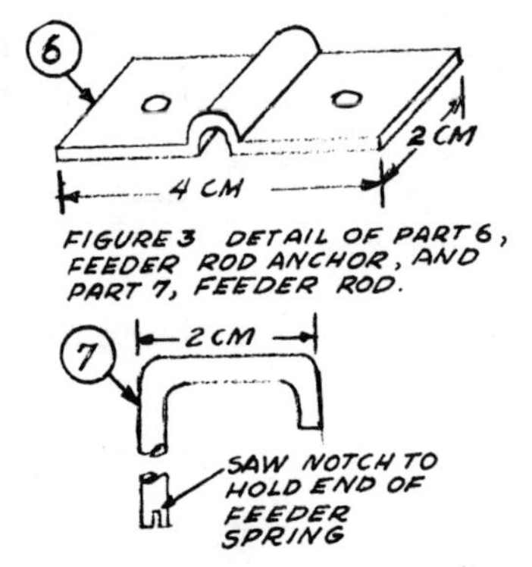

FIGURE 3 DETAIL OF PART 6, FEEDER ROD ANCHOR, AND PART 7, FEEDER ROD.

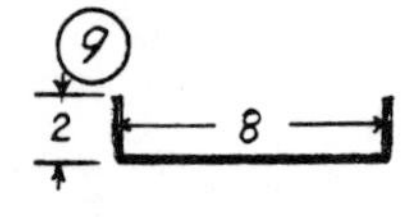


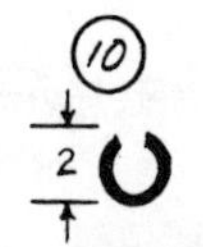


FIGURE 4 CROSS-SECTION OF PART 9, BELLOWS, AND PART 10, BRACE.

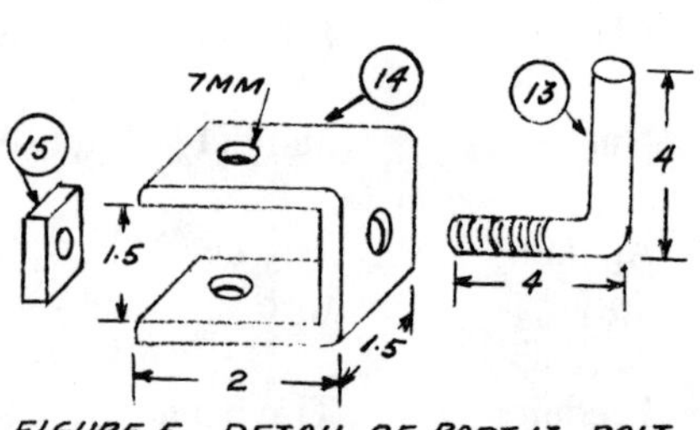

FIGURE 5. DETAIL OF PART 13, BOLT, PART 14, CLAMP AND PART 15, NUT.

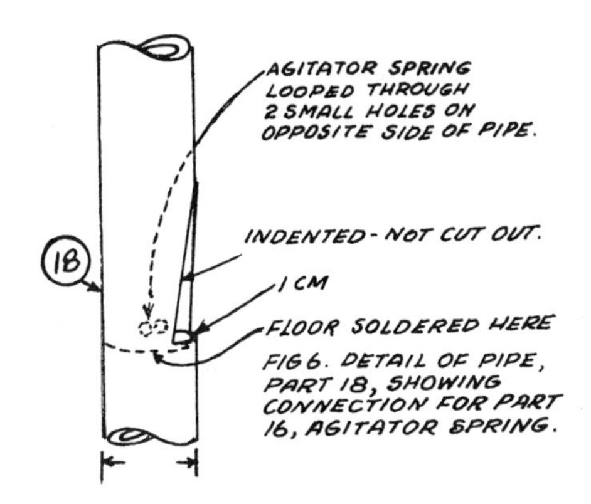


FIG 6. DETAIL OF PIPE, PART 18, SHOWING CONNECTION FOR PART 16, AGITATOR SPRING.

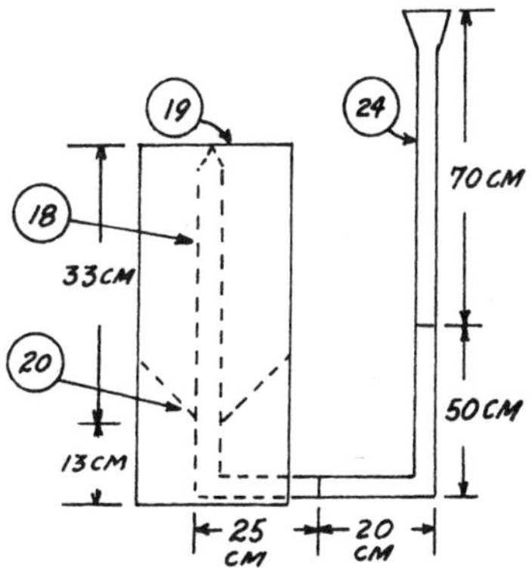


FIGURE 7, DETAIL OF PART 19 HOPPER AND PART 24, PIPE. BROKEN LINES SHOW PART 18, PIPE, AND PART 20, FLOOR.

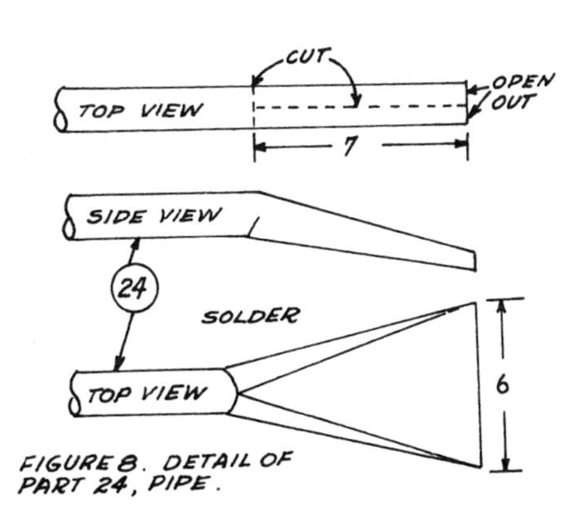


FIGURE 8. DETAIL OF PART 24, PIPE.

## How the Duster Operates

In operating the duster, the rod (23) is used to pump the inner-tube bellows, which pivots about point A (see Figure 1).

Air is admitted to the bellows through valve (4), also made of innertube rubber, and passes down the pipe (18). A measured amount of dust is injected into pipe (18) at point B. The feed mechanism consists of a 6mm (1/4") rod (7) covered by a spring (17). As the bellows is worked up and down, the rod and spring go in and out of the hole (at point B) in the delivery pipe (18). The dust lodges between the loops of the spring and is carried into the pipe. The amount of dust delivered is controlled by stretching the spring on the rod so that there is more space between the loops. The greater the space between the loops, the greater the amount of dust carried into the pipe. An easily adjustable clamp (13) and (14) is provided on the rod to regulate the amount of dust applied to the plants. The air-dust mixture is blown out the delivery pipe at (24).

The bellows of the duster is made from truck innertube rubber. There are several sizes of innertubes. If the size shown in the list of parts is not used, the diameter of the hopper must be adjusted to the size of the tubes available. The hopper is made from galvanized tin, from 24 to 28 gauge.

In the illustrations, the feeder rod (7) is shown as being straight. However, it is necessary to bend the rod to allow it to work in and out of the hole in the delivery pipe without binding.

To fill the duster, slip the bellows off of the top of the hopper. The hopper must not be filled above the top of the delivery pipe. The top of the delivery pipe (18) is cut so as to prevent dust from spilling in the tube during filling, and to provide a means for fastening it to the hopper (19).

## Adjusting the Duster

To increase the amount of dust being applied:

- Slip the bellows (8) off of the top of the hopper (19).
- Loosen the bolt (13).
- Pull up on the clamp (14) stretching the spring (17).
- Tighten the bolt (13).
- Replace the bellows and test the amount of dust delivered to see if it is satisfactory.

To decrease the amount of dust, the procedure is the same except that the clamp is pushed down on the rod.

## Filling the Duster

Before filling the duster, make sure that all lumps of dust have been broken up. Putting the dust through a piece of window screen is a good way to break up the lumps. This will also remove any foreign matter.

**Source:**

Dale Fritz, VITA Volunteer, Schenectady, New York

## Making Springs for the Duster

This method for winding springs can be used to make springs of any size. Figures 1 and 2 show spring winders for springs that will be the right size for use in the Backpack Crop Duster described in the preceding entry.

**Tools and Materials**

Drill
Drill bit: 2mm (1/12")
Drill bit: 6mm (1/4")
Drill bit: 12.5mm (1/2")
Wood: 10cm x 10cm x 1m (4" x 4" x 39")
Metal rod: 6mm (1/4") by 1m (39") long
Metal pipe: 12.5mm (1/2") by 30cm (12") long
4 small nails
Steel spring wire

A good source of spring wire is from the bead of an old tire. The rubber should not be burned off as this destroys the spring-strength of the wire.

One winder is made of the 6mm (1/4") rod. The other winder is made from the 12.5mm (1/2") pipe with a section of the rod used as a handle. Cut a piece of the 6mm (1/4") rod about 30cm long. Bend to form handle shown in Figure 1; set aside. Bend remaining piece as shown in Figure 2.

Drill a 6mm (1/4") hole in one end of the wood block and a 12.5mm (1/2") hole in the other end. Drill a 2mm (1/12") hole through the longer section of 6mm (1/4") rod and through the 12.5mm (1/2") pipe to insert the end of the wire. Drill a 6mm (1/4") hole through the 12.5mm (1/2") pipe to hold the length of the rod to be used as a winding handle. Drive two nails close together, about 1.5mm to 2mm (1/12" to 1/16") from each hole in the wood block. Put the pieces together as shown in Figures 1 and 2.

The wire is fed through the nail wire guide and then through the 1/12 inch hole in the rod or pipe spool. The spool is then turned in a clockwise direction until the desired length of spring is wound. The springs for the backpack duster are 9mm (11/32") from the 6mm (1/4") spool and 3.5cm (1 3/8") from the 12.5mm (1/2") spool.

**Source:**

Dale Fritz, VITA Volunteer, Schenectady, New York

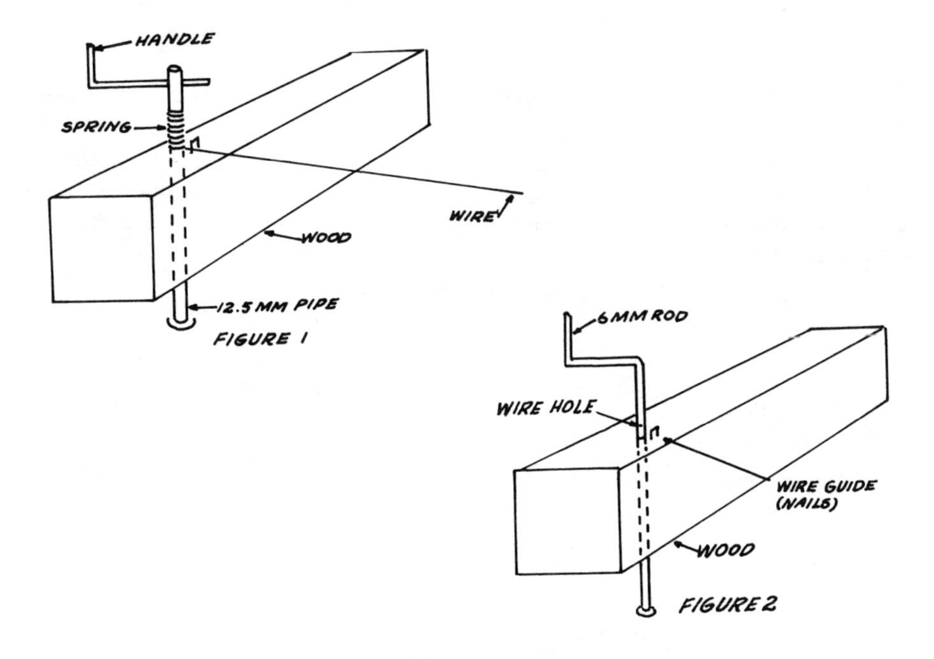

# Poultry Raising

## BROODER WITH CORRAL FOR 200 CHICKS

This chick brooder (see Figure 1) is hinged for easy access to corral and brooder. The brooder has been used successfully in Ecuador and elsewhere to raise broilers for a cash crop.

The brooder is heated by a regular electric light bulb, placed under the brooder floor. Depending on the temperature rise required, the wattage of the light bulb will have to be chosen by experimentation. The metal floor and roof prevent predators such as rats from entering the brooder. If electric power is not available, an excavation can be made for a lantern. Be sure the lantern has adequate ventilation.

FIGURE 1

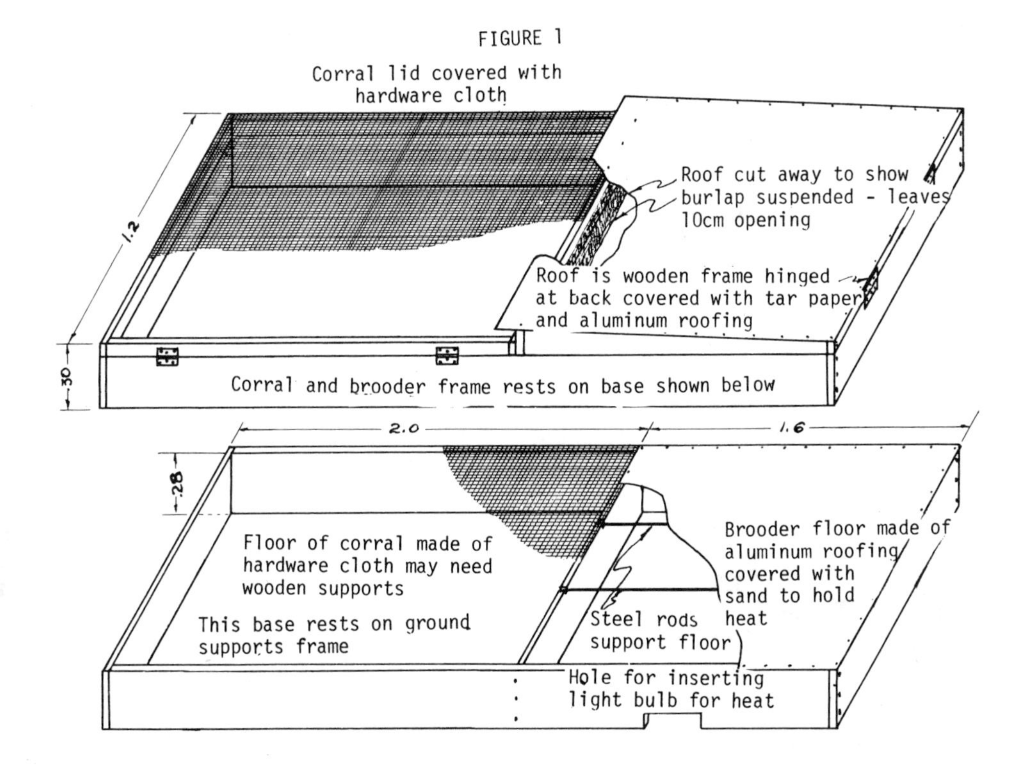

**Tools and Materials**

Small carpentry tools
Hardware cloth 1.2 x 2m (4' x 6' 6 3/4"), 2 pieces this size needed.
Aluminum roofing:
 1 piece: 1.2m x 1.6m (4' x 5'3")
 1 piece: 1.2m x 1.7m (4' x 5'7")
Wood, approximately 30cm x 2cm x 20m (1' x 3/4" x 65'8")
Steel rod 1cm (3/8") diameter x 3.2m (10' 6")
4 hinges about 8cm (3 1/8") long
Woodscrews for hinges
2 buckets clean dry sand
Nails, tacks, staples

**Source:**

Kreps, George. Article in *Rural Missions,* #122, Agricultural Missions, Inc.

# KEROSENE LAMP BROODER FOR 75 TO 100 CHICKS

This brooder has been used by more than 300 farmers in eastern Nigeria.

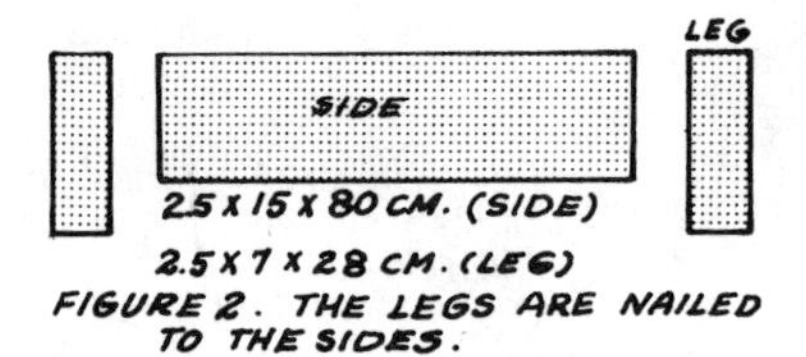

FIGURE 2. THE LEGS ARE NAILED TO THE SIDES.

Nail legs to side (see Figure 2). If desired, make the height of the brooder adjustable by drilling a row of holes in each leg and bolting the legs to the sides.

Assemble and nail top support rails 1cm (3/8") below the upper edge of the sides (see Figure 3).

Make the top of plywood, sheet metal, or wooden boards so that the top fits inside the frame and rests on the support rails (see Figure 4). The hole in the center of the top is for ventilation. A swinging metal cover regulates the size of the opening.

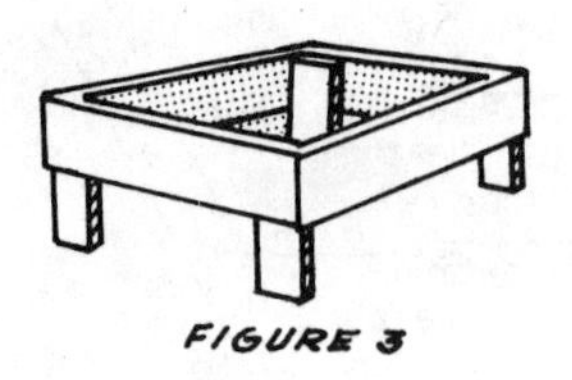
FIGURE 3

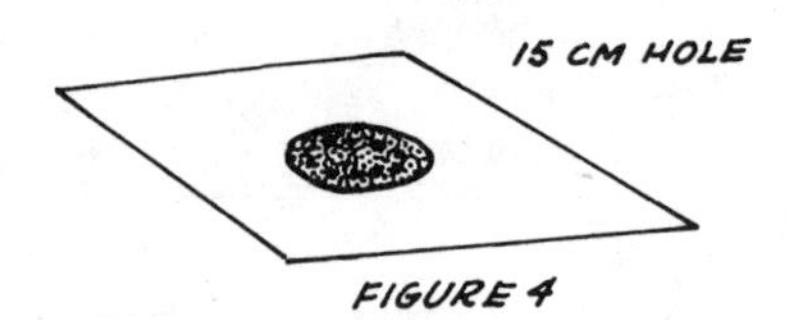

FIGURE 4

A bush or hurricane lamp is placed inside wire mesh or a perforated tin can to protect the chicks and to help radiate the heat (see Figure 5).

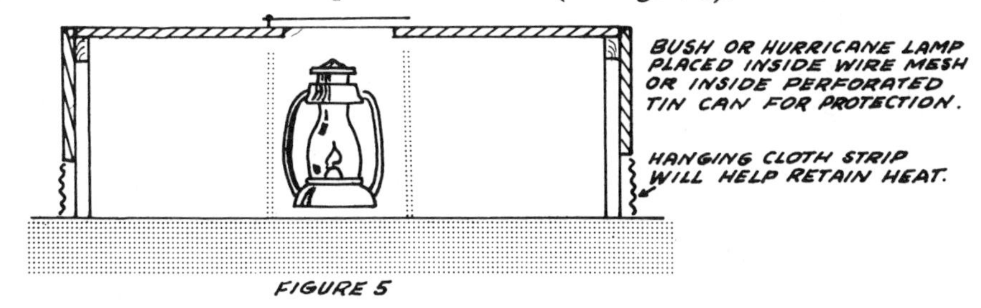


FIGURE 5

The dimensions given in the illustrations can be altered slightly to use available materials.

The wicks of the lanterns should be cleaned daily to cut down on soot.

**Source:**

W. H. McCluskey, Poultry Science Department, Oregon State University, Corvallis.

# BROODER FOR 300 CHICKS

This brooder (see Figure 6) is similar to the other two brooders. It can be used with either lanterns or electric light bulbs. If lanterns are used, their wicks should be cleaned daily. Construction details are given in Figure 7.

**Source:**

Stopper, W.W. "Brooder for 300 Chicks". New Delhi: U.S. Technical Cooperation Mission to India. (Mimeographed).

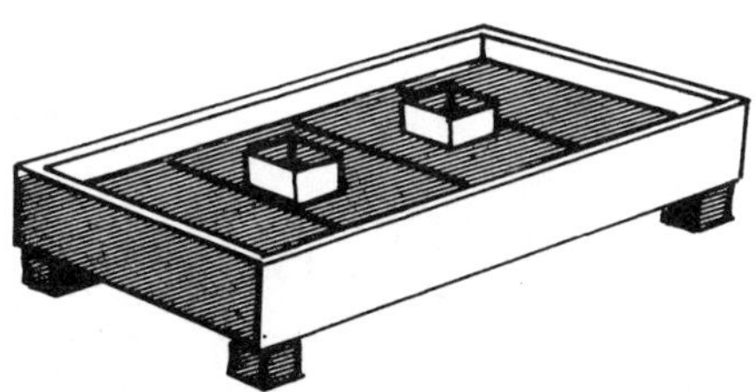

FIGURE 6 BROODER FOR 300 CHICKS.

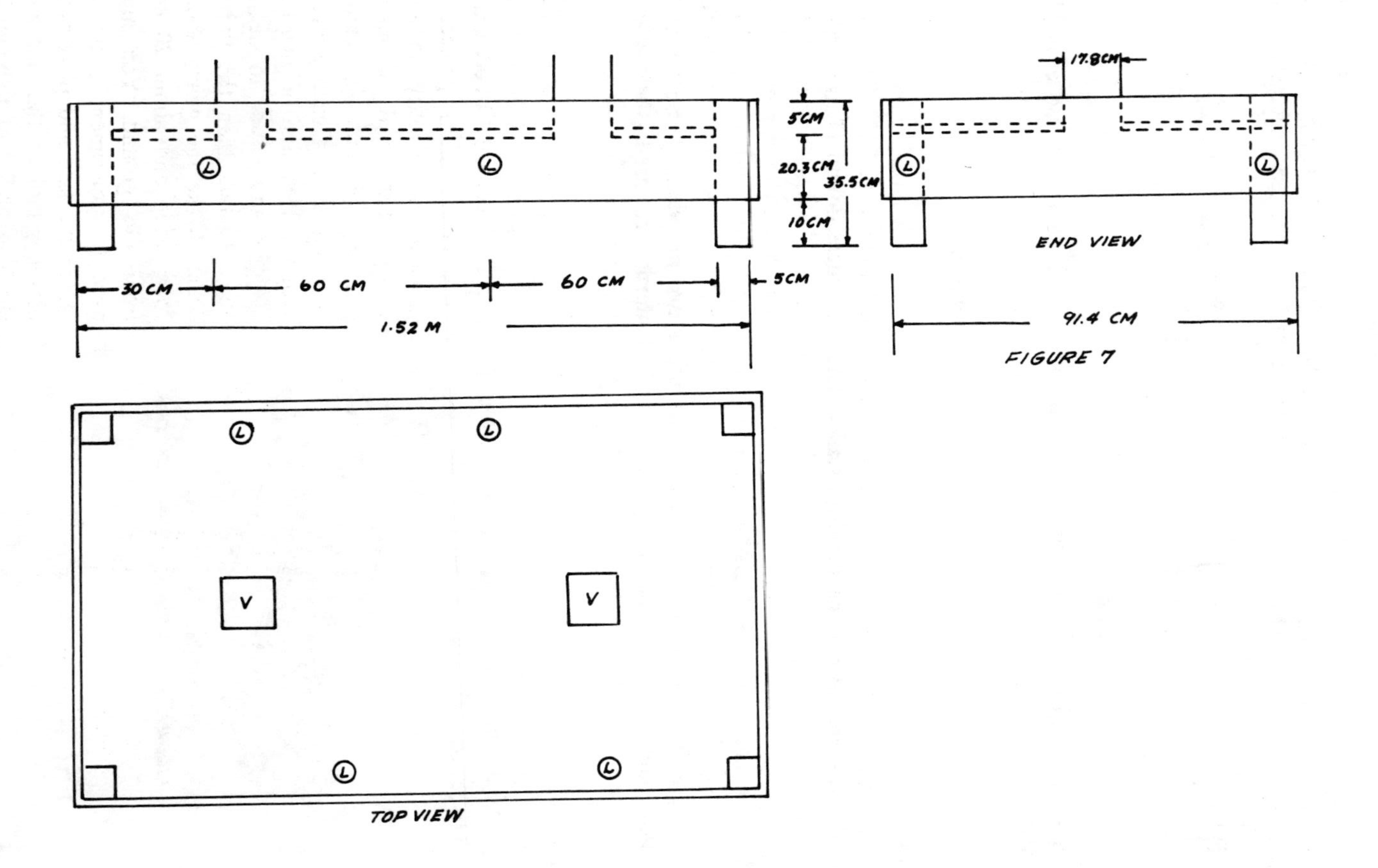
L
L
5CM
20.3CM
35.5CM
10CM
30 CM
60 CM
60 CM
5CM
1.52 M
17.8CM
L
L
END VIEW
91.4 CM
L
L
V
V
L
L
TOP VIEW

FIGURE 7

# BAMBOO POULTRY HOUSE

This bamboo poultry house has a thatch roof and slat walls to provide good ventilation. The elevated slat floor keeps chickens clean and healthy while the egg catch and feed troughs simplify maintenance. It has been used successfully in the Philippines and Liberia.

**Tools and Materials**

Bamboo
Nails
Thatching materials
Small tools

## House

The house is built on a frame of small poles, with floor poles raised about 1m (3') from the ground. (See section on construction with bamboo, p. 302.) The floor poles are covered with large bamboo stalks, split into strips 38mm (1 1/2") wide, spaced 38mm (1 1/2") apart. Floors so constructed have several advantages: better ventilation, no problem of wet moldy litter during rainy reason or dry dusty litter during dry season; droppings fall between split reeds to ground away from chickens. This eliminates parasites and diseases normally passed from hen to hen through droppings remaining warm and moist in litter. However, it has been suggested that wide spacing of floor and wall slats might invite marauders such as weasels and snakes.

Metal shields on all the support poles will keep rats and other pests from climbing (see Figure 1a). Be sure you don't inadvertently leave a hoe or other tool leaning against the house, or the rats will climb that. (Note: A VITA grain storage project in Central African Republic has had good results protecting granaries–not poultry houses–with a flat band of metal [Figure 1b] that is simply wrapped around each granary support. This kind of guard is cheaper and easier to install and maintain than the flared collar. Make the guard about 25cm wide and about 20 cm from the ground. You may have to experiment a bit to match the size and placement of the guard to the size and climbing ability of the rats in your neighborhood.)

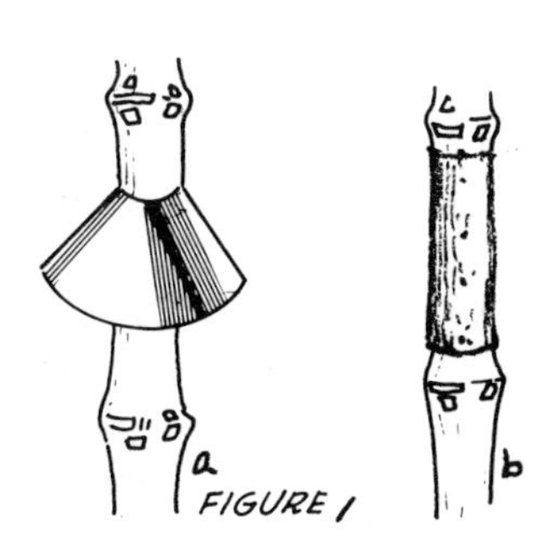

FIGURE 1

Walls are made from vertical strips of bamboo 38mm (1 1/2") wide, spaced 6cm to 8cm (2 1/2" to 3") apart. This also allows ample ventilation, needed to furnish oxygen to the chickens and to allow evaporation of excess mosture produced in the droppings. In the tropics the problem is to keep chickens cool, not warm. Using a closed or tight-walled poultry house with a solid floor would keep them too warm and result in lowered production and increased respiratory problems. Shade over and around these houses is very important. If the ground around the houses is not shaded, heat will bounce into the houses.

## Roof

The roof must protect the chickens from the weather. In Liberia thatch roofing keeps the birds cool, but it must be replaced more often than most other materials. Since it is cheap and readily available to the small farmer or rural family, it is most likely to be used. Aluminum, which reflects the heat of the sun, and asbestos, an efficient insulator, are desirable roofing materials in the tropics. Zinc, which is commonly used to roof houses in Liberia, is undesirable for chicken houses because it is an efficient conductor of heat.

Whatever the roofing material, the roof must have an overhang of 1m (3') on all sides to prevent rain from blowing inside the house. It may be desirable to slope the overhang toward the ground.

## Feeders

Feeders and waterers are made from 10cm to 12.5cm (4" to 5") diameter bamboo of the desired length (see Figure 2). A node or joint must be left intact in each end of the bamboo section to keep the feed or water in. A section 7.5cm to 10cm (3" to 4") wide around half the circumference of the bamboo, except for 7.5cm (3") sections on the ends, is removed to make a kind of trough. All nodes between the ends are removed. These feeders must be fastened at the base, to keep them from rolling.

FIGURE 2. EGG CATCH AND FEED TROUGH IN BAMBOO POULTRY HOUSE.

The feeders are fastened to the outside of the walls about 15cm (6") above floor level. The hens place their heads through the bamboo strips to feed or drink, thus conserving floor space for additional chickens.

## Nests

The demonstration nests are 38cm (15") long, 30cm (12") wide, and 35.5cm (14") high (see Figure 3). The strips used on the floor of the nest are about 13mm (1/2") wide, spaced 13mm (1/2") apart, and must be very smooth. The floor slopes 13mm (1/2") from front to back, so that when the eggs are laid they will roll to the back of the nest. An opening 5cm (2") high at the back of the nest allows the eggs to roll out of the nest into an egg catch (see Figure 1). This type of nest results in cleaner eggs and fewer broken eggs. It also yields better quality eggs because they begin to cool as soon as they roll out of the nest. In addition, the eggs are outside the nest where egg eating hens cannot reach them. Placing the egg catch so it protrudes outside the wall of the house allows the eggs to be gathered from outside. Placing the nests 1 meter (3') above the floor conserves floor space and permits more laying hens to be placed in the laying house. One nest is put in for every five hens.

FIGURE 3. NESTS IN BAMBOO POULTRY HOUSE.

In laying houses, nests are also constructed of split bamboo for unobstructed ventilation. Conventional lumber nests are hotter; this may cause hens to lay eggs on the floor instead of in the nests. This means more dirty eggs, more broken eggs, and more likelihood of the hens eating the broken eggs. The only way to cure a hen of eating eggs once the habit is formed is to kill her. In addition, as the hens enter the nests they sit on eggs laid previously by other hens, keeping them warm. The quality of eggs deteriorates very fast under these conditions.

**Source:**

USAID, Monrovia, Liberia, described in OTS Information Kit, vol. I, No. 5, May 1961.

# POULTRY FEED FORMULAS

| Ingredients—**Ceylon** | Percentage required for Layers |
|---|---|
| Sorghum | 42.0 |
| Rice bran | 19.5 |
| Fish meal | 8.5 |
| Coconut meal | 18.5 |
| Gingelly *(sesamus indicum)*cake | 2.0 |
| Cowpeas | 3.0 |
| Shell grit | 6.5 |
| Salt | 0.5 |
| TOTAL | 100.5 |
| Added per 100.5 kg: | |
| Potassium Iodide (g) | 0.145 |
| Choline Chloride (21.7%)(g) | 540 |

| Ingredients—**Congo** | Percentage required for Layers |
|---|---|
| Maize, ground | 20.0 |
| Millet, ground | 18.0 |
| Rice, husk, ground | 10.0 |
| Fish meal | 4.0 |
| Meat meal | 5.0 |
| Groundnut cake meal | 25.0 |
| Lucerne meal | 12.0 |
| Dicalcium phosphate | 2.0 |
| Oyster shells | 3.0 |
| Salt | 1.0 |
| TOTAL | 100.0 |

| Ingredients—**Uruguay** | Percentage required for Layers |
|---|---|
| Ground maize | 40.0 |
| Ground wheat | 5.0 |
| Sorghum | 3.0 |
| Ground barley | 20.0 |
| Bran | 10.0 |
| Meat meal | 7.0 |
| Ground sunflower cake | 10.0 |
| Oyster shells | 4.0 |
| Salt | 1.0 |
| TOTAL | 100.0 |

Compiled by Harlan Attfield, from *Poultry Feeding in Tropical and Subtropical Countries*, Food and Agriculture Organization of the United Nations, Rome.

# Intensive Gardening

Intensively cultivated vegetable gardens can supply a great deal of a family's food from very little land. However, to maintain their productivity, these gardens require a lot of fertilizer and some special techniques, which are discussed below.

As one crop is finished, another is put in its place throughout the growing season. Without additional fertilizer the soil would soon be worn out. Cost of the garden can be kept low by using compost and a crop rotation system that also includes poultry or other livestock, which can give a steady supply of manure. This virtually eliminates fertilizer costs. The best way to ensure a large supply of manure is to keep the animals in a pen, barn, or corral, especially at night.

## THE SOIL

Fertile soil includes organic matter and minerals. The best soil is loose and has a crumbly texture that breaks easily into small pieces a few millimeters in diameter. The deeper the crumb structure exists in the soil the better.

If the soil is compacted or dense, it can be loosened by first plowing or tilling to break up the soil. Tilling also controls weeds. This work can be done with a pick and shovel, a hoe, or a heavy fork. A small tractor, or animal drawn tools, may be helpful in a very large garden.

The soil can be improved by: 1) adding manure or compost, or by returning to the soil plant materials that you or your animals do not eat, 2) rotating crops, 3) working the soil only when it is dry enough. Test for dryness by taking a handful of soil and squeezing it. If it sticks together tightly, it is still too wet to work.

## THE GROWING BEDS

Make planting beds no wider than you can reach to the middle of for planting, weeding, and harvesting. In that way you won't have to step on the beds and compact the soil. One meter (three feet) is a typical width. Lay the beds across any slope to slow water runoff and reduce erosion. The soil may be raised in long mounds so that it will warm more readily and be less subject to flooding. Edge the mounds with stone, brick, concrete block, heavy boards, or other material to hold the soil in place. This is not essential, but makes the garden easier to care for in the long run.

Leave a footpath between the beds that is wide enough to walk in and to allow some space for the tops of the growing plants. You will want to be able to work between the beds without damaging plants. Build a secure fence around the garden to keep out chickens, rabbits, cattle, and other animals.

If there is a stream or a tubewell nearby, the garden can be watered by running water in furrows between the beds, or by hand watering. Widely spaced individual plants, such as tomato, pepper, or eggplant, can be watered by burying a jar with a tiny hole near the bottom in the ground near the plant (Figure 1). The jar is filled with water, which seeps out to be used by the plant as needed. This is quite a bit of work, but can be very effective in very dry areas. Bury the jar when you set out the plant so you don't disturb the roots later. Check the water level in the jar about once a week, oftener if need be.

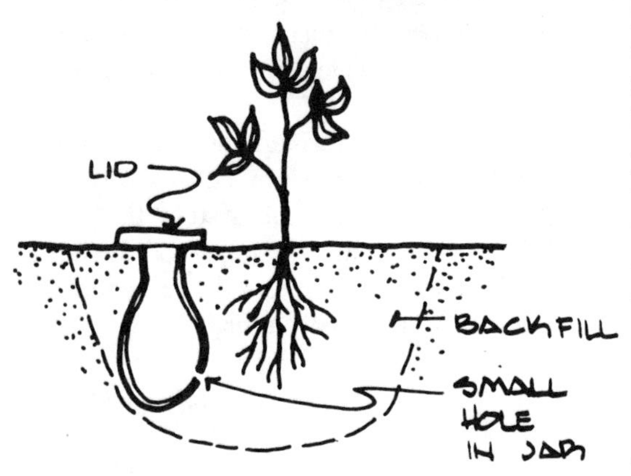


*Figure 1: Automatic watering using a buried jar of water.*

## FERTILIZING THE SOIL

Growing plants take nutrients from the soil, which must be replaced or crop yields will slowly diminish, and intensive cultivation uses up nutrients rapidly. The major nutrients are nitrogen, phosphorus, potassium, and calcium. These can be bought as chemical fertilizers, but are also found in plant matter and manure.

An inexpensive way to enrich the soil is to use compost from a compost pit or crib that is located near the garden (Figure 2). Pile the materials into layers as shown. Keep moist. Turn and mix every week or so as they decay. When the compost gets to be dark and crumbly, it is ready for the garden. Composting will not usually supply all the fertilization needed, but will add nutrients to the garden soil and improve soil texture.

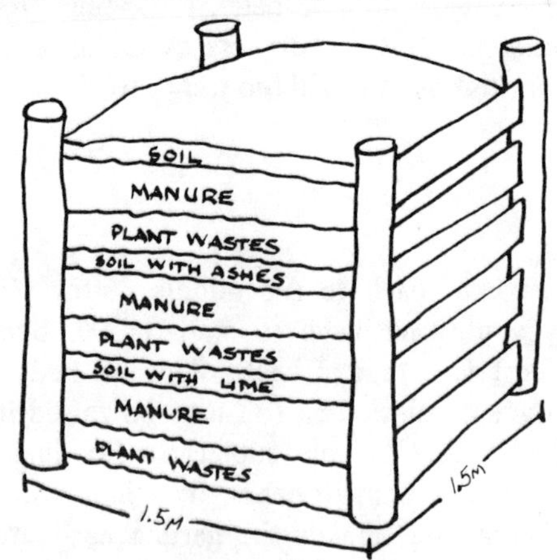


*Figure 2: Compost Pile*

The simplest way to fertilize and improve soil texture at the same time is to use animal manure. If you use fresh manure, spread it over the garden at the end of the growing season and work it into the soil. During the growing season, it is best to use only seasoned manure.

If only fresh manure is available, a small amount of it may be used to make a weak "tea" that can be poured around the growing plants. To make the "tea" put a shovelful of fresh manure into a bucket of water and let it stand for about a week. Dilute the liquid until it is the color of weak tea and use it to water your plants about once a week.

## SELECTION OF CROPS

Select crops that suit the climate and your family's tastes. If you want to grow vegetables to sell, consider community tastes as well. Try to choose an assortment that will give you something fresh from the garden throughout the season. Unless you have some way to preserve the produce, don't plant more than you can eat, give away, or sell fresh. But do plant vegetables you like a lot or want in quantity at intervals of a couple of weeks so that you will be able to harvest them over a long period. Keep in mind that in a well-fertilized garden plants can be more closely spaced and will yield a larger harvest for the space.

Some crops may be planted directly in the beds while others are best started in a seed box and later transplanted into the garden beds. The table below gives a partial listing of both types of vegetables.

**Seeds To Plant and Seedlings To Transplant**

| Vegetables that should be Transplanted | Vegetable Seeds to Plant Directly in the Garden | |
|---|---|---|
| Broccoli | Black Colocasia (roots) | Okra |
| Cabbage | Beet | Onion |
| Cauliflower | Bitter gourd | Pigeon pea |
| Chinese cabbage | Carrot | Pointed gourd |
| Eggplant | Collard | Potato (tuber) |
| Indian spinach | Coriander | Radish |
| Lettuce | Cowpea | Red Amaranth |
| Mustard | Cucumber | Soybean |
| Pepper | Field bean | Sweet corn |
| Spinach | French bean | Sweet potato (cuttings) |
| Tomato | Green Amaranth | Sweet pumpkin |
| | Jute | Sword bean |
| | Kohlrabi | Turnip |

It is a good idea to rotate the vegetables in the beds each season. That is, plant one type of vegetable one season, another type the next season, and so on. Each type or family of vegetable is subject to similar pests and soil diseases. Planting a different type vegetable in the beds each season helps prevent the build up of these pests and diseases and gives the soil a rest.

There are four basic families of vegetables--root vegetables, leafy vegetables, legumes, and fruiting vegetables--so the rotation would span four seasons.

Peas, beans, and such are legumes, which means that they can make their own nitrogen plant food and so enrich the soil. Plant vegetables that need a lot of nitrogen in the bed when the legumes are finished. Root vegetables are grown primarily for their thick fleshy roots--radish, carrot, onion, beet. The leaves of some root vegetables, like beet, are often eaten as greens. Fruiting crops include peppers, eggplant, tomato, and white potato.

Leafy vegetables--cabbages of various kinds, lettuces, spinach, collard--are grown for their leaves, which are rich in vitamins and minerals. Some leafy vegetables tolerate cold weather better than others and some do well when it is hot, so it is possible to have some kind of fresh greens from the garden almost all year round. When you are planning your crop rotation, include broccoli and cauliflower in the leafy group, even though you don't eat the leaves, because they are attacked by some of the same pests as the leafy vegetables.

Plan the beds so that as one crop is finished another takes its place (with the addition of a little compost or seasoned manure). Save space by planting vines like beans and cucumbers on trellises at the edge of the garden, situated so they don't shade other crops. Stake tomatoes, peppers, etc., with posts of bamboo or whatever is available to keep the fruit from rotting on the ground.

## MULCH

Cover the soil around seedlings with a thick layer of grass clippings, leaves, straw, or other material. Some people use black plastic, which is expensive, or even layers of newspaper. The idea is to keep the soil from drying out so fast and to keep weeds from sprouting. Mulching may seem like a lot of extra work in the beginning, but it saves a lot of work over the season. It also saves water, and the organic mulches, like grass and straw, enrich the soil as they decay.

**Sources:**

Paul J. Abrahams. VITA Volunteer, Atlanta, Georgia
J.W. and J.B. Fitts, VITA Volunteers, North Carolina
Harlan H.D. Hatfield, VITA Volunteer, Bend, Oregon
James M. Corven, VITA Volunteer, Washington, D.C.

# Silage for Dairy Cows

The small dairy farmer who maintains five or six cows on two or three hectares (four or five acres) of fodder and pasture grass is usually faced with a serious decline in milk production during dry or cold periods. The decline in milk production is nearly always the result of the seasonal scarcity of fresh, succulent, nutritious feed. Without good feed, cows are obliged to eat dry, strawy, weedy grass, which not only lacks nutritive value, but often causes digestive troubles, constipation, and difficult birth. These troubles can be dealt with easily and cheaply; good health and a high level of production can be maintained–by the use of silage.

FIGURE 1. TILE-BLOCK TOWER SILO WITH TILE CHUTE.

Silage can be stored in permanent or temporary silos. Permanent silos can be either upright tower-shaped structures (see Figure 1) or horizontal, like the trench silo (see Figures 2, 3, and 4). Upright stack silos (see Figure 5) and fence silos are examples of temporary silos. The use of successive rings of fencing is becoming widespread; these silos can be lined with plastic or paper or they can be unlined. Many farmers have saved the money needed for permanent silos by using temporary silos for several years.

Losses of silage vary with the type of silo, the crop ensiled, its stage of maturity and moisture content, fineness of chopping, and the extent to which air and water have been excluded from the silage. Losses run from 5 to 20 percent in permanent upright silos; from 10 to 30 percent in permanent horizontal silos; from 15 to 50 percent in temporary trench, fence, and stack silos.

A silo should be located near the barn to keep to a minimum the time and labor involved in feeding.

Detailed instructions on silo building are given "Farm Silos," Miscellaneous Publication No. 810, Agricultural Research Service, U.S. Department of Agriculture, 1967 (revised).

FIGURE 2

It is not worthwhile to make a silo of less than four tons capacity, except under very special conditions. Spoilage in smaller silos is often excessive. A cow of average size not provided with any other fodder will consume about 23kg (50 pounds) of silage in 24 hours; on this basis a farmer knowing the number of cows to be provided for and the approximate length of the period during which silage is to be used, may estimate the quantity needed; for example:

| | |
|---|---|
| 20 cows @23kg (50 pounds) per day for 90 days | 41,400kg ( 90,000 pounds) |
| 5 heifers @14kg (30 pounds) per day for 90 days | 6,300kg ( 13,500 pounds) |
| 5 calves @7kg (15 pounds) per day for 90 days | 3,150kg ( 6,750 pounds) |
| | 50,850kg (110,250 pounds) |
| 51 metric tons | (56 short tons) |

The bare requirements would be 51 metric tons (56 short tons) of silage, and an allowance for wastage should be added. The tables may be used to estimate the dimensions of a silo.

A silo of ten tons capacity or less should be filled in two operations, that is, on two separate days with two or three days between operations. Similarly, a large silo should be filled in proportionate operations, though this is not so essential as with the smaller size. Table 1 gives trench silo capacities.

| Dimensions in Meters (Feet) | | | Approximate Kilograms (Pounds) of Silage Per 30cm (1') of Length |
|---|---|---|---|
| Top Width | Bottom Width | Depth | |
| 2.4( 8) | 1.8( 6) | 1.8(6) | 756 (1680) |
| 3 (10) | 2.1( 7) | 1.8(6) | 918 (2040) |
| 3.7(12) | 2.4( 8) | 1.8(6) | 1080 (2400) |
| 2.4( 8) | 1.8( 6) | 2.1(7) | 882 (1960) |
| 3 (10) | 2.1( 7) | 2.1(7) | 1071 (2380) |
| 3.7(12) | 2.4( 8) | 2.1(7) | 1260 (2800) |
| 3 (10) | 1.8( 6) | 2.4(8) | 1152 (2560) |
| 3.7(12) | 2.4( 8) | 2.4(8) | 1440 (3200) |
| 4.3(14) | 3 (10) | 2.4(8) | 1728 (3840) |

Material for silage varies considerably. Corn, guinea corn, sugar cane leaves, uba cane leaves, napier grass, guatemala grass may be used singly or in mixtures; the important point to be borne in mind is that the material should be young, fresh, and green. Uba and sugar cane should be cut before the stem is formed; guinea grass should be cut before flowering and seeding takes place; napier, guatemala, and elephant should be cut while the stems are still tender and green. If only fresh, leafy growth described above is used, there is no need for chopping the material as it is brought to the silo. It should be scattered thinly over the entire surface of the silo, and should be constantly trampled to cause consolidation. Trampling close to the walls is especially important.

Silage that is considerably more nutritious than grass silage can be produced by combining fresh young leguminous fodders with grass when filling of the silo. Cow peas, edua peas, soya beans, Bengal beans, and St. Vincent plum fodders have been used with success at the level of 20-25 percent of the total bulk. This material must be chopped.

The use of molasses is recommended in all silos, for increased palatability, increased nutritive value, and in the case of young grasses, or silage with leguminous mixtures, as an aid to the essential fermentation. Molasses should be used at the rate of 10kg per metric ton (20 pounds per ton) of grass material, as follows: if the material is wet with rain or dew, add two parts of water to one of molasses before application; if the material is dry, add four parts of water to one of molasses. As each layer of material, several centimeters or a few inches thick, is laid down, sprinkle on the molasses-water mixture, unless a blower with a continuous molasses sprayer attached is used. In leguminous mixtures 25 percent more molasses should be used.

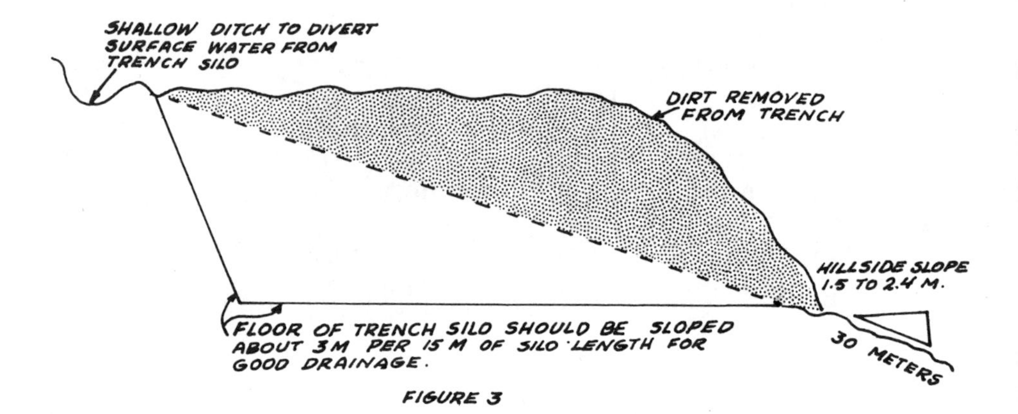


FIGURE 3

FIGURE 4

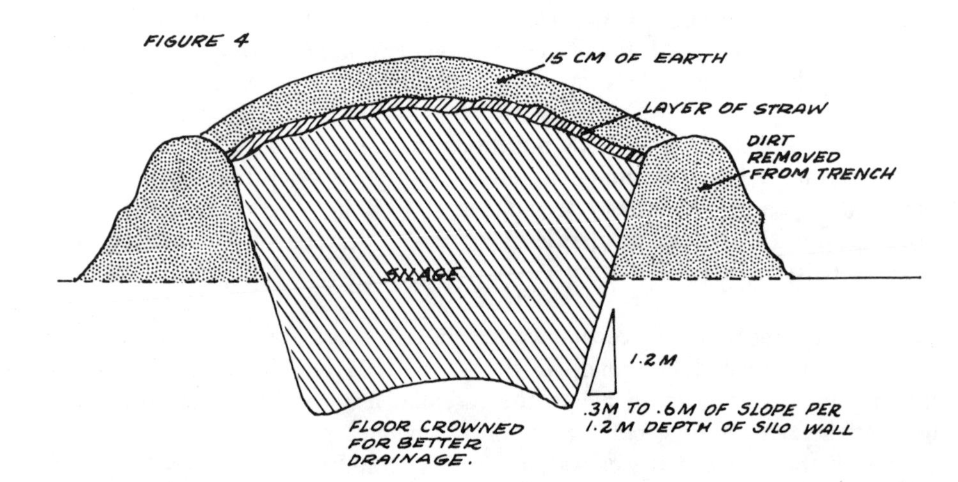

| Inside Diameter of Silo in Meters | Depth of Silage in Meters | | | | | | | | | | | |
|---|---|---|---|---|---|---|---|---|---|---|---|---|
| | 2.4 | 3 | 3.7 | 4.3 | 4.9 | 5.5 | 6.1 | 6.7 | 7.3 | 7.9 | 8.5 | 9.1 |
| 3 | 9.9 | 12.6 | 15.3 | 18 | 20.7 | 23.4 | 25.2 | 28.8 | 31.5 | 35.1 | 37.8 | 42.3 |
| 3.7 | 14.4 | 18 | 21.6 | 26.1 | 29.7 | 34.2 | 36 | 40.5 | 45 | 49.5 | 54.9 | 60.3 |
| 4.3 | 18.9 | 24.3 | 29.7 | 35.1 | 40.5 | 45.9 | 48.6 | 54.9 | 61.2 | 67.5 | 74.7 | 81.9 |
| 4.9 | 25.2 | 31.5 | 38.7 | 45.9 | 53.1 | 61.3 | 63.9 | 72 | 81 | 88.2 | | |

Table 2. Number of metric tons of silage in a vertical silo.

When it is not possible to obtain young, fresh material, and older material must be used, then chopping is essential. Once the material has been chopped the remaining operations are similar to those described above, with the exception that only 6kg of molasses need be used per metric ton (12 pounds per ton) of grass material plus 35 percent more if legumes are included.

After a silo has been filled level with the top and has been thoroughly trampled, the silage will settle gradually over a period of several days, bringing the need for refilling once or perhaps twice to compensate for shrinkage. After the final refill a thick layer of dried grass should be laid over the silage and trampled down; finally, a few heavy logs laid over the dried layer will assist consolidation. A pointed roof over the silo with eaves reaching down below the rim will shed rain water.

Silage made in the spring of the year when grass is young and nutritious will keep perfectly until the winter or drought period comes; then it is possible to supply cows with feed every bit as nutritious and as palatable as fresh grass in the natural state. It is true that some cows do not take naturally and readily to silage, but they may be taught to consume it with relish.

FIGURE 5.

When a silo is opened to feed cows, logs and the dried grass layer should be removed. It is commonly found that a layer of silage several centimeters (a few inches) thick from the top downward will have spoiled--turned black or slimy with white streaks of fungus here and there. This should be thrown away.

The color of the good silage exposed below may be green, yellow-green, or brownish-green, and it will have a strong pleasant smell; there will be no sliminess or streaks of fungus. The silage may be fed at will to cattle, care being taken only that each day's supply should be removed from the whole surface of the silage rather than from one spot; in this way an even surface will be maintained and no one section will be over-exposed to air. After each day's supply has been taken out, the surface of the silage should be covered with old bags to prevent drying out; if it should become necessary to interrupt the feeding of silage for more than a day or two, then the silage must be sealed off as it was when the silo was first filled.

**WARNING – GAS DANGER IN SILOS**

Suffocating and, in some cases, poisonous gas may be present around silos. Suffocating gas from fermenting silage, mostly carbon dioxide, forms in all silos shortly after filling begins and continues until fermentation stops. Poisonous gas, when present, is nitrogen dioxide. Its color and density vary with temperature. At room temperature it is orange yellow and 2 1/2 times as heavy as air. As the temperature rises, its color becomes darker and its density becomes lighter. The gas, being heavier than air, collects and remains in any depression or enclosed space when there is no strong, free movement of air. Danger of nitrogen dioxide gas occurs only during filling and for about a week after.

Many lives have been lost because of carelessness in entering a silo where there may be danger of gas. Gas is a particular hazard in below-ground silos. To stir the air in a silo, tie a rope to a basket, a blanket, a large piece of canvas, or a tree branch and then drop the article into the silo and raise it a number of times with the rope.

**Source:**

*The Farmer's Guide*. Marvin D. Van Peursem, VITA Volunteer, Newton, Iowa. Kingston, Jamaica: Jamaica Agricultural Society, 1962.

# Food Processing and Preservation

Mary Anne Schlosser

# Storing Food at Home

You work hard when you grow food and prepare it to eat. Buying food takes money that you have worked hard to earn. You do not want to waste it. To keep food clean and safe in the home you must have good storage space, suitable containers, and a way to keep foods cool and dry.

> **IMPORTANT**
>
> Only water that is pure enough to drink should be used for washing or cooking food. If the purity of water is in doubt, it should be boiled for 10 minutes or disinfected. See section on water purification, p. 138, for proper disinfection procedures.

## HOW TO CARE FOR VARIOUS KINDS OF FOOD

Different kinds of food need special care. Treating each food properly will make it keep longer.

### Dairy Foods

Fresh milk is safe if it is boiled. If you do not have refrigeration, boiled milk will keep longer than milk that has been pasteurized. Cream will keep longer if it is boiled.

After milk and cream are boiled, then cooled, store them in clean containers. These foods will keep longer if stored in a refrigerator, ice chest (see p. 290), or evaporative cooler (see p. 28). If refrigeration is not available store them in the coolest place you can find.

FIGURE 1

Use boiled water to reconstitute canned, evaporated, condensed, or dried milk or add water and boil for 10 minutes. Unsafe milk should not be used for any purpose.

Cooked foods using milk or cream spoil very quickly. Use them immediately in hot climates. **Do not store.**

Dried milk in its original container will keep for several months in a cupboard or on open shelves. Close the container tightly after using. The milk will take up moisture and become lumpy if exposed to air. Then it is hard to mix with water and food. A glass jar with a tight lid, or a tin can with a press-in lid, are recommended to store dry milk powder after the package has been opened.

After dried milk has had safe water added to it, store it the same as fresh fluid milk.

Canned evaporated milk may be stored at room temperature until opened. Before opening shake the can to mix thoroughly. After opening, cover tightly and store the same as fresh fluid milk.

Canned sweetened condensed milk may be stored in the cupboard or on open shelves. After the can has been opened it can be stored in the same place as the unopened can but it needs protection from ants and other insects. Sweetened condensed milk does not require refrigeration unless it has been diluted with water.

Butter should be kept in a cool place, in a covered container.

Keep hard cheese in a cool place. Wrap tightly in a clean cloth or paper to keep out air. Put in a box or metal container if possible. Before using, trim away any mold that forms on the surface.

Soft cheeses should be stored in a tightly covered container in a refrigerator or other cool place.

## Fresh Meat, Fish, Poultry

The moist surfaces of dressed meats, poultry, and fish attract bacteria that cause spoilage. Keep these foods clean, cold, and dry. They should be allowed some air when stored. Wrap loosely with a clean cloth or paper. Wipe or scrape off any dirt before wrapping.

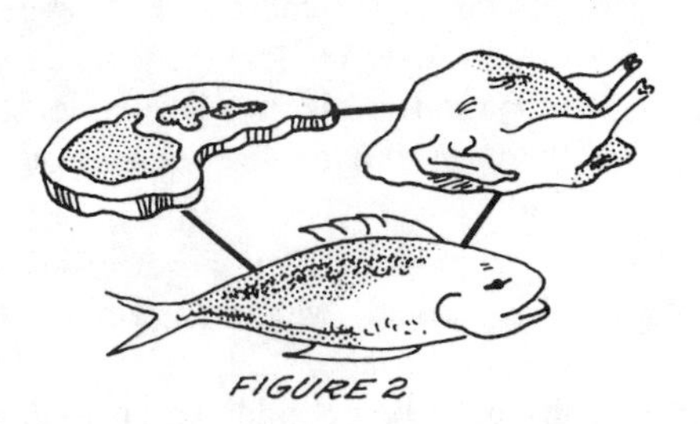

FIGURE 2

These foods spoil very quickly. They should not be kept long in warm, moist climates.

Rubbing cured or smoked meats with dry baking soda may help prevent molding. If meat is attacked by insects or shows spoilage, cut out the bad part.

# Eggs

Sort eggs as soon as they are brought from the poultry yard or market. Cracked ones should be removed and cooked for immediate use. Spoiled eggs should be thrown away. Rough handling and high temperatures shorten eggs' keeping quality.

FIGURE 3

Keep eggs in a covered container in a cool, dry, clean place. Eggs keep fresh longer if stored in an airtight container.

Don't wash the eggs unless you want to sell them. Water removes the thin film on the shell that protects the egg. This film helps to stop evaporation, the entrance of harmful bacteria, and the absorption of odors. Do wash eggs just before using them. Wash with cooled boiled water.

# Fresh Fruits and Vegetables

Fresh fruits and vegetables need to be kept clean and in a cool place with good air circulation and out of direct sunlight. Such conditions help to prevent spoilage. Avoid breaking or cutting the skin.

Sort fruits and vegetables before storing. Use bruised ones immediately, throw away decayed or spoiled ones. Ripe fruits and vegetables should be used in two or three days. Allow them to ripen in the open air out of the sun. Wash fruits and vegetables in clean water before using them.

Fruits and vegetables stored in boxes, baskets, barrels, and bins should be sorted frequently to remove decayed or spoiled ones. Some fruits such as oranges and apples may be wrapped in separate papers. The wrappers help to keep the fruit from bruising each other and also help to prevent mold.

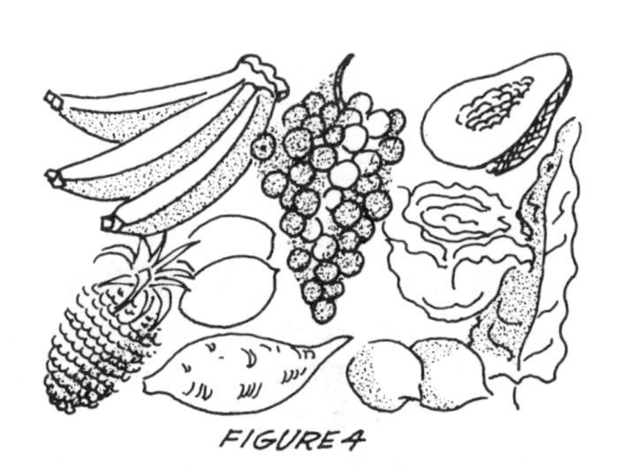

FIGURE 4

Soft fruits and vegetables such as berries, peaches, papayas, figs, tomatoes, and plums should be spread out on clean wrapping paper or in shallow pans or platters rather than deep containers.

Potatoes and other starchy tubers should be sound, dry, and free from soil, cuts, and bruises when put into storage. Wet tubers rot more quickly than dry tubers. Store potatoes in a dark place because light promotes the formation of green skin and the poisonous glycoalkaloid called solanine in the potato.

Potatoes keep better if cured within 1-3 days after harvest. The easiest way to cure potatoes is to keep them in a container with restricted ventilation (to establish a high relative humidity of about 85 percent) for about 15 days at 15°C (60°F), or 10 days at 20°C (68°F), *or* 6 days at 25°C (77°F). After curing, fully open the container to allow free air movement and store in a cool, dark place.

## Fats and Oils

Keep all fats cool, covered, and in lightproof containers. Heat, light, and air help to make fats rancid. Use no iron, copper, or copper alloy vessels or equipment to store or handle fats and oils because traces of iron or copper make them turn rancid quickly.

Fats and oils should be kept dry with no moisture mixed with them. Mold on the surface of fats shows moisture is present. Remove the mold carefully. If possible, heat the fat to drive off the moisture.

Foods like nuts and chocolate, which have some fat, may get rancid. Nuts keep best when left in shells. Keep these foods cool, clean, and dry in light-proof containers.

Peanuts that are much darker in color than the rest of the batch should be thrown out. They are probably contaminated with aflatoxin, which causes cancer of the liver.

## Baked Goods

Cool bread, cakes, pies, cookies, and other baked goods rapidly after they are taken from the oven. Be sure the place is free from dust and insects. Wrap bread with a clean cloth or paper when cool.

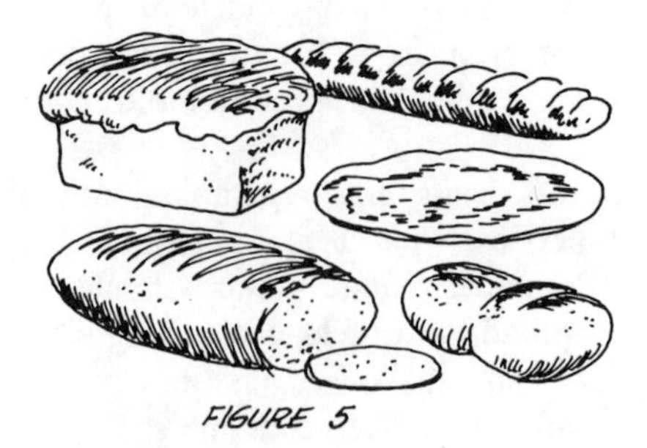

FIGURE 5

Stored baked goods in a clean tin box or other suitable container off the floor.

Molds grow on bread. Scald and air the bread box at least once a week. In hot humid weather do not shut the bread box tightly when it is filled with fresh bread.

Store crackers, crisp cookies, pretzels, and other crisp baked goods in airtight containers to retain crispness. A tin can with a press-in lid is ideal. If not available use a sealed plastic bag made from thick plastic.

## Dried Foods

Dried meats and dried fruits and vegetables may be kept in closely woven cloth bags if the bags of food are kept in a cool, dry place. If these dried foods are hung in a damp place they are likely to mold.

Properly dried foods are best stored in airtight containers if you live in a humid climate. A tin can with a press-in lid or a large glass jar with a tightly fitting lid will prevent moisture pickup from the humid atmosphere. Look at the product occasionally and check that it is in good condition. If there is any sign of mold it means the food is not dry enough.

Open bags of dried foods should be kept in a pottery or metal container. Seal the container tightly to keep out insects and rodents.

## Canned Goods

Canned foods should be kept in a clean, dry, cool place. Destroy any swelled or leaking cans. Do not eat or even taste the food in swelled or leaking cans. Don't even open the can. Dispose of it.

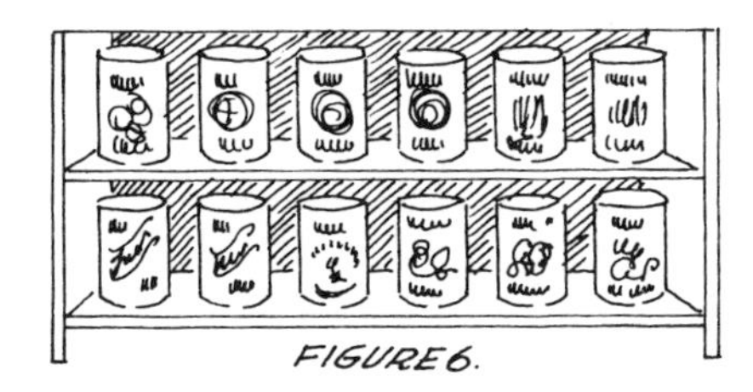
FIGURE 6.

The outside of the cans will become rusty if they are stored in a damp place or in humid atmosphere. The contents of rusty cans are safe to eat provided there are no holes, leaks, or bulges in the cans and the contents appear normal when can is opened.

## Leftover Cooked Foods

Moist cooked foods, particularly those made with milk, eggs, meat, or fish, spoil easily. Leftover cooked foods should be cooled quickly. Store in refrigerator, ice chest, or evaporative cooler. Use at the next meal if not refrigerated.

# FOOD SPOILAGE

## When is Food Spoiled?

Food generally shows when it is spoiled. Check it often. It may have an unpleasant appearance, taste, or smell. Look for these signs of food spoilage:

- slime on the surface of meats and other moist foods
- bad odors
- sour taste in bland foods
- gas bubbles, or foaming
- discoloration
- liquid that has become cloudy, thick, or slimy
- texture becomes very soft
- signs of mold growth

It is important to destroy spoiled foods as soon as they are found. Throw away any food that has a bad smell. Chopped meat, eggs, and sea food usually spoil rapidly. Watch grains for signs of weevils. Look for insects and mold in dried foods. Destroy the part that has insects or mold at once.

## Why Food Spoils

Foods may be spoiled by:

- bacteria, molds, and yeasts
- parasites of meat animals
- insects and rodents
- warm air, freezing temperatures, light
- too little or too much moisture
- storing too long

**Dirt and careless handling** increase food spoilage. Good care of food in the home can help avoid waste. Keep food in a clean and safe place. **Bacteria** are living things so small you can't seem them. Many are harmful. They live almost everywhere. Sometimes food is made unsafe because bacteria causing disease have gotten into it. Food can carry these and many other diseases:

- amoebic dysenteries
- typhoid
- botulism
- tuberculosis
- diphtheria
- salmonellosis

People may appear healthy and still carry these disease bacteria in their bodies. When they handle food, the bacteria may be passed on to the food. Then the food is unsafe for others.

**Bacteria** need water to live. Removing water prevents their growth. Foods are dried to preserve them. Then they are kept dry. Some foods that are dried are meat, fish, beans, peas, grapes, figs, currants, cereal grains, flour, spaghetti, noodles and other pasta products, dates. They are dried in the sun or smoked over a fire.

Bacteria, molds, and yeast in foods may be destroyed by heating and some chemical preservatives. They cannot grow in properly dried foods. They grow more slowly at refrigerator temperature than at room temperature.

Molds can be harmful. They grow where it is damp. Molds look like delicate velvety or powdery growths of various colors spread through food.

If meat, cheese, or jam have mold on the surface, cut away the moldy part. The food that is left may be eaten.

**Parasites**, such as tapeworm and trichina, live in meat animals. The tiny larvae of these parasites may be in the lean meat. They are waiting to complete their development in the human body or some other place.

Thorough cooking of meat is the best way to destroy these parasites. Preservatives such as salt and smoke do not destroy them. There is great danger in eating uncooked or lightly cooked sausages, for example, even though they have been smoked.

Many chemical substances either destroy certain harmful bacteria or prevent their growth. For food, two of the simplest to use are common salt and sugar. Salt is used for meat and vegetables. Sugar is used to preserve fruits. Sugar and salt have to be used at a high level to be effective.

**Insects and rodents** eat some food and damage several times as much as they eat with urine, feces, and hairs. They may also leave dangerous bacteria on them.

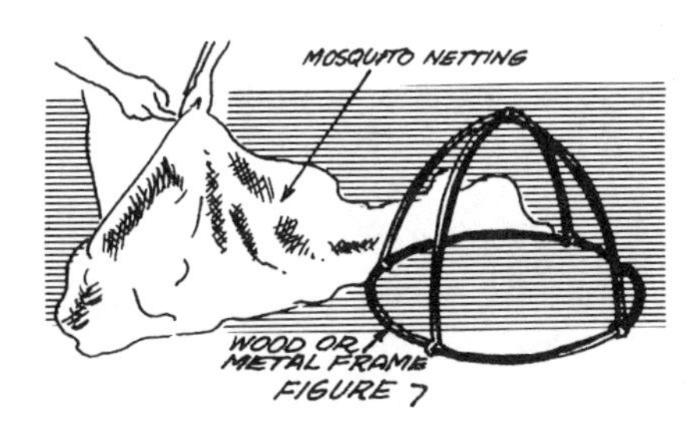

FIGURE 7

The house fly spreads typhoid fever, cholera, dysentery, tuberculosis, and many other diseases. Keep flies away from foods. A cloth net fastened to a simple wire frame keeps flies out of contact with food (Figure 7).

The "fly specks" often found on food or dishes may have disease germs and mice destroy many types of food.

To help keep insect and rodent pests out of food:

- o keep food covered or in closed containers
- o get rid of garbage and trash
- o keep the storage area clean

Poisoned bait, powders, or sprays may be necessary to rid storage areas of household insects and rodents. Ask your health department, sanitation, or other official what pesticide to use, where to get it, and how to use it. These people have special training on how to control household pests. They can help you.

Use pesticides with care. They are POISONOUS to people and animals. Keep them out of reach of children. Never store insecticides in the same place you store food. Always wash off any dust, spray, or solution that gets on you. When spraying, remove dishes, pots and pans, other cooking utensils, and food from the room. If you have a cupboard with solid, tight fitting doors store the dishes and cooking equipment there while spraying. Never use oil spray or solutions near a fire.

Rats and mice can be caught in traps or killed with poison bait. Destroy or block up all places where they are likely to nest and breed. Rodents cannot chew through metal, glass, or pottery containers so try to use containers made from these materials for storage of food.

**Temperature** affects food. Fruits ripen more quickly, vegetables become old and wilt more quickly, and nuts, fats, and oils become rancid more quickly as the temperature increases. Insects, bacteria, molds, and yeasts grow more quickly at higher temperature. Therefore, store food in a cool place. Do not store food near a hot stove.

Food in direct sunlight gets hotter and spoils more quickly than food in the shade. Food should never be left in direct sunlight unless it has been put there for a limited time to dry it or to drive out insects.

Freezing temperatures can ruin the texture and flavor of some foods. Frozen potatoes, for example, are watery and have an unpleasant flavor. Frozen and thawed foods are safe to eat but may have an off flavor or bad texture.

**Moisture** in the air is necessary where green leafy vegetables are stored. If there is not enough moisture in the air, the moisture from these vegetables will evaporate into the air. Then they become wilted or limp and look bad even though they are still safe to eat. These vegetables keep best when stored in a sealed plastic bag or box and kept in a refrigerator, ice chest, or evaporative cooler.

# CONTAINERS FOR FOOD

It is very important to have good containers for storing food. Some foods must be stored in containers with tight fitting covers. Generally each food is best stored in a separate container. Label food containers to save time and avoid mistakes.

## Types of Containers

Dry foods should be stored in glass, pottery, wooden, or tin or other metal containers. The type of container will depend on the food to be stored and whether the container can be washed. Dry tin quickly to avoid rust.

For moist and watery foods the choice of containers is more limited. Leakage must be avoided. You must consider the effect acids in watery foods have on the container, especially metals. A container that can be washed and aired before fresh supplies are stored in it is best.

FIGURE 8

Pottery jars are good for storing many kinds of food. Jars that are glazed on the inside are best. They can be washed easily. If the jars do not have a tight fitting cover, make one. Use a plate, saucer, or piece of metal. A good cover helps to keep out insects and rodents.

Glass jars with tight lids are also good for storing many foods. Foods that are affected by light should not be stored in glass jars unless the jars can be stored in a dark place. Glass jars can be used again. Wash them in hot soapy water. Rinse them with hot water that has been boiled for 10 minutes. Dry them in the sun if possible.

Bottles are good for storing liquids and some dry foods. In many countries people preserve fruit and vegetable juices in bottles.

Coconuts, gourds, and calabashes may be used for storing some dry foods for a short time. Covers can be made of closely woven materials. Insects tend to eat away the soft lining of these containers, so they are not good for storing meal and flour for long. Wash these containers often to keep out weevils. Dry in the sun.

A simple cupboard can be made from a wooden box with shelves. The door is made of chicken wire so air can circulate. Use it to store root vegetables and some fruits.

FIGURE 9. IN ETHIOPIA COVERED BASKETS ARE HUNG FROM THE RAFTERS TO STORE DRIED FRUITS AND VEGETABLES AND BREAD.

Tin cans of all sizes are good for storing foods. Sometimes the lids of cans containing food have been removed with a hand or mechanical can opener. Then the lid does not fit. If you use these cans to store food, make a cover out of a plate, saucer, or a piece of metal.

Use a food cover to keep out flies and other insects when you store food on a table in an uncovered container. You can make a food cover out of mosquito netting and a metal or wooden frame (see Figure 7). Store foods this way for a short time only.

A bread box may be made of metal or wood. Punch holes in each end for air circulation.

Open baskets are good for storing fresh fruits and vegetables for short periods. A tight cover is not needed for these foods.

## Care of Food Containers

Food containers must be kept clean. Wash and dry containers before fresh supplies are stored in them.

Water for washing containers should be clean and hot. Use soap or detergent. Rinse the containers carefully with clear clean water. Dry them in the sun if you can.

Do not store food in containers that have held kerosene, gasoline, heavy oil, chemicals, or pesticides.

Containers holding food that does not need to be kept cool may be stored on shelves or on a table.

## THE STORAGE AREA

A good storage area is:

- clean and neat
- cool and dry
- well ventilated
- free of rodents and insects

You may store food in the kitchen in cupboards on open shelves, or in a closet with shelves. Sometimes a separate room next to the kitchen, called a pantry, is used for storing food. Also cellars, caves, and outdoor pits are used in some parts of the world for food storage.

FIGURE 10. IN THIS PHILIPPINE HOME SOME FOODS ARE STORED ON OPEN SHELVES. OTHER FOODS ARE STORED IN CUPBOARDS WITH VENTED DOORS SO AIR CAN CIRCULATE.

## Good Ventilation

Ventilation is important for good food storage. Good circulation is needed around food to carry off odors and to keep the right temperature and the right amount of moisture.

## Keep the Storage Area Cool and Dry

Many fresh fruits soon spoil in a warm place. Then they are unsafe to eat. Cooking oils, table fats, and other foods with fat in them may get a stronger flavor if stored in a warm place. A dry storage area helps to prevent mold on foods such as bread, cheese, and berries. It also prevents rust on tin cans in which food may be canned or stored.

## Keep the Storage Area Clean

There is no substitute for cleanliness. Scrub shelves, cupboards, and floors often. Paint, whitewash, or line shelves with clean paper. Clean the walls, then paint or whitewash them. Keeping the storage area clean helps to keep away household pests.

Remember, cleaning rcmoves insecticides. Apply them again after you clean, not before.

# Keeping Foods Cool

Some foods are quite perishable. They are:

- o fresh meat, fish, poultry
- o some fresh fruits and vegetables
- o milk, butter, margarine, cream
- o leftover cooked foods

In a warm climate it is best to buy these foods in small quantities and use them quickly rather than store them. If you have to store these foods, keep them as cool as possible. This is one way to keep them fresh and prevent spoilage.

There are several ways to keep foods cool. Some ways work better than others:

1. Mechanical refrigerators are the most effective in cooling and preserving foods, but are expensive and require an outside fuel source.

2. Ice chests come next; if ice is available they are quite effective.

3. Evaporative coolers follow ice chests.

4. Window boxes are the poorest devices.

5. In some situations, it is possible to enclose food in watertight containers and place in a cool stream or spring.

6. Keep food in shade, out of the sun, if no other means is available to protect it.

Obviously there is direct relationship between effectiveness and price. Each family should install the best cooling system it can afford; that is, option 1 is better than option 2, but 2 is better than 3, etc.

The information given in this section will help you to choose a practical way to keep foods cool given your particular situation.

# EVAPORATIVE FOOD COOLER

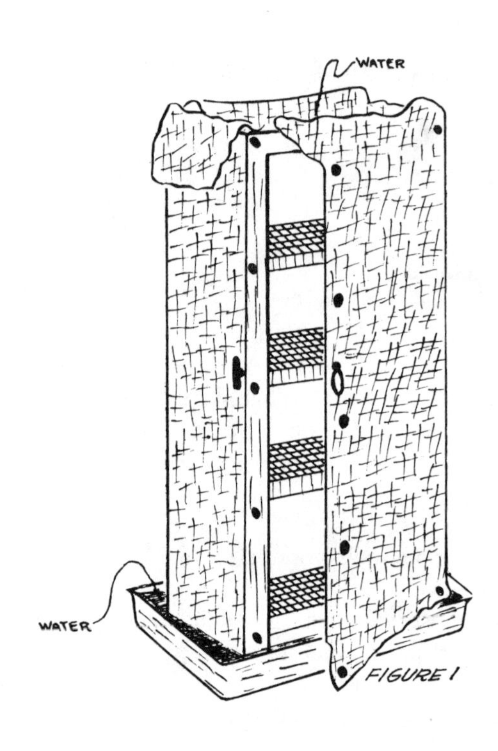


The evaporative food cooler is cooled by the evaporation of water from its cloth cover. The cloth is moistened as capillary action moves water from the pans through it.

If the climate is dry and the cooler is kept in a breezy spot in the shade, it will cool food considerably below the prevailing temperature. **To be safe,** the cooler must be kept clean. The cooler's cloth cover keeps flying insects out. The water-filled lower pan discourages roaches and other crawling insects.

It should be emphasized that coolers based on the principle of evaporation of water need readily available water of reasonably good quality and a low humidity environment. **These coolers do not cool in a humid climate.**

**Tools and Materials**

Saw
Hammer
Nails, tacks
Burlap or other cloth: 2m x 2m (78 3/4' x 78 3/4')
Wood for frame: 3cm x 3cm x 13m (1 1/4' x 1 1/4" x 42.7')
Pan: 10cm (4") deep, 24cm x 30cm (9 7/16" x 11 13/16") for top
Screen, hardware cloth, or galvanized iron: 2m x 2m (78 3/4" x 78 3/4") (non-rusting)
Hinges: 2 pair
Pan larger than 30cm x 36cm (11 13/16" x 14 3/16") for legs to stand in
Paint for wooden and metal parts
Buttons or lacing material for cover

Make the wooden frame to fit the upper pan (see Figure 2). This might be the bottom of a discarded 20-liter (5-gallon) oil can. The lip of the pan fits over the top of the frame to keep the pan from falling into the compartment. Hinge the door carefully so that it swings easily, and make a simple wooden or thong latch. Paint or oil all the wooden parts. The upper and lower pans should also be painted to prevent rust. Cover the shelves (see Figure 3) and frame with screening or hardware cloth and tack it in place.

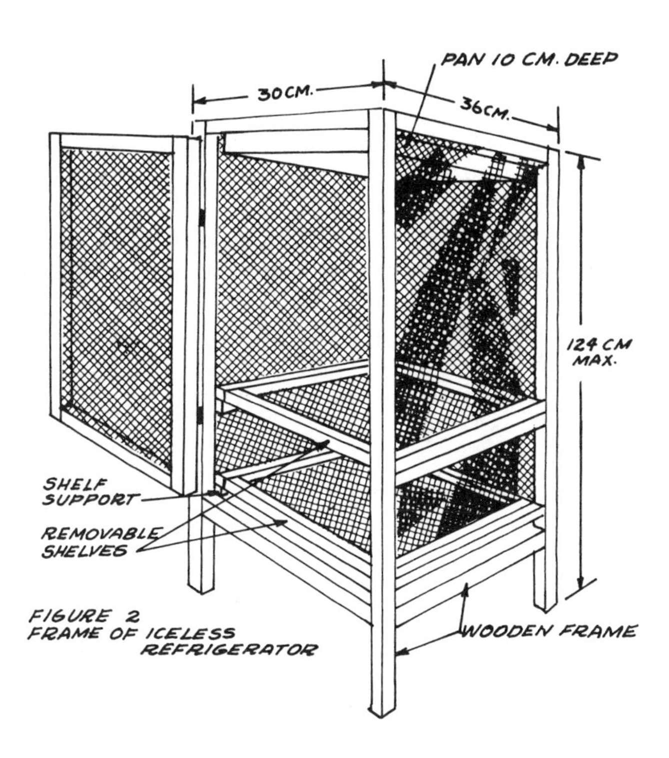


FIGURE 2
FRAME OF ICELESS
REFRIGERATOR

The frame can be strengthened by putting the screen on diagonally, although this will take more material than applying it with the wires parallel to the frame. Make the shelves adjustable by providing several shelf supports. Flatten the pointed ends of the nails slightly to keep the wood from splitting when it is fastened.

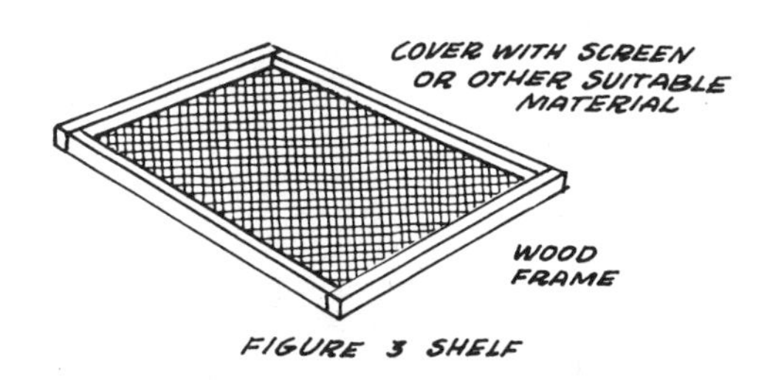


FIGURE 3 SHELF

Make two covers of canton flannel, jute burlap (not sisal or henequin burlap), or heavy-grade absorbent coarse cloth to fit the frame. Wash and sun one cover while using the other. On the front, fasten the cover to the door instead of the frame. Allow a wide hem to overlap the door closing. To form wicks that will carry water from the pans into the cover, the top and bottom of the frame and door covers should extend into the upper and lower pans. If the cloth cover does not stay moist, extra pieces of cloth can be placed at the top of the frame to serve as additional wicks.

# ICELESS COOLER

A second type of cooler may be made from a basket with a loose fitting cover. It may be made of bamboo or other slender wood with open weave. The size depends upon the family's needs. In addition to the basket, you will need a container to set the basket in. This may be square or round, of earthenware or metal. A clean oil drum could be used. This container should be about 30cm (12") high and wider than the basket. Other materials include bricks or stones and soft jute burlap.
To build the cooler (see Figure 4):

- o Select a cool place in the kitchen away from the stove for your cooler.
- o Place the outer container here.
- o Arrange the bricks or stones in the container so the basket will balance evenly on them.
- o Sew burlap around the rim of the basket. Let it hang loose around the bottom and extend into the earthenware or metal container.
- o Sew burlap loosely over the cover of the basket.

Set the basket on the bricks. Place food in the basket. Cover. Put water in the bottom of the container. Wet the cover of the basket the first time the basket is used. Later do this just occasionally. The basket itself should not be in the water. The burlap cover should hang down into the water.

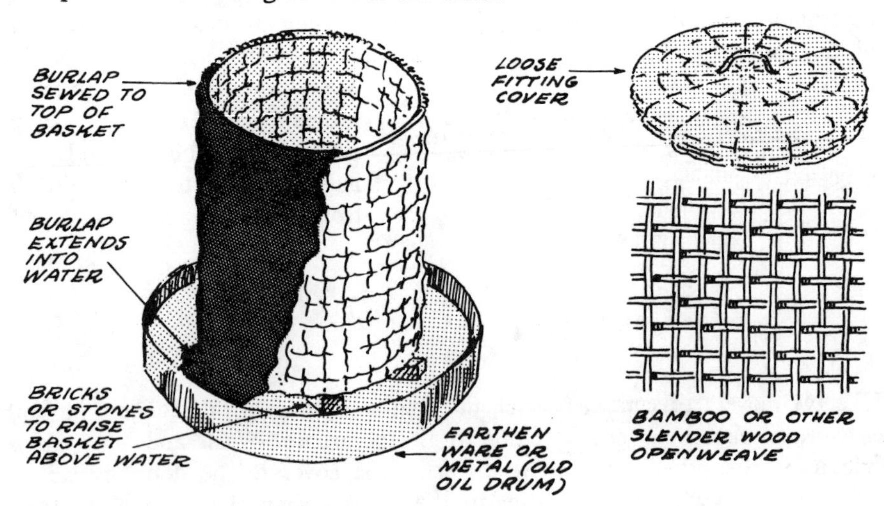


FIGURE 4

# WINDOW BOX

In some countries window boxes are used to store foods during the cool months of the year. They must have good ventilation and tight covers to keep out rain or snow. An ordinary light wooden box may be used or you can make one.

To install a window box:

- o Fit the box to the outside of the window. The window is the door. Select the window that is in the shade longest during the day. Keep the window closed when the box is not actually in use. This will keep the box from getting too warm and the room from getting too cold.
- o Put a shelf on the window sill. Support the shelf with wooden braces.
- o Set the box on the shelf. Fasten the box to the window case with screws or nails.
- o Fit a sloping top over the box to shed the rain.
- o Make holes in the end of the box so air can circulate. Screen the holes.
- o Shelves may be made of heavy screening, poultry wire, or wood.
- o Rest the shelves on cleats fastened to the sides of the box.
- o Paint the box inside and out. It will be easier to keep clean. Wash the inside with soap and water from time to time.
- o Food placed in the box should be in clean covered containers.

A similar food storage closet may also be built on the outside of the house. You can make it open into a room by a special door through the wall.

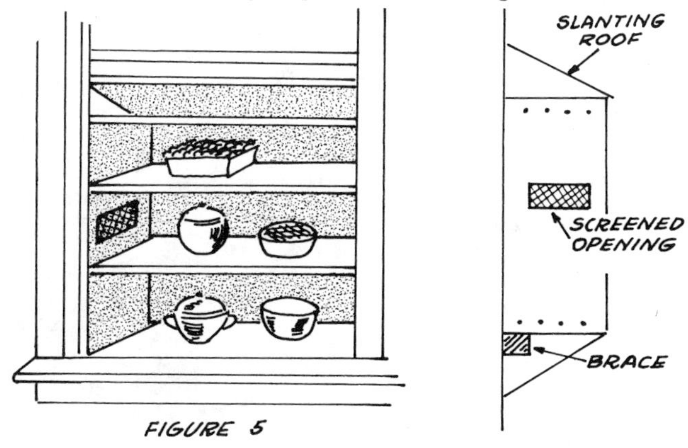


FIGURE 5

# OTHER WAYS TO KEEP FOOD COOL

A **mechanical refrigerator** is ideal for storing perishable foods. However, refrigerators are not available in all parts of the world and are often very expensive to buy and operate. Where a refrigerator is used, it needs special care.

Clean and defrost it regularly. To do this, turn it off. Allow the ice to melt. Wash the inside of the refrigerator thoroughly, using warm water and soap. Pay special attention to the corners.

An **ice chest** can be made at home. Line a wooden packing case with galvanized iron. You will need to put insulation between the wooden box and the iron to keep out heat. Use sawdust, cork, or similar material. Be sure to insulate the top and bottom as well as the sides. Make a hole at the bottom for water to drain out as the ice melts. Keep the ice chest clean. Wash it with soap and water often.

To pack the chest, allow at least one fourth to one third of the volume of the chest for the ice. Block ice lasts longer than chopped or crushed ice. Keep the packed chest out of the sun and away from sources of heat. Cool cooked foods to room temperature before placing them in the chest.

A **wooden keg** lined with cement makes a good food cooler. You may store leafy vegetables such as spinach and lettuce here. The vegetables can be kept in a strong paper or plastic bag. Hang the bags on a hook screwed into a cover of the keg. Fill the bottom with water.

On some farms cold water pumped from deep wells may first be used to cool foods, by running it through a suitable storage box. Also, a house or box may be built over a spring or brook to keep foods cool.

Special wells or caves are sometimes built for cool storage of foods.

FIGURE 6 THIS IS A SIMPLE COUNTRY ICELESS COOLER USED IN INDIA. YOU CAN MAKE IT EASILY WITH TWO DIFFERENT SIZED POTTERY JARS. PUT WATER BETWEEN THE JARS. COVER THE TOP WITH PALM, BANANA OR OTHER LARGE LEAVES.

# Storing Vegetables and Fruits For Winter Use

In some countries the climate is too cold to grow foods the year around. Farmers and gardeners in many parts of the world have found good ways to store some vegetables and fruits.

Some of their methods may be ones you will want to study and tell others about. An agricultural adviser can help you decide which type of storage is best for your climate and the foods grown in your area. Storage methods described here are practical only in areas where outdoor winter temperatures average -1°C (30°F) or lower. They do not work when the climate is warm all year long.

**Some vegetables**, like tomatoes, can be planted late in the season so that they can be picked just before frost. If picked when white or turning red, tomatoes will ripen in a warm room. To store them for longer periods, they can be packed in boxes of sawdust; when they are to be used, the boxes are opened and the tomatoes are put in a warm room to ripen.

**Dry bean seeds** can be kept for winter use by picking the pods as soon as they are mature and spreading them in a warm, dry place until dry. The beans are then shelled, stored in bags, and hung in a cool, dry, ventilated place until needed. Cellars are usually too damp for storing dry beans. Dry beans of all kinds, soybeans, and peas can be stored this way. Keep the beans as dry as possible.

**Root crops** such as beets, carrots, celery, kohlrabi, turnips, winter radish, and horseradish are not stored until late fall. When the soil is dry, the roots are pulled and the tops are removed. Cone-shaped pits make good storage places for root crops in areas where they can be kept from freezing. Turnips may be left in the garden until later than most crops but are hurt by alternate thawing and freezing. Parsnips may be left in the ground until needed as freezing does not hurt them, but put a few in storage for use when the ground is frozen.

**Sweet potatoes** store best in a warm, moderately dry place. A small supply can be placed near a cooking stove or a warm chimney or some other place where the temperature will stay around 12°C to 15°C (55°F to 60°F).

**Late maturing pumpkin and squash** can be kept in rows out of doors until late winter. They can also be kept on shelves in an area with a temperature ranging from 12°C to 15°C (55°F to 60°F).

Some helpful pointers on storing fruits and vegetables:

- Different vegetables and fruits need different storage conditions and methods
- Anything showing decay or injury should not be stored.
- Vegetables and fruits will dry out unless the storage place is damp and the temperature low but not freezing.
- Ventilation not only changes air and removes odors, it also helps maintain desirable temperature and humidity.
- Windows and ventilators should be kept open when temperature is not freezing.
- Walls and ceilings should be insulated so moisture will not condense and drop on stored foods.

The following sections show how to build some kinds of storage facilities.

## POST PLANK CELLAR

This type of storage cellar is low in cost, but does not last long because the wood will decay. (See Figure 1). If creosote or other waterproofing material is available, paint the wood with it to slow down decay.

- Dig a hole big enough to hold the foods to be stored and 120cm (4') deep.
- Keep the soil piled nearby to use to cover the roof and bank the sides.

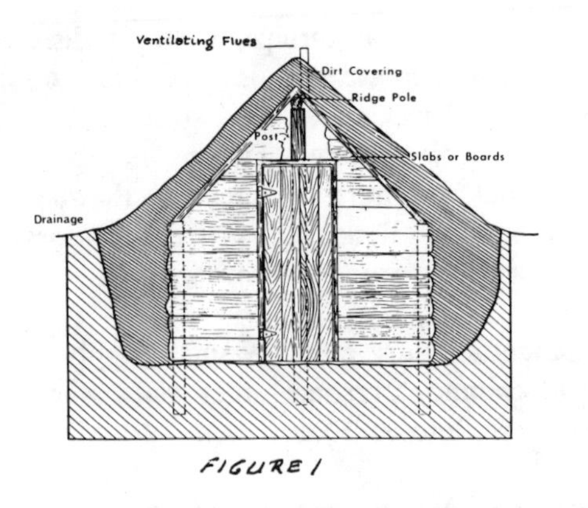

FIGURE 1

- Set two rows of posts of the same height in the bottom of the pit near the side walls.
- Set a middle row of posts about 150cm (5') higher than the outside posts. Put a ridge pole on the center row. Lay planks on the two outside rows.
- Next place a roof of planks.

- o Close the ends and cover the whole cellar except the door with soil. The door may be made of planks or other durable material. The thickness of the cover depends upon the climate.

- o Be sure that water drains away from the cellar. Extend a pipe from the storage area up through the dirt for ventilation.

## CABBAGE PITS

A good way to store cabbage, collards, and other greens is in a pit made of stakes and poles covered with straw (see Figure 2).

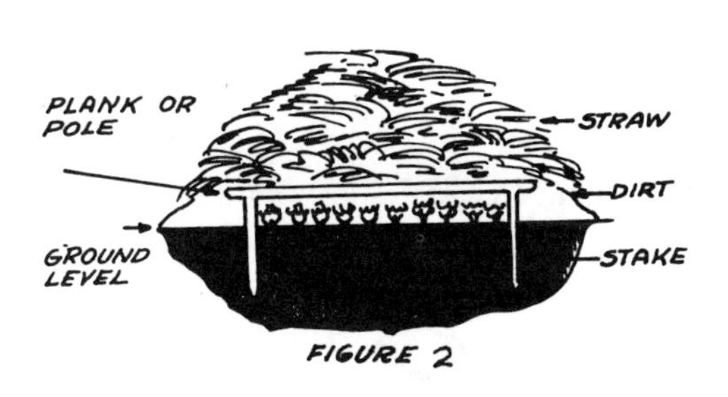

FIGURE 2

- o Dig a trench long enough to hold the number of cabbages to be stored.

- o Pull the plants by the roots and set them side by side in the trench.

- o Pack soil around the roots.

- o Build a frame about 60cm (2') high around the bed. This may be of boards, poles, or stakes driven into the ground.

- o Bank soil around the frame.

- o Place poles across the top to hold a covering of straw, hay, leaves, or corn fodder.

Cabbages can also be stored above ground in an area protected by drains from excess moisture (see Figure 3). Cabbage plants are pulled out by the roots, placed head down in the storage area and covered with soil. The advantage of this method of storage is that you can remove a few heads of cabbage without disturbing the rest of the pit.

## STORAGE CONES

- o Build the cones either on the surface of the ground, or in a hole 15cm to 20cm (8" to 10") deep in a well-drained location.

- o Spread a layer of clean straw, leaves, or similar material on the ground.

- o Stack the food to be stored on the litter in a cone-shaped pile.

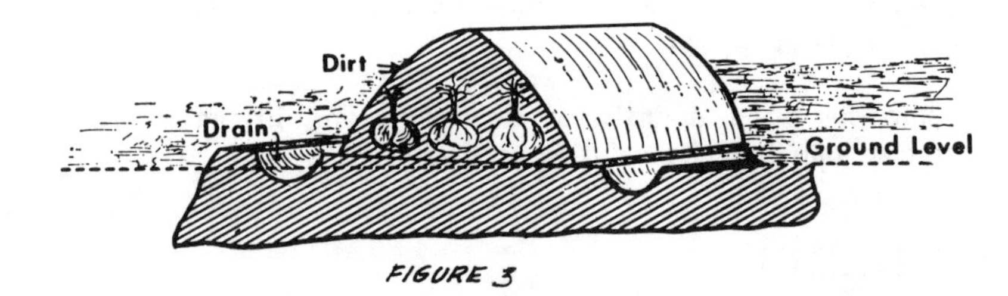


FIGURE 3

- Cover the food with more straw, leaves, etc.

- Cover the entire pile with 7cm to 10cm (3" to 4") of soil.

- Firm the soil with the back of a shovel to make it waterproof. More soil may be needed in very cold weather.

- Dig a shallow drainage ditch around the cone to carry away water.

- Ventilation or air circulation is necessary.

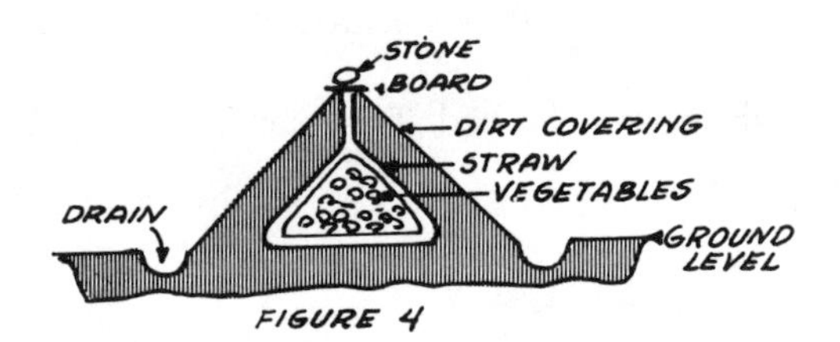


FIGURE 4

**Small cones** with 100 to 150 liters (a few bushels) of vegetables will get enough air if the straw between the vegetables and soil extends through the soil at the top opening. To keep out rain, cover the top with a board or piece of sheet metal held with a stone.

**Large cones** - Place two or three rough boards or stakes up through the center of the pile of vegetables to form a flue. Cap the flue with two boards nailed together at right angles.

- Opening the cone - Once the cone is opened it is best to remove all the food at once. It is better to make several small cones rather than one large one, and place small amounts of vegetables in each cone. When several kinds of vegetables are stored in the same cone, separate them with straw or leaves.

- Cones should be made in a different place every year to avoid decay from spoiled food left in an old cone.

# Fish Preservation

Fish can be an important source of protein, and more and more people are adding fish to their diets. Whether fish are caught from the sea or raised in a pond, a problem many people face is that they have more fish on hand at one time then they can eat or sell fresh.

If the proper equipment and a reliable supply of energy are available, fish can be kept for long periods by canning or freezing. Without these resources, salting and/or smoking are good low-cost choices for preserving fish.

Whichever method is chosen, quality and cleanliness are especially important:

- The quality of the fish to be preserved–the fish must be top quality; salting and smoking will not help poor quality, old, or rotting fish; and

- Cleanlines in all operations–all water used must be unpolluted; all waste must be removed from working and drying areas; whatever comes in contact with the fish, including all the equipment, must be kept clean.

## SALTING FISH

Salting, one of the oldest methods of preserving food, is an art as well as a science. The process of salting fish is influenced by weather, size and species of fish, and the quality of salt used. Therefore, experience is needed to adapt the process outlined here to your situation. Start by salting small lots of different varieties of the available fish. By salting small amounts of fish at first, you will learn how much time is required for each step. Salted fish, if properly packed to protect it from excessive moisture, will not spoil.

One word of caution: Start by salting non-fatty, white-meated varieties of fish. The salting of fatty fish brings up problems of rancidity, rusting, and spoilage that can be handled better after you have experience in salting.

The process of salting fish has four operations:

- Preparing the fish
- Salting
- Washing and drying to remove excess salt
- Air drying

IMPORTANT POINTS TO REMEMBER

- Use only top quality fish
- Work cleanly
- Work fast
- Keep the brine saturated–when in doubt, add more salt.
- Try to follow local custom in style and length of cure
- All water used must be unpolluted

**Tools and Materials**

A clean sharp knife

Salt: the amount varies with local conditions, but figure about 1 part salt (by weight) to 5 parts of raw, prepared fish. Use good quality salt. Salt that is dirty, discolored, or has a bitter taste is unsuitable for salting fish.

Clean containers for washing fish

Clean, flat working surfaces; such as tables

Clean containers for removing waste

Waterproof vats: one or more, depending on the amount of fish to be salted. The dimensions are not too important; a good size is 183cm x 152cm and 91cm deep (6' x 5' x 3'). But fish can be salted in a container as small as a wide-mouthed glass jar. Metals other than stainless steel should not be used. Wooden boxes will work because moisture will swell the wood and seal it effectively.

Clean boards and weights (for pressing).

Clean slats or lines for hanging fish (see Figures 3 and 4).

Portable thatch-roof shelters or small roofed sheds (see Figure 5).

# Preparing the Fish

Fish should be gutted and beheaded as soon as possible after catching.

Remove the head by cutting it off on a slanted line following the gills. Sharks can be beheaded at the last line of gill slits. (Only the "wings" of rays or skates are usually considered edible.) Fish that weigh 250gm (1/2 pound) do not have to be beheaded but they should generally be gutted. Local custom will determine whether or not they should be beheaded.

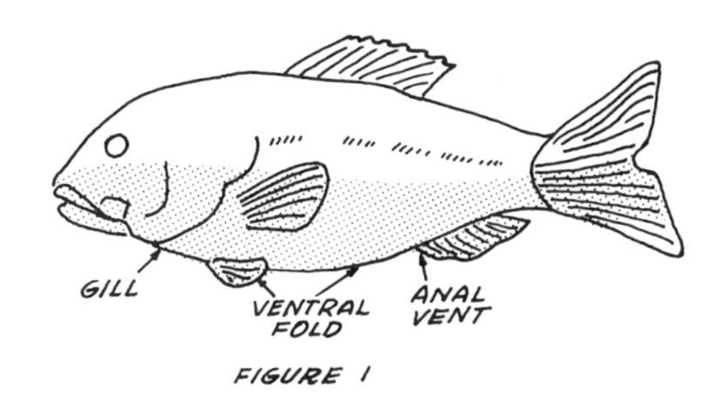

FIGURE 1

In gutting a fish, cut from the gill cavity along the ventral fold to the anal vent (see Figure 1). All the guts must be removed. It is also good commercial practice to remove the black membrane located in the visceral cavity (the hollow in the body of the fish which contains the guts) of many species.

The next step is to bleed the fish. All species of fish must be thoroughly bled: if the head has not been removed, cut the throat; remove the gills and all blood vessels. Blood clots can cause discoloration, as well as bacterial infection that would make the fish unfit for eating.

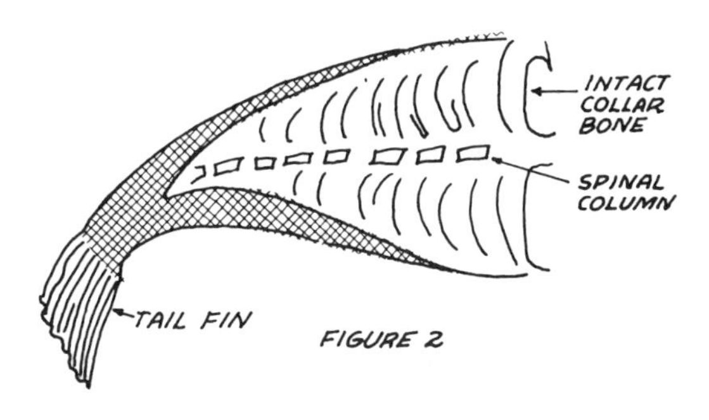

FIGURE 2

Cut the fish according to local custom. As a rule of thumb: under 0.5kg (1 pound), the fish may be left whole; from 0.5kg to 5kg (1 to 10 pounds) it should be split in half from head to tail (see Figure 2); over 5kg (10 pounds), split the fish in two again from head to tail. The collarbone behind the gills should be left intact when a fish is split in half.

## Salting

To salt fish, follow these steps carefully:

- Sprinkle a thin layer of salt in a waterproof vat. Use just enough to cover the bottom completely.
- Place a layer of fish, FLESH side up, with enough room for each fish to avoid overlapping. Try for a neat pattern, alternating head to tail and tail to head.
- Cover the fish with salt - a thin layer, but with no open spaces.
- Continue to layer the fish flesh side up, up to two or three layers from the top of the vat.
- Reverse the fish, packing them SKIN side up to the top of the vat, alternating with layers of salt. The top layer must be salt.

- o The salt will extract moisture from the fish, forming a brine. Use boards and weights to keep all the fish under the salt.

- o The brine must be kept saturated (90 on a Salinometer, or when no more salt can be dissolved) at all times. As moisture is extracted, more salt must be added to keep the brine saturated. With too little salt the fish will spoil.

As moisture is extracted from the fish, the level of fish in the vat will fall. More fish can be added, skin side up, alternating a layer of fish with a layer of salt, the top layer always being salt. Continue to add salt to keep the brine saturated. The fish are "struck through," or thoroughly salted, in 12 to 15 days in warm weather. In cold weather, the fish should stay in the brine for 21 days or more; in the tropics, 15 days may be a good limit. The higher the temperature, the quicker the fish will be struck through. When properly salted, the flesh of the fish is translucent but the eyes are opaque and no longer translucent. The flesh is firm but yields to gentle pressure. It has a whitish salt cover. An odor of fish and brine should prevail. There should be no spoilage odors.

## Washing and Drying to Remove Excess Salt

- o When the fish are struck through, remove them from the vat and wash in unpolluted sea water or fresh brine to remove excess salt.

- o Then place the fish on flat surfaces, using any arrangement of boards and weights to press them as flat as possible:

  - to remove excess moisture; and

  - to make the fish thinner, which will reduce the length of the air-drying process and improve the appearance of the fish for marketing.

## Air Drying

The final drying can be done either by sunlight and natural air currents or by artificial heat and air currents generated by fans. In most areas, in the proper season, drying can be done outdoors in the sun and fresh air. Choose an open area to get the most sunlight and wind. Avoid swampy areas, locations near human or animal waste, and, especially, fly-breeding areas.

When freshly salted fish is first brought out to dry, there is danger of sunburn. If fish is exposed at this stage to the direct rays of the sun, it may harden on the outside and turn yellow. This will keep the inside from drying properly. To avoid this, keep the fish under shade or semi-shade for the first day.

After the first day, expose the fish to as much sunlight and wind as possible. One method is to lay the fish on triangular slats--so that the fish rests on the least possible amount of surface–fresh side facing the sun (see Figure 3). Another method is to hang the fish by the tail (see Figure 4).

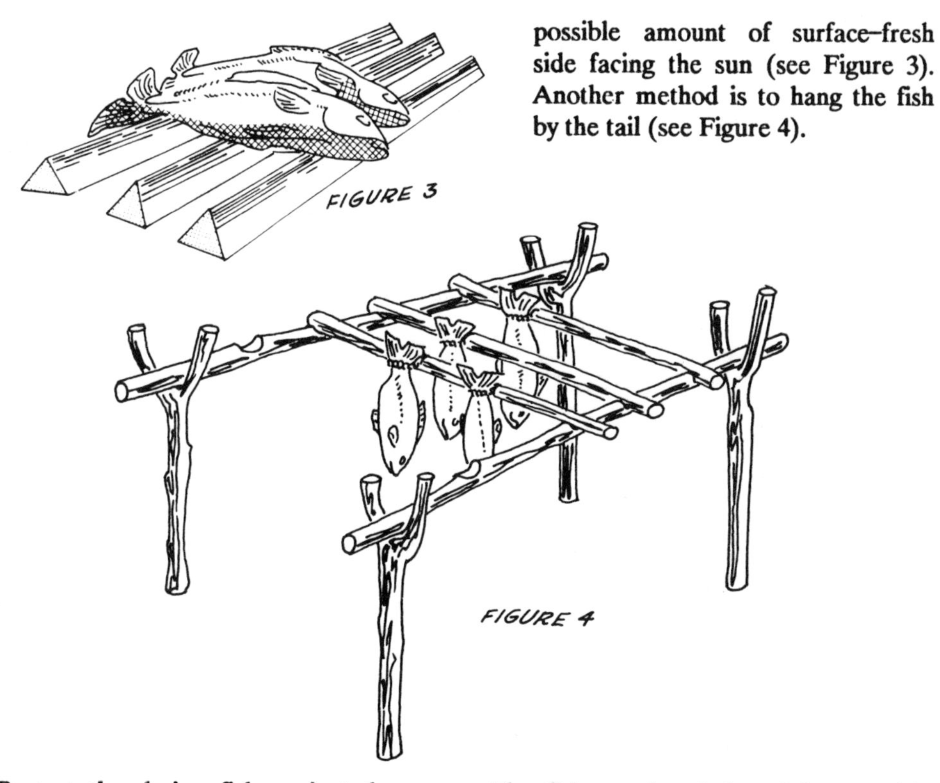

FIGURE 3

FIGURE 4

Protect the drying fish against dampness. The fish can be sheltered by portable thatch roofs (see Figure 5) or moved into small roofed sheds built nearby for protection from rainfall and night-time dampness. The fish should be free of discoloration, mold, or other defects. Split fish should not have ragged edges.

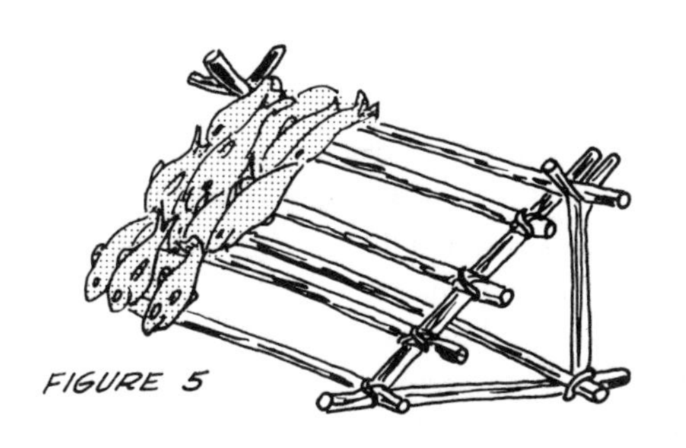

FIGURE 5

Generally, six warm days with winds of more than 5km (3 miles) per hour should dry the fish enough to prevent spoiling in storage or shipping, provided the fish is properly packed to protect it from excessive moisture.

## Using Salted Fish

Salted fish is usually soaked overnight, with at least one change of water, to remove most of the salt before it is eaten. The longer it is soaked, the more salt is removed. Then it is used in the same way as fresh fish, except that it is not good for frying.

**Source:**

Daniel Casper, Product Manager, Seabrook Farms, Co., Seabrook, New Jersey

# SMOKING FISH

Smoked fish does not last as long as salted fish, and must be refrigerated, frozen, or canned if it to be stored for any length of time. Smoked fish are prepared in a smokehouse, which is simply a shed or box over a fire that is controlled so that it produces smoke instead of flames. The fish are hung inside the smokehouse so that they are surrounded by smoke. It takes about six hours to smoke fish for eating or storage.

Prepare the fish as you would for salting. Bleed and gut the fish and split them from head to tail. Wash the fish in fresh, clean water. Place in a salt water brine for about one hour. Remove the fish from the brine and wash again in clean fresh water. Drain, and hang in a cool breezy place for about an hour.

Build a fire in the smokehouse. When the fire is burning properly–that is, producing lots of smoke–place the fish on hooks and hang or tie them in the top of the smokehouse. Make sure the fish are placed securely so they will not fall. Watch the fire carefully to make sure that it is smoking the fish and not burning them–and also to be sure that the smokehouse itself doesn't catch on fire.

FIGURE 6

After the fish are smoked for about six hours they can be eaten immediately, stored in jars (to be canned), or frozen or refrigerated until they are eaten.

Smoked fish do not last as long as salted fish, so do not smoke all of the fish unless it will be used soon after harvest.

Source:

Chakroff, Marilyn. *Freshwater Fish Pond Culture and Management.* Arlington, Virginia: Volunteers in Technical Assistance, 1978.

Carruthers, Richard T. *Understanding Fish Processing and Preservation.* Arlington, Virginia: Volunteers in Technical Assistance, 1986.

# Construction

VITA Archives

# Concrete Construction

## OVERVIEW

Concrete is a strong and inexpensive construction material when it is properly prepared and used. This introduction explains the importance of a good mixture and describes the materials used in the mixture. Following this are entries on:

- Calculating amounts of materials for concrete
- Mixing concrete by machine or by hand
- Testing concrete mixtures
- Making forms for concrete
- Placing concrete in forms
- Curing concrete
- Making quick-setting concrete
- Useful sources of information on concrete

Concrete is made by combining the proper proportions of cement, water, fine aggregate (sand), and coarse aggregate (gravel). A chemical reaction, hydration, takes place between the water and cement, causing the concrete to harden or set rapidly at first, then more slowly over a long period of time.

## Importance of a Good Mixture

After concrete has set, there is no simple non-destructive test to find out how strong it is. Therefore, the entire responsibility for making concrete as strong as a particular job demands rests with the supervisor and the people who prepare, measure, and mix the ingredients, place them in the forms, and watch over the concrete while it hardens.

FIGURE 1.

The most important factor in making strong concrete is the amount of water used. Beginners are likely to use too much. In general, the lower the ratio of water to cement, the stronger the concrete will be.

The proper proportioning of all materials is essential. The section on "Calculating Amounts of Materials for Concrete" provides the necessary information.

## Aggregates: Gravel and Sand

To make strong concrete, the coarse aggregate (gravel) and fine aggregate (sand) must be the right size, have the right shape, and be properly graded.

Coarse aggregate sizes can vary from 0.5cm (1/4") to 4 or 5cm (1 1/2" or 2") in diameter. The maximum size depends on the nature of the work. In general, the largest particles should not be more than one-fourth the thickness of the smallest dimension of the section. Sand can vary from sizes smaller than 0.5cm down to, but not including, silty material.

Very sharp, rough, or flat aggregate should not be used in concrete. The best aggregate is cubical material (from a rock crusher) or rounded gravel (from a stream bed or beach).

Proper grading means that there are not too many grains or pebbles of any one size. To visualize this, think of a large pile of stones all 5cm (2") in diameter. There would be spaces between these stones where smaller pebbles would fit. We could add to the pile just enough smaller stones to fill the largest spaces. Now the spaces would be smaller yet, and even smaller pebbles could fill these holes; and so forth. Carried to an extreme, the pile would become nearly solid rock, and only a very small amount of cement would be needed to fill the remaining spaces and hold the concrete together. The resulting concrete would be very dense, strong, and economical.

It is extremely important that the aggregate and sand be clean. Silt, clay, or bits of organic matter will ruin concrete if too much is present. A very simple test for cleanliness makes use of a clear wide-mouth jar. Fill the jar to a depth of 5cm (2") with the fine aggregate (sand) and then add water until the jar is three quarters full. Shake the mixture vigorously for a minute. The last few shakes should be sideways to let the sand level off. Then let it stand for three hours. If there is silt in the sand, it will form a distinct layer above the sand. If the layer of very fine material is more than 3mm (1/8") deep, the concrete will be weak.

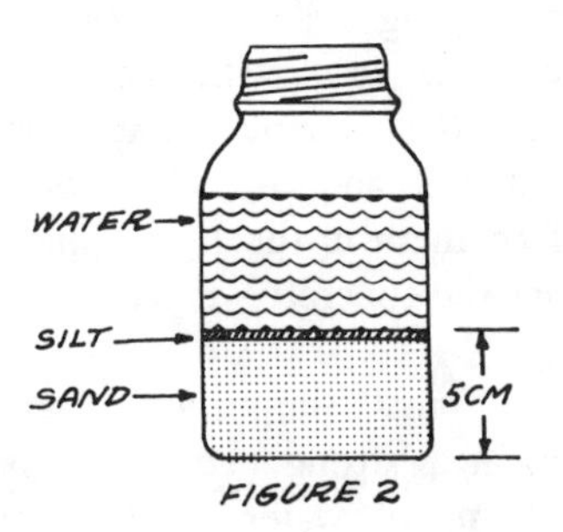

FIGURE 2

If there is too much fine or silty material, another source of sand should be found. If this is impractical, it is possible to remove the fine particles. This can be done by putting the sand in a container like a drum. Cover the sand with water, stir or agitate vigorously, let it stand for a minute, then pour off the liquid. A few such treatments will remove most of the fine and organic matter.

In very dry climates, the sand may be perfectly dry. Very dry sand will pack into a much smaller volume than sand that is moist. If 2 buckets of water are added to 20 buckets of bone dry sand, you can carry away about 27 buckets of damp sand. If your sand is completely dry, add some water to it.

Another point to consider in selecting an aggregate is its strength. About the only simple test is to break some of the stones with a hammer. If the effort required to break the majority of stones is greater than the effort required to break a piece of concrete of about the same size, the aggreage will make strong concrete. If the stone breaks easily, the concrete made of these stones will be no stronger than the stones themselves.

## Water

The water used to prepare concrete must be clean, and free of organic matter. Water acceptable for drinking is preferable. Any clear, fresh water is acceptable. Salt water may be used if fresh water is not readily available, but it will reduce the strength of concrete about 15 percent.

If you must use dirty or muddy water, let the water settle in a large pan or tank to remove most of the dirt.

Cement for concrete, if it is a U.S. brand, comes in 42.6kg (94 pound) sacks, and is 28.4 liters (exactly 1 cubic foot) in volume. It must be kept perfectly dry prior to use, or the chemical action will begin and the cement will be ruined.

Mixing the materials, getting them in place rapidly, tamping or compacting to a dense mixture, and proper curing are important parts of the construction process. These will be discussed in the sections on mixing and curing concrete.

Concrete reinforced with steel rods is used for structures such as large buildings and bridges. Proper design of reinforced concrete and placement of steel reinforcing is a complex procedure that requires the help of a trained engineer.

# CALCULATING MATERIALS FOR CONCRETE

Three methods are given here for finding the correct proportions of cement, water, and aggregate for concrete:

- A "Concrete Calculator" fold-out chart
- Using water to estimate proportions
- A "rule of thumb"

# Using the Concrete Calculator

The amounts of materials needed for a concrete construction job can be estimated quickly and accurately with the "Concrete Calculator" chart. The chart is given in both English (Chart A) and metric (Chart B) units.

To use one of the charts, you must know:

- o The area of concrete needed in square meters or square feet.
- o The thickness of concrete needed in centimeters (inches).
- o The kind of work to be done (see below).
- o The wetness of the sand (see below).

To use the calculator, follow these steps

- o Make a light pencil mark on Scale 1, representing the area of concrete needed. If the volume is less than 400 liters or 15 cubic feet, multiply it by a convenient factor (for example, 10); then, when you find the amounts of materials the chart says to use, divide them by the same factor to get the actual amounts needed.

- o Make a similar mark on Scale 2, the slanted scale indicating thickness.

- o Draw a straight line through the two marks intersecting Scale 3 to find the volume of concrete needed.

  (If the shape of the area is complex, measure it in sections, add up the volumes of all the parts and mark the total volume on Scale 3.)

- o Mark the kind of work on Scale 4. A line through the marks on Scales 3 and 4 to Scale 5 will give the amount of fine aggregate needed.

- o Continue on a zig-zag course as shown in the KEY to calculate the rest of the materials.

- o Add 10 percent to the amounts indicated by the chart to allow for wastage and miscalculation.

- o If the mix is too wet or too stiff, see page 312 for instructions on adjusting it.

Materials can be measured in buckets. Most buckets are rated by the number of gallons they hold. To convert to liters, multiply gallons by 3.785. To convert to cubic feet: 1 cubic foot = 7.5 gallons. A 4-gallon bucket would hold 15.15 liters or 0.533 cubic feet.

# NOTES

# NOTES

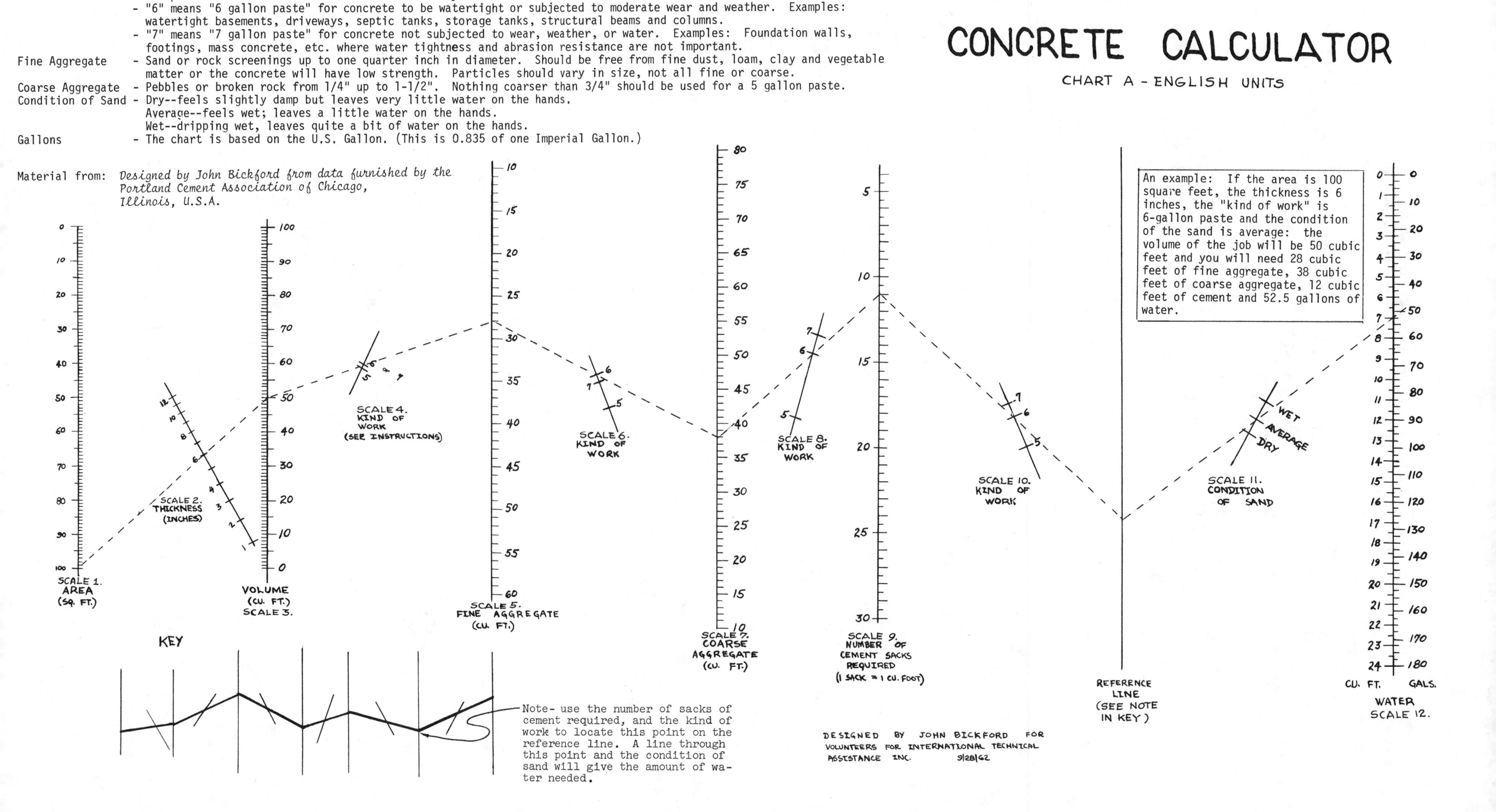
Kind of work - "5" means "5 gallon paste" which is concrete subjected to severe wear, weather, or weak acid and alkali solutions. Examples would be the floor of a commercial dairy.
- "6" means "6 gallon paste" for concrete to be watertight or subjected to moderate wear and weather. Examples: watertight basements, driveways, septic tanks, storage tanks, structural beams and columns.
- "7" means "7 gallon paste" for concrete not subjected to wear, weather, or water. Examples: Foundation walls, footings, mass concrete, etc. where water tightness and abrasion resistance are not important.
Fine Aggregate - Sand or rock screenings up to one quarter inch in diameter. Should be free from fine dust, loam, clay and vegetable matter or the concrete will have low strength. Particles should vary in size, not all fine or coarse.
Coarse Aggregate - Pebbles or broken rock from 1/4" up to 1-1/2". Nothing coarser than 3/4" should be used for a 5 gallon paste.
Condition of Sand - Dry--feels slightly damp but leaves very little water on the hands.
Average--feels wet; leaves a little water on the hands.
Wet--dripping wet, leaves quite a bit of water on the hands.
Gallons - The chart is based on the U.S. Gallon. (This is 0.835 of one Imperial Gallon.)
Material from: Designed by John Bickford from data furnished by the Portland Cement Association of Chicago, Illinois, U.S.A.
CONCRETE CALCULATOR
CHART A - ENGLISH UNITS
An example: If the area is 100 square feet, the thickness is 6 inches, the "kind of work" is 6-gallon paste and the condition of the sand is average: the volume of the job will be 50 cubic feet and you will need 28 cubic feet of fine aggregate, 38 cubic feet of coarse aggregate, 12 cubic feet of cement and 52.5 gallons of water.
SCALE 1. AREA (SQ. FT.)
SCALE 2. THICKNESS (INCHES)
VOLUME (CU. FT.) SCALE 3.
SCALE 4. KIND OF WORK (SEE INSTRUCTIONS)
SCALE 5. FINE AGGREGATE (CU. FT.)
SCALE 6. KIND OF WORK
SCALE 7. COARSE AGGREGATE (CU. FT.)
SCALE 8. KIND OF WORK
SCALE 9. NUMBER OF CEMENT SACKS REQUIRED (1 SACK = 1 CU. FOOT)
SCALE 10. KIND OF WORK
REFERENCE LINE (SEE NOTE IN KEY)
SCALE 11. CONDITION OF SAND
WET
AVERAGE
DRY
CU. FT.
GALS.
WATER SCALE 12.
KEY
Note- use the number of sacks of cement required, and the kind of work to locate this point on the reference line. A line through this point and the condition of sand will give the amount of water needed.
DESIGNED BY JOHN BICKFORD FOR VOLUNTEERS FOR INTERNATIONAL TECHNICAL ASSISTANCE INC. 9/28/62

# CONCRETE CALCULATOR

CHART B METRIC UNITS

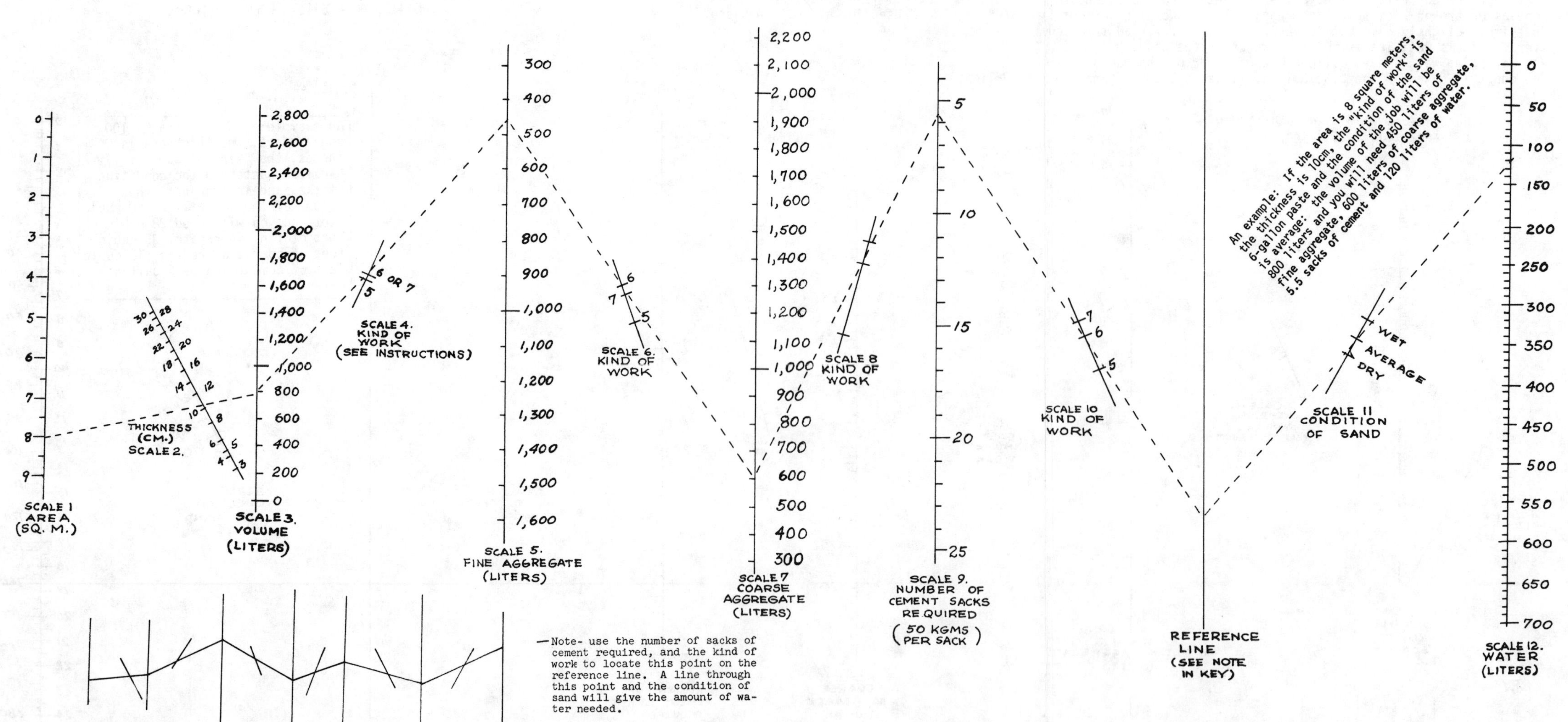

The definitions used in the chart are:

**Kind of Work:**

**"5"** means "5-gallon paste" (5 gallons or 19 liters of water to one sack of cement), for concrete subjected to severe wear, weather, or weak acid and alkali solutions. An example is the floor of a commercial dairy.

**"6"** means "6-gallon paste," for concrete that is to be watertight or subjected to moderate wear or weather. Examples: watertight basements, driveways, septic tanks, storage tanks, structural beams and columns.

**"7"** means "7-gallon paste," for concrete not subjected to wear, weather, or water. Examples: Foundation walls, footings, and mass concrete where water tightness and abrasion resistance are not important.

**Fine Aggregate:**

Sand or rock screenings up to 0.5cm (1/4") in diameter. It should be free from fine dust, loam, clay, and organic matter or the concrete will be weak. The particles should vary in size.

**Coarse Aggregate:**

Pebbles or broken rock from 0.5cm (1/4") up to 4 or 5cm (1 1/2" or 2"). Nothing larger than 2cm (3/4") should be used with a 5-gallon paste.

**Condition of Sand:**

**Dry:** feels slightly damp but leaves very little water on the hands.
**Average:** feels wet, leaves a little water on the hands.
**Wet:** dripping wet, leaves a lot of water on the hands.

**Gallons:** Chart A is based on the U.S. gallon (0.835 Imperial Gallon).

## Using the Water Displacement Method

The "Concrete Calculator" chart assumes that the aggregate is well graded. When the aggregate is not well graded, an alternate method can be used to find the correct proportions for a concrete mixture. The advantage of this method is that only a small sample of the ungraded aggregate needs to be divided into coarse and fine particles.

Well-graded aggregate seldom occurs naturally. Some "pre-mix" processing would be needed to grade it.

Remember that when you make concrete, you are filling the spaces in the aggregate with cement mortar or paste. The amount of cement paste needed can be found by adding water to a known volume of aggregate. To do this:

1. Divide a sample of the aggregate into coarse and fine particles by sifting it through a 0.5cm (1/4") screen.

2. Fill a pail with the coarse aggregate (dry).

3. Fill the pail with water. The amount of water used equals the amount of fine aggregate and cement paste needed to fill the spaces.

4. Into another pail, put an amount of fine aggregate equal to the volume of water used in Step 3.

5. Fill the pail with enough water to bring the water level to the top of the fine aggregate. The volume of water used equals the volume of cement paste needed to fill the remaining spaces.

   Add about 10 percent to this volume to allow for waste and to make the mix more workable.

6. To find the correct ratios of materials, divide the volume of cement paste needed into the volumes of fine and coarse aggregates.

7. Add these two ratios to get the ratio for ungraded aggregate. For example: If you are using a 19-liter (5-gallon) pail, and it takes 12.8 liters (3.4 gallons) of water to fill the pail in Step 3, put 12.8 liters (3.4 gallons) of fine aggregate in the second pail (Step 4). If Step 5 takes 6.4 liters (1.7 gallons) of water, this is the volume of cement paste needed. Divide this volume into the volumes of fine and coarse aggregates to get the ratios of materials:

$$\frac{\text{19 liters (coarse aggregate)}}{\text{6.4 liters (cement paste)}} = 3$$

$$\frac{\text{12.8 liters (fine aggregate)}}{\text{6.4 liters (cement paste)}} = 2$$

The sum of the two ratios is 5, so the ratio of ingredients in this case is 1:5, or 1 part cement paste to 5 parts ungraded aggregate, by volume.

To find the ratio of water to cement, see "Kind of Work" page 309. For directions on adjusting a mixture that is either too wet or too stiff, see page 312.

## Using "Rule of Thumb" Proportions

For a variety of small concrete construction tasks and for repair and patch-work, the following simple "rule of thumb" can be used as a simple guideline.

Use the ratio 1:2:3, by volume, to proportion the cement and aggregate and use a water-cement ratio of 6 gallons water to 1 sack of cement. That is, for every sack of cement (28.4 liters or 1 cubic foot) used, add 56.8 liters (2 cubic feet) of fine aggregate and 85.2 liters (3 cubic feet) of coarse aggregate. Add 28.7 liters (6 gallons) of water for each sack of cement.

A home-made box of 28.4-liter (1-cubic foot) volume will help in proportioning the mixture. The volume of concrete produced by a one-sack batch using the proportions given above will be about 142 liters (5 cubic feet).

The most common mistakes made by inexperienced persons are using too much cement, which increases the cost, and using too much water, which produces weak concrete.

# MIXING CONCRETE

Concrete must be thoroughly mixed to yield the strongest product. For machine mixing, allow 5 to 6 minutes after all the materials are in the drum. First, put about 10 percent of the mixing water in the drum. Then add water uniformly with the dry materials, leaving another 10 percent to be added after the dry materials are in the drum.

## Making a Mixing Boat or Floor

On many self-help projects, the amount of concrete needed may be small or it may be difficult to get a mechanical mixer. Concrete can be mixed by hand; if a few precautions are taken, it can be as strong as concrete mixed in a machine.

**Tools and Materials**

Lumber, 2 pieces: 183cm x 91.5cm x 5cm (6' x 3' x 2")
Galvanized sheet metal: 183cm x 91.5cm (6' x 3')
Nails, Saw, Hammer

Or:

Concrete for a mixing floor: about 284 liters (10 cubic feet) of concrete is needed for a 244 cm (8') diameter mixing floor that is 5cm (2") thick with a 10cm (4") high rim
Shovel

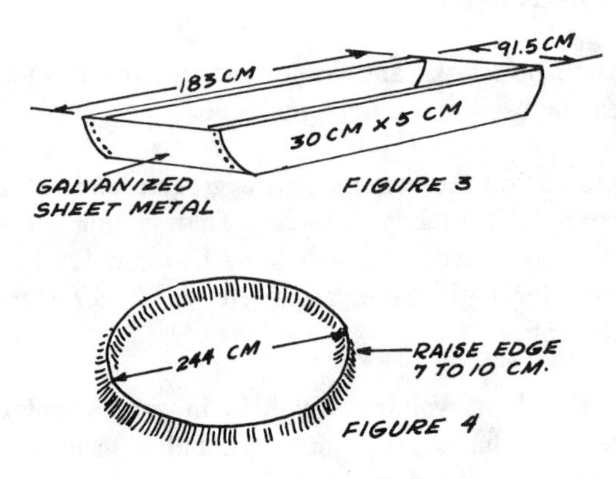


The first requirement for mixing by hand is a mixing area that is both clean and water-tight. This can be a wood and metal mixing boat (see Figure 3) or a simple round concrete floor (see Figure 4).

The ends of the wood and metal mixing boat are curved to make it easier to empty. The raised edge of the concrete mixing floor prevents loss of water while the concrete is being mixed.

The procedure is:

- Spread the fine aggregate evenly over the mixing area.
- Spread the cement evenly over the fine aggregate and mix these materials by turning them with a shovel until the color is uniform.
- Spread this mixture out evenly and spread the coarse aggregate on it and mix thoroughly again.
- Form a hollow in the middle of the mixture and slowly add the correct amount of water and, again, mix thoroughly.

The mixture should be placed in the forms within 20 minutes after it is completely mixed.

When work is finished for the day, be sure to rinse concrete from the mixing area and the tools to keep them from rusting and to prevent cement from caking on them. Smooth shiny tools and boat surfaces make mixing surprisingly easier. The tools will also last much longer. Try to keep from getting wet concrete on your skin because it is caustic. If you do, wash it off as soon as possible.

A workable mix should be smooth and plastic–neither so wet that it will run nor so stiff that it will crumble.

If the mix is too wet, add small amounts of sand and gravel, **in the proper proportion**, until the mix is workable.

If the mix is too stiff, add small amounts of water and cement, **maintaining the proper water-cement ratio**, until the mix is workable.

Note the amounts of materials added so that you will have the correct proportions for subsequent batches.

If a concrete mix is too stiff, it will be difficult to place in the forms. If it is not stiff enough, the mix probably does not have enough aggregate, which is uneconomical.

## Slump Tests

### *Slump Cone*

A "slump cone" is a simple device for testing a concrete mixture to see that it has the right proportion of materials.

**Tools and Materials**

Heavy galvanized iron sheet: 35.5cm x 63.5cm (14 1/8" x 25 1/2")
Iron strap: 3mm x 2.5cm x 7.5cm (1/8" x 1" x 3") 4 pieces
16 Iron rivets: 3mm in diameter and 6mm long
Wooden dowel: 16mm in diameter and 61cm long

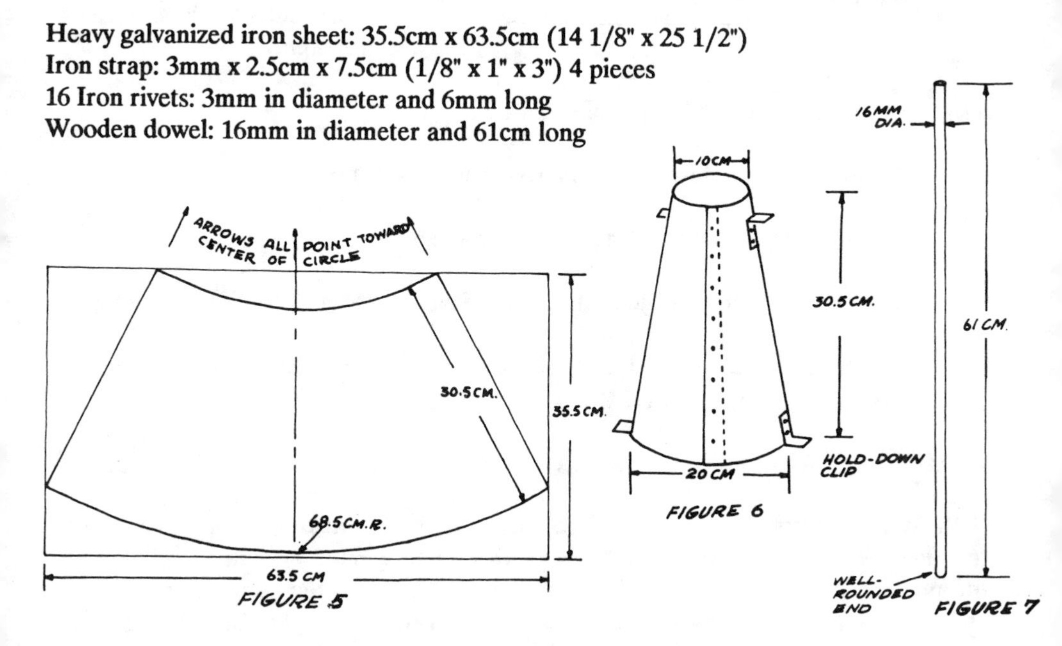


FIGURE 5

FIGURE 6

FIGURE 7

### *Testing*

To perform the test:

- Dampen the slump cone and set it on a flat, moist, non-absorbent surface. Stand on the clips at the bottom of the cone to hold it down.

- Fill the cone in three layers approximately equal in volume. Because the diameter at the bottom of the cone is large, the first layer should fill the cone to about one-fourth its height.

o Stroke each layer 25 times with the wooden dowel.

o After the top layer has been stroked with the dowel, smooth the surface of the concrete so the cone is filled exactly.

o Carefully lift the cone off the concrete.

o Place the empty cone alongside the concrete. Measure the difference between the height of the cone and the height of the concrete. This difference is the slump.

Suggested slumps for various types of construction are:

o Reinforced walls and footings: 5cm to 13cm (2" to 5")

o Unreinforced walls and footings: 2.5cm to 10cm (1" to 4")

o Thin reinforced walls, columns and slabs: 7.5cm to 15cm (3" to 6")

o Pavements, walkways, culverts, drainage structures, and heavy mass concrete: 2.5cm to 7.5cm (1" to 3")

### *Correcting the Mixture*

If the slump is not within the desired range, or if the mixture is obviously either too fluid or too stiff, the proportions of the mixture must be changed. To make the mixture more fluid and increase the slump, increase the proportion of water and cement without **changing the water-cement ratio.** To make the mixture stiffer and decrease the slump, increase the proportion of the aggregates **without changing the fine aggregate-coarse aggreagate ratio.** Do not add just water to make the mix more fluid; this will weaken the concrete.

## MAKING FORMS FOR CONCRETE

Fresh concrete is heavy and plastic. Forms for holding it in place until it hardens must be well braced and should have a smooth inside surface. Cracks, knots, or other imperfections in the forms may be permanently reproduced in the concrete surface.

Wood is commonly used for forms, because of its light weight and strength. Since cracks between boards can mar the concrete surface, plywood, which has a special high-density overlay surface, is often used. The finish on plywood provides a smooth casting surface and makes it easier to remove the forms for reuse.

If unsurfaced wood is used for forms, oil or grease the inside surface to make removal of the forms easier and to prevent the wood from drawing too much water from the concrete. Do not oil or grease the wood if the concrete surface will be painted or stuccoed.

Forms for flat work, such as pavements, may be 5cm x 10cm (2" x 4") or 5cm x 15cm (2" x 6") lumber, the size depending on the thickness of the slab. Stakes spaced 122cm (4') apart hold the forms in place.

Figures 9 and 10 show forms for straight-wall construction. To prevent the forms from bulging, opposite studs should be tied together with 10- to 12-gauge wire, which should be twisted to draw the form walls tight against wooden spacer blocks. (The blocks are removed as the concrete is placed.)

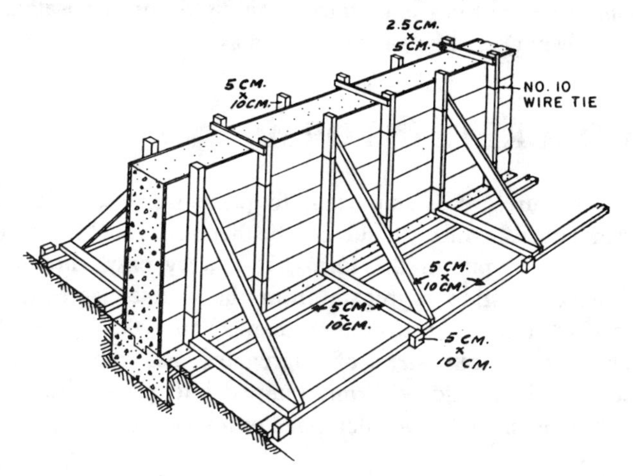


FIGURE 9--Forms for a straight wall on level ground.

The ties should be spaced about 76cm (2 1/2') vertically on the studs. When the forms are removed, clip the wires close to the concrete and punch them back. Pit holes caused by punching back the wires should be pointed up with mortar.

Forms should be easy to fill with concrete and easy to remove once the concrete has hardened. Screws or double headed nails which can be taken out easily can be a great help in removing wooden forms without damaging the concrete.

Forms are sometimes made of other materials. For example, metal forming is more economical for repeated work, such as curbs, slip forming for monolithic concrete tanks or silos, and reinforced concrete floors for multistory buildings.

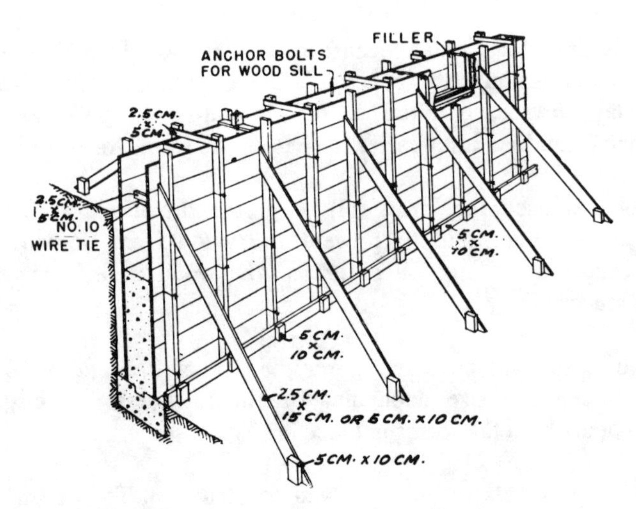


FIGURE 10--Forms for a basement or cellar wall. The earth can be used as the outside form if sufficiently firm.

The finest natural finish on a concrete surface can be obtained by casting on polyethylene. Sometimes polyethylene forms are used for decorative work, or a kraft paper with a polyethylene film surface is used as form liner.

## PLACING CONCRETE IN FORMS

To make strong structures, it is important to place fresh concrete in the forms correctly. The wet concrete mix should not be handled roughly when it is being carried to the forms and put in the forms. It is very easy, through joggling or throwing, to separate the fine aggregate from the coarse aggregate. Do not let concrete drop freely for a distance greater than 90 to 120cm (3' to 4'). Concrete is strongest when the various sizes of aggregates and cement paste are well mixed. The concrete mix should be firmly tamped into place with a thin iron rod (about 2cm or 3/4" in diameter), a wooden pole, or a shovel.

## CURING CONCRETE

When the forms are filled, the hard work is done, but the process is not finished. The concrete must be protected until it reaches the required strength. It starts to harden almost immediately once the water is added, but the hardening action may not be complete for several years.

The early stage of curing is extremely critical. Special steps should be taken to keep the concrete wet. In temperate climates, the mixture should be kept wet for at least 7 days; in tropical and subtropical climates, it should be kept wet for at least 11 days. Once concrete dries, it will stop hardening; after this happens, re-wetting will **NOT** re-start the hardening process.

Newly-laid concrete should be protected from the sun and from drying wind. Large areas such as floors or walls that are exposed to the sun or wind should be protected with some sort of covering. Protective covers often used are: canvas, empty cement bags, burlap, palm leaves, straw, and wet sand. The covering should also be kept wet so that it will not absorb water from the concrete.

Concrete is strong enough for light loads after 7 days. In most cases, forms can be removed from standing structures like bridges and walls after 4 or 5 days, but if they are left in place they will help to keep the concrete from drying out. In small ground-supported structures such as street drains, the forms can be removed within 6 hours of completion provided this is done carefully. Plans will usually say if forms should be left in place longer.

Concrete is usually expected to reach the strength for which it was designed after 28 days. Concrete that is moist cured for a month is about twice as strong as concrete that cures in the open air.

# QUICK-SETTING CONCRETE

Quick-setting concrete is often useful; for example, when repeated castings are needed from the same mold. A concrete mixture that contains calcium chloride as an accelerator will set about twice as fast as a mixture that does not. The mixed batch must be put into the forms faster, but since quick-setting batches are usually small, this is not a problem. Calcium chloride does not lessen the strength of fully-cured concrete.

No more than 1kg (2 pounds) of calcium chloride should be used per sack of cement. It should be used only if it is in its original containers, which should be moisture-proof bags or sacks or air-tight steel drums.

To add the calcium chloride, mix up a solution containing 1/2kg per liter (1 pound per quart) of water. Use this solution as part of the mixing water at a ratio of 2 liters (2 quarts) per sack of cement (42.6kg or 94 pounds). Solid (dry) calcium chloride must never be added to the concrete mix; only use it in solution.

**Sources:**

VITA Volunteers:

John Bickford, Connecticut; Robert D. Cremer, New York; Kenneth D. Hahn, California; R. B. Heckler, Florida

*A Building Guide for Self-Help Projects,* Accra, Ghana: Department of Social Welfare and Community Development, 1961.

*Design and Control of Concrete Mixtures,* Chicago: Portland Cement Association

*Use of Concrete on the Farm,* Farmers' Bulletin No. 2203, Washington, D.C.: U.S. Department of Agriculture, 1965.

**Other Useful Publications:**

*Basics of Concrete,* Ideas and Methods Exchange No. 49, Washington, D.C.: U.S. Department of Housing and Urban Development, Division of International Affairs

*Concrete Technology: Student Manual,* Albany, New York: Delmar Publishers

Hobbs, Wesley. *Making Quality Concrete for Agricultural and Home Structures,* University, Addis Ababa, Ethiopia: Haile Sellassie

**Useful sources of information on concrete, including how-to-do-it manuals:**

Portland Cement Institute
18 Kew Road
Richmond
Johannesburg, South Africa

Instituto del Cemento Portland Argentino
San Martin 1137
Buenos Aires, Argentina

Cement and Concrete Association of Australia
147-151 Walker Street
North Sydney, Australia, N.S.W.

Associacao Brasileria de Cimento Portland
Caixa Postal 30886
Sao Paulo, Brazil

Cement and Concrete Association
52 Grosvenor Gardens
London, S.W. 1, England

The Concrete Association of India
P.O. Box 138
Bombay 1, India

Portland Cement Association
33 West Grand Avenue
Chicago, Illinois 60610 USA

# Bamboo Construction

Bamboo is one of the oldest materials people have used to increase their comfort and well-being. In today's world of plastics and steel, besides continuing to make its traditional contributions, bamboo is growing in importance. Outstanding varieties of bamboo from throughout the world are being tested to find out how they can contribute to local economies.

FIGURE 1--From supporting posts to rafters and sheathing, this cottage in the Ecuadoran lowlands is made entirely of native bamboo, Guadua angustifolia. The posts may serve for five years; the siding may remain in serviceable condition for decades.

As the best species are identified and disseminated, their use will help to improve the lives of many. With a few plants of superior bamboos in the backyard, a family will have at hand the wherewithal to fence the garden, build a pigpen or chicken coop, or add a room to the house. The family will also be able to increase its daily income by making baskets or other specialties for sale or exchange.

Bamboos are prominent elements in the natural vegetation of many parts of the tropical, subtropical and mild temperate regions of the world, from sea level to altitudes or more than 13,000 feet (4000m). People have widened the distribution of many species of bamboo, but some of the more valuable species have not been distributed as much as they could be.

Bamboo can be prepared for use in construction with simple tools. Once prepared, bamboo can be used extensively in the construction of houses: in making foundations, frames, floors, walls, partitions, ceilings, doors, windows, roofs, pipes, and troughs. For further detail, see *Bamboo as a Building Material,* by F. A. McClure.

The entries that follow explain:

- o Splitting and preserving bamboo
- o Bamboo joints
- o Making bamboo board
- o Bamboo walls, partitions, and ceilings

# PREPARING BAMBOO

## Splitting Bamboo

To prepare bamboo for use in construction, the culms (stems) must be carefully split.

**Tools and Materials**

Iron or hardwood bars, 2.5cm (1") thick
Ax
Steel wedges
Wooden posts
Splitting knives (Figure 4)

Several devices can be used for splitting culms. When bamboo is split the edges of the bamboo strips can be razor-sharp; they should be handled carefully.

### *Splitting Small Culms*

Small culms can be split to make withes (strips) for weaving and lashing:

- o Use a splitting knife with a short handle and broad blade to make four cuts, at equal distances from each other, in the upper end of the culm (Figure 2).

- o Split the culm the rest of the way by driving a hardwood cross along the cuts (Figure 3).

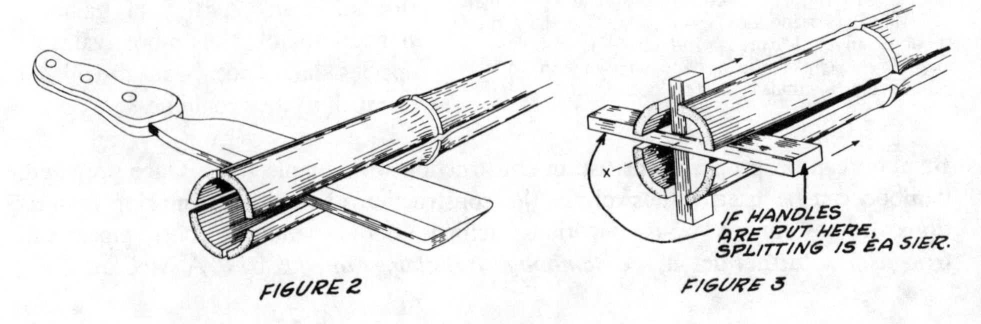


FIGURE 2 FIGURE 3

- o Using a long-handled knife (see Figure 4), cut each strip in half (see Figure 5). A strip of bamboo can be held on the blade to make it thicker and speed up the work.

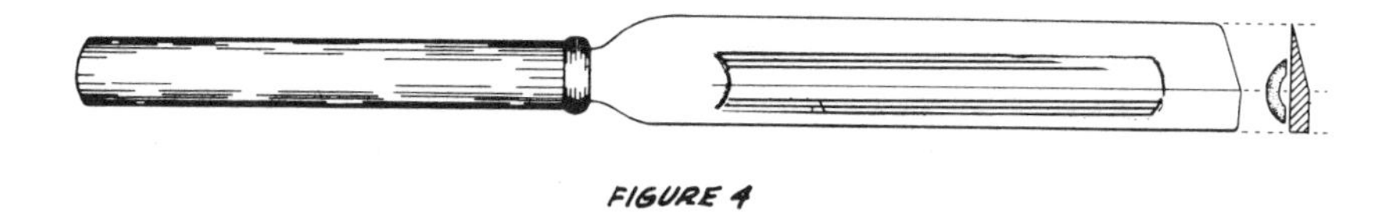

FIGURE 4

- o Use the same knife to split the soft, pithy inner strip from the hard outer strip (see Figure 6). The inner strip is usually discarded.

FIGURE 5

FIGURE 6

## *Splitting Heavy Culms*

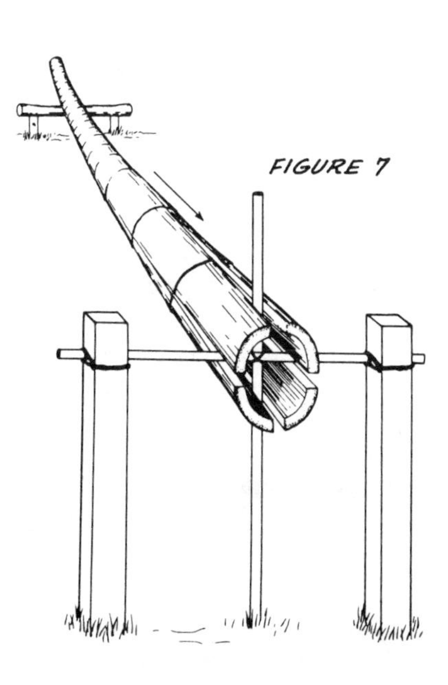

FIGURE 7

- o Build a cross of iron or hardwood bars about 2.5cm (1") thick, and place it on firmly set posts about 10cm (4") thick and 90cm (3') high (see Figure 7).

- o At the top end of the culm, use an ax to make two pairs of breaches at right angles to each other (see Figure 7).

- o Hold the breaches open with steel wedges placed a short distance from the end of the culm, until the culm is on the cross as shown in Figure 7.

- o Push and pull the culm until the cross splits the whole culm.

- To split the culms again after they are split into four strips, use a simple steel wedge mounted on a post or block of wood (see Figure 8).

- Paired wedges solidly mounted on a block or heavy bench can be used to split strips into three narrower strips (see Figure 9).

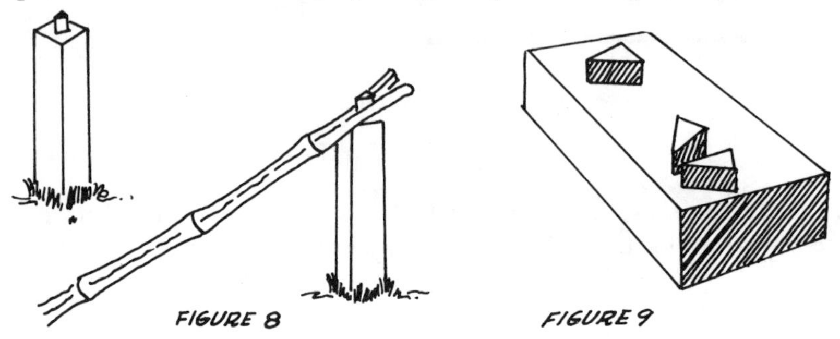

FIGURE 8 FIGURE 9

## Bamboo Preservation

Most bamboos are subject to attack by rot fungi and wood-eating insects. Bamboos with higher moisture and starch content seem to be more prone to attack, and insect pests may be more of a problem in some seasons than others. So bamboo should be cut if possible in the bugs'"off-season." There are many methods for making bamboo more resistant to attack. A simple method that combines proper curing and the use of a pesticide (insecticide and/or fungicide) is described here.

If bamboo is to be used to hold food or water, the only treatment recommended is immersion of green bamboo in a borax-boric acid solution (see Bamboo Piping).

**Tools and Materials**

Machete and hacksaw for felling and trimming bamboo culms
Pesticide–choice depends on the insect or fungus pests that are prevalent in your area. Consult your local extension agent or farmers in the neighborhood about the type and its use. Follow directions carefully.
Talc–to mix with dry pesticide according to package instructions. If talc is not available, other dry dusty materials such as finely powdered dried clay is used.
Dusting bag (made from cloth with an open weave)

Bamboo should not be cut before it is mature. This is usually the end of the third season. Freshly-cut bamboo culms should be dried for 4 to 8 weeks before being used in building.

A clump-curing process tested by the U.S. Department of Agriculture Federal Experiment Station in Puerto Rico helps to reduce attack by insects and rot fungi. The steps are:

- o Cut the bamboo off at the base, but keep it upright in the clump.

- o Dust the fresh-cut lower end of the culm at once by patting it with a dusting bag filled with the pesticide-talc mixture. An alternative method of dusting is to dip the ends of the culms into a tray containing the mixture.

- o To keep the bamboo from being stained or rotted by fungi, raise each culm off the ground by putting a block of stone, brick, or wood under it.

- o Leave the culms in this position for 4 to 8 weeks, depending on whether the weather is dry or damp.

The culms should be as dry as possible before being placed near buildings, where wood eating insects usually are.

- o When the culms have dried as much as conditions will permit, take them down and trim them. Dust **all** cut surfaces immediately with the pesticide-talc mixture.

- o Finish the seasoning in a well-aired shelter where the culms are not exposed to rain and dew. Rain will stain the culms when they become dry.

  This method will prevent damage by wood-eating insects while the culms are drying.

If the bamboo is to be stored for a long time, stacks and storage shelves should be sprayed every six months with the appropriate pesticide mixed in water or light oil. Local conditions may shorten or lengthen the time between sprayings.

In both storage and use, bamboo culms are best preserved when they are protected against rain in a well-ventilated place where they do not touch the ground.

# BAMBOO JOINTS

A number of methods of joining bamboo for making implements or for construction are shown in Figures 10 and 11.

**Tools and Materials**

Bamboo
Lashing material: cord or wire
Machete, hacksaw, knife, drill, and other bamboo working tools

Bamboo is useful for heavy construction because it is strong for its weight. This is because it is hollow with the strongest fibers on the outside where they give

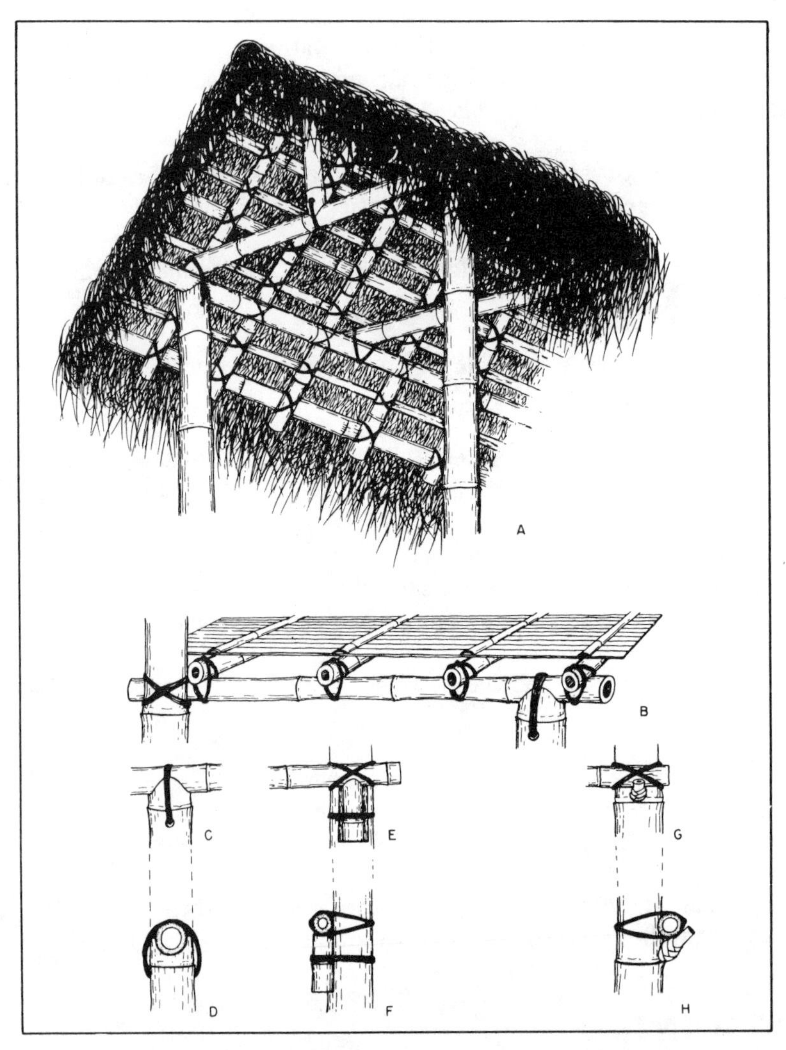


FIGURE 10--Details of bamboo construction: A, fitting and binding culms at joints in roof and frame; B, fitting and securing bamboo boards of floor; C and D, saddle joint; E and F, use of inset block to support horizontal load-bearing elements; G and H, use of stump of branch at node of post to support horizontal load-bearing elements.

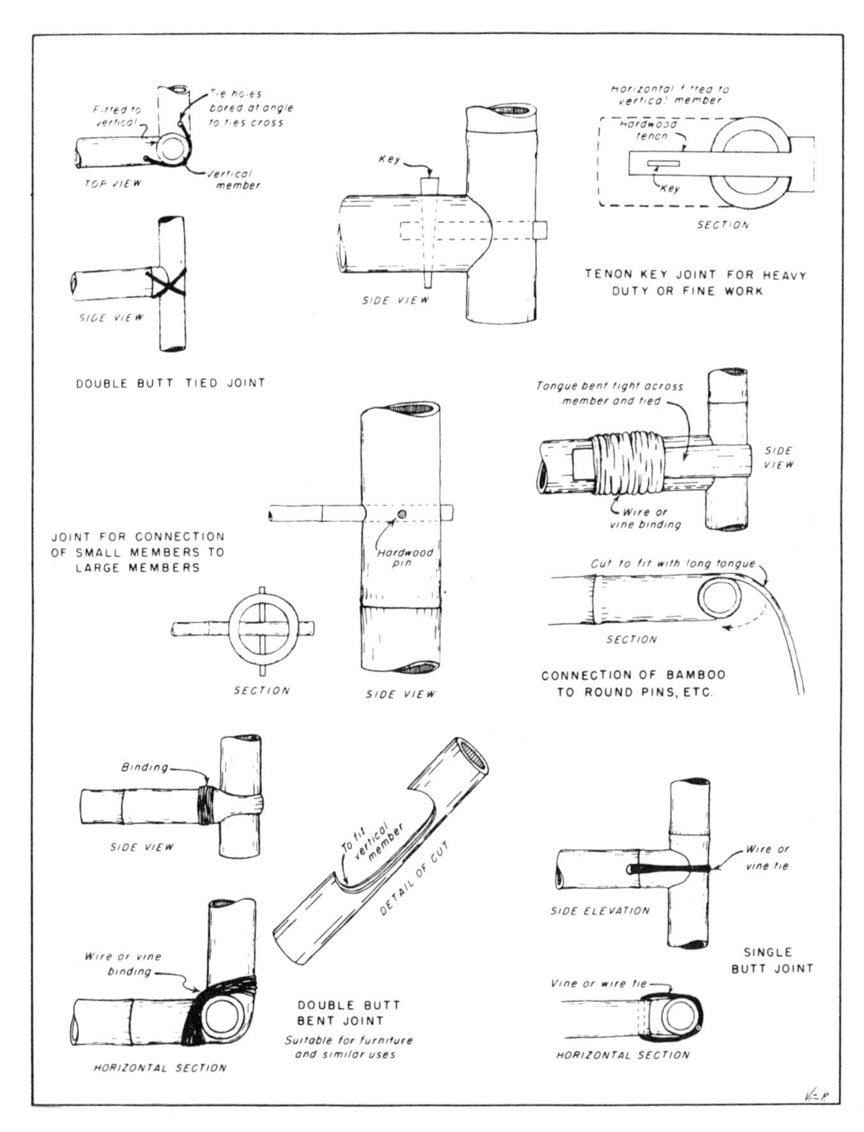


FIGURE 11--Joints used in building with bamboo.

the greatest strength and produce a hard attractive surface. Bamboo has solid diaphragms across each joint or node, which prevent buckling and allow the bamboo to bend considerably before breaking.

Any cut in the bamboo, such as a notch or mortise, weakens it; therefore, mortise and tenon joints should not be used with bamboo. However, notches or saddle-like cuts can be made at the upper ends of posts which hold cross pieces (see Figure 10, C an D).

Bamboo parts are usually lashed together because nails will split most culms. The withes (strips) for lashing are often split from bamboo and sometimes from rattan. When all local bamboo yields brittle withes, lashing must be done with bark, vines or galvanized iron wire.

In bending bamboo–for example, for the "Double Butt Bent Joint" in Figure 11–you can help to keep the bamboo from splitting by boiling or steaming it and bending it while it is hot.

Local artisans often know the best species of bamboo and they have frequently worked out practical methods for making joints.

# BAMBOO BOARDS

Bamboo culms can be split and flattened to form boards for use in sheathing, walls, or floors.

**Tools and Materials**

Machete
Ax–lightweight, with a wedge-shaped head
Spud–a long-handled shovel-like implement with a broad blade set at an angle to work parallel to the surface of the board.
Large bamboo culms

Not all of the tools listed above are necessary, but they speed up the work when a large quantity is being produced.

- Remove the thick-walled lower part of the culm.

- Use an ax with a well-greased blade to split each node of the culm in several places (see Figure 12). This should be done carefully to avoid injuring one's feet.

- Spread the culm wide open with one long split.

- Remove the pith at the joints with a machete, adz, or spud (see Figure 13).
- Store the boards as shown in Figure 14.

FIGURE 12--An ax with a well-greased bit is used in Ecuador for making bamboo boards. Each node is split in several places; then with one long split, the culm is spread wide open. Not used for boards is the thick-walled basal part of the culm.

FIGURE 13--Final step in making a bamboo board--removing diaphragm fragments from the newly opened culm. It may be done with a machete, as here, or with an adze or a long-handled, shovellike curved spud.

FIGURE 14--Bamboo boards stand ready for use. The making of these boards is a well-developed trade in both Ecuador and Colombia.

# BAMBOO WALLS, PARTITIONS, AND CEILINGS

Bamboo buildings can be built to meet a variety of requirements for strength, light, ventilation, and protection against wind and rain. A few of the methods of building with bamboo are described here.

The parts of a building that are not usually made from bamboo are the foundation and the frame.

Both split and unsplit bamboo culms are used in building. They can be used either horizontally or vertically. Culms exposed to the weather, however, will last longer if they are vertical because they will dry better after rain.

**Tools and Materials**

Local bamboos
Bamboo-working tools, such as machete, hacksaw, chisel, drill
Lashing material: wire or cord
Nails
Barbed wire
Plaster or stucco

## Walls

A method commonly used in Ecuador for making walls is to lash wide bamboo strips or thin bamboo culms, horizontally and at close intervals, to both sides of hardwood or bamboo uprights. The spaces between the strips or culms are filled with mud alone or with mud and stones.

In Peru, flexible bamboo strips are woven together and then plastered on one or both sides with mud.

An attractive but weaker wall can be built by using bamboo boards, stretched laterally as they are attached, as a base for plaster or stucco. Barbed wire can be nailed to the surface to provide a better bond for the stucco. The exterior can be made very attractive by whitening it with lime or cement.

## Partitions

Partitions are usually much lighter and weaker than walls. Often they are no more than a matting woven from thin bamboo strips and held in place by a light framework of bamboo poles. Bamboo matting is often used to finish ceilings and both interior and exterior walls; bamboos with thin-walled, tough culms are usually used for this.

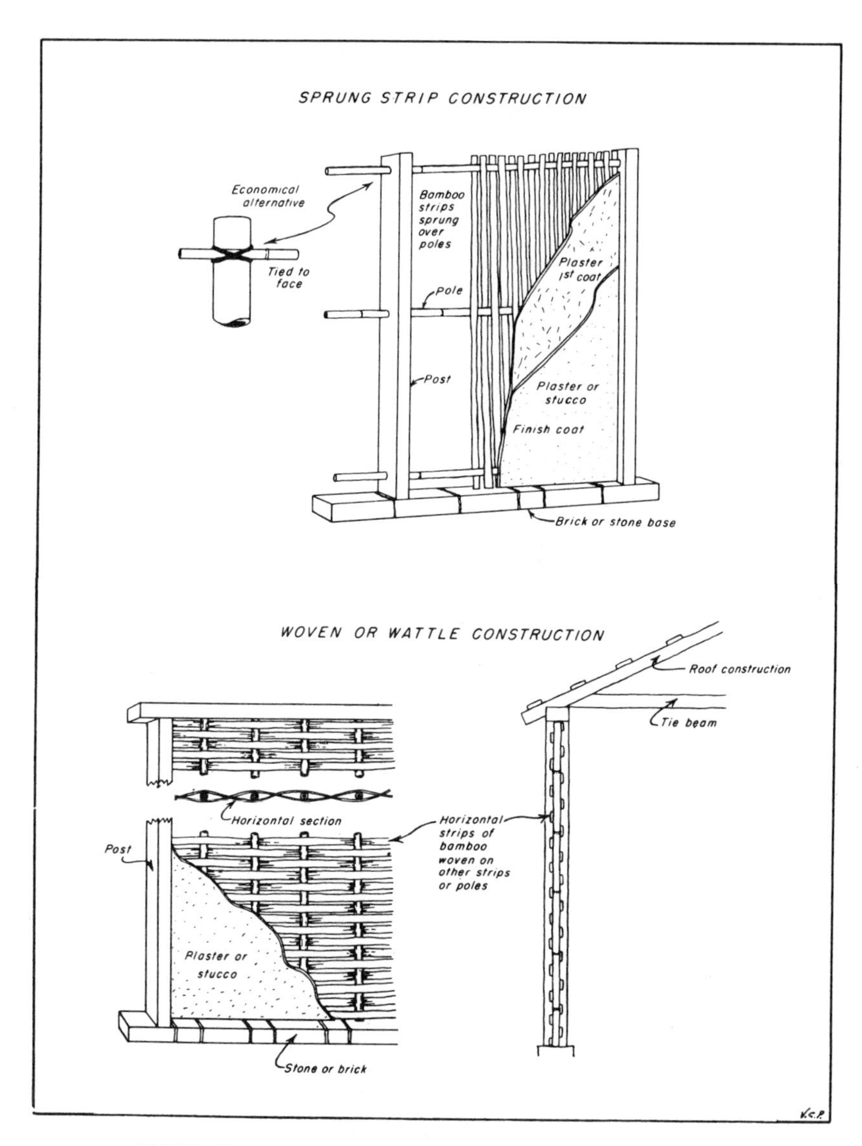


FIGURE 15--Types of wall construction used with bamboo.

# Ceilings

Ceilings can be built with small, unsplit culms placed close together or with a lattice of lath-like strips split from larger culms. There should be some space to let smoke from kitchen fires escape.

**Source:**

McClure, F.A. *Bamboo as a Building Material.* Washington, D.C.: Foreign Agricultural Service, U. S. Department of Agriculture, 1953; reprinted 1963 by Office of International Housing, Department of Housing and Urban Development.

**Sources of information of bamboo are:**

Forestry Division
Joint Commission on Rural Reconstruction
37 Nan Hai Road
Taipei, Taiwan

Forest Research Institute
P.O. New Forest
Dehru Dun, India

Tropical Development & Research Institute
56-62 Grays Inn Road
London, WC 1
England

Federal Experiment Station in Puerto Rico
U.S. Department of Agriculture
Mayaguez, Puerto Rico

# Stabilized Earth Construction

## OVERVIEW

Soil is a universal building material and is one of the oldest known to humanity. Simple soils (without additives), or soils improved by adding stabilizing materials such as bitumen or cement, are suitable for homes, schools, roads, and other construction.

For construction purposes, soil is usually formed into blocks. Two general types of blocks are described here: adobe block and stabilized earth block formed under great pressure. Adobe blocks are made from moistened soil that may be mixed with straw or other stabilizers. They are formed without pressure and usually cured in the sun. Stabilized earth blocks (sometimes called rammed earth blocks) are made from soil mixed with stabilizing material such as Portland cement, formed into blocks under high pressure, and cured in the shade.

Low cost is a primary advantage of soil block construction. An overall cost reduction of about 50 percent over conventional construction can be realized. Other advantages are that building materials are usually readily available and little skill and training are required for their use. The material is culturally acceptable in nearly all countries, including the United States.

## SOIL CHARACTERISTICS

The composition of soil varies from one region to another, and with soil depth. In any one area, it may be desirable to mix soils from several locations or depths to obtain a composition more suitable for construction.

The primary components of soil that are of importance in construction are sand, clay, and silt. (Organic materials are also found in surface soil. These tend to reduce the quality of the blocks.) The fraction of clay in the soil is important because it acts to bind the larger soil particles together but the clay content should not exceed one-third. Above that, deep cracks and weakening of the dried blocks are likely to occur. Silt, which is usually found mixed with the sand, should not exceed one third because silt is vulnerable to erosion from wind and rain.

Proportions of sand, silt, and clay vary widely. One of the few soil block standards that exist is California's Uniform Building Code Specification, which recommends 55 to 75 percent sand, and 25 to 45 percent clay and silt. A good mixture for most blocks might be:

sand.... 65 percent
clay.... 20 percent
silt.... 15 percent

To assure that the composition to be used is suitable for construction, several test blocks should be produced using various mixtures. After curing, the test blocks should be hard and resist a scratch or prick from a knife. Striking two compressed/stabilized blocks together should produce a click sound. The blocks should sustain a drop of two feet (.6 meter) without breaking. If the block crumbles or breaks, the sand or organic content is probably too high, and clay should be added to the mix. On the other hand, if large cracks appear during curing, the clay content is probably too high and sand should be added to the mix.

# TESTING THE SOIL

Soil tests should be made before any block production is started. If the testing is not done first, a great deal of time and money may be wasted in the production of unusable blocks. The agricultural departments of most countries can provide laboratory tests at modest costs. If field tests must be made instead, some simple methods to determine the soil's suitability can be tried.

## Composition Test

- Pass the soil through a 1/4" (6mm) screen to remove stones and other large particles.
- Pour the screened soil into a wide mouth jar until it is half full.
- Fill the jar with water. (You may add two tablespoons of salt to make the soil settle faster)
- Cover the jar tightly, and shake vigorously for two minutes.
- Let settle for at least 30 minutes.

The small gravel and sand will settle rapidly to the bottom of the jar. The clay and silt will settle more slowly. After 30 minutes, the jar should look like the drawing in Figure 1c. Hold a scale vertically on the side of the jar to measure the amounts of sand, silt, and clay. Record the sample number and the amounts. Then convert the amounts to percentages.

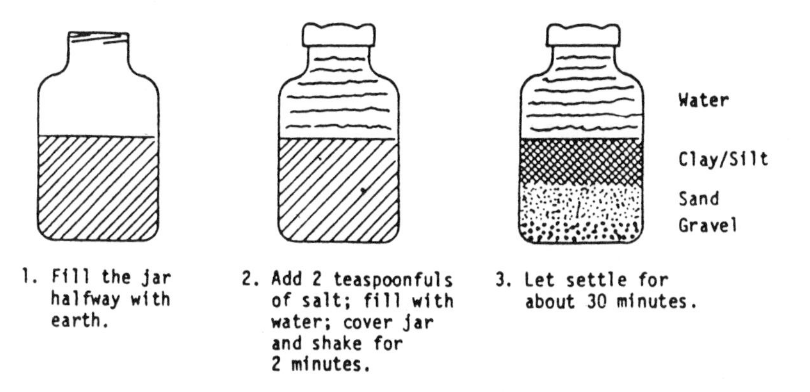


Figure 1. Soil Particle Composition Test

## Compaction Test

In addition to the soil composition test, a compaction test should be done to determine the packing quality of the clay, which depends on the percentage of clay in the sample and the quality of the clay itself. A simple field test can be done as follows:

- Take a handful of dry, screened earth and add some water to it until it is damp enough to form a ball when squeezed in the hand, but not so damp that it leaves more than a slight trace of water in the hand when squeezed.

- Drop the ball from a height of about 3 feet (1 m) onto hard ground. If the ball breaks into a few small pieces, the packing quality is good to fair. If it disintegrates the quality is poor and a soil mix with more clay should be prepared and tested.

## Shrinkage Test

If stabilizing material such as Portland cement is to be added to the soil, a shrinkage test of the soil should also be made. This test will indicate the suitability of the soil and also the best cement-to-soil ratio to use. It measures the shrinkage of soil that contains no stabilizer. As shown in Figure 2, the box should have these inside measurements: 24" x 1-1/2" (4 cm x 4 cm x 60 cm).

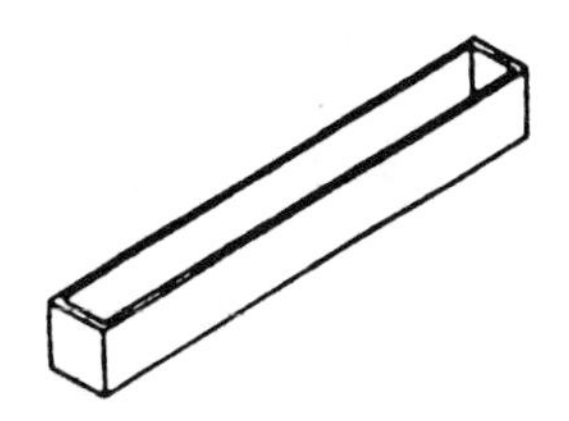

Figure 2. Box for Box Test

To test soil with this method:

- o Oil or grease the inside surface of the box thoroughly.
- o Pack the box well with moist soil (previously passed through a 6mm to 10mm (1/4" to 3/8") mesh screen. The soil should be moistened to pack well, but it should not be muddy.
- o Tamp, especially at the corners.
- o Smooth off the surface with a stick.
- o Place the box in the sun for three days or in the shade for seven days. It should be protected from rain.

Measure the contraction (shrinkage) by pushing the dried sample to one end of the box.

| **Shrinkage** | **Cement to Soil Ratio** |
|---|---|
| Not over 1/2" (15 mm) | 1 part to 18 parts |
| Between 1/2" and 1" (15 mm - 30 mm) | 1 part to 16 parts |
| Between 1" and 1-1/2" (30 mm - 45 mm) | 1 part to 14 parts |
| Between 1-1/2" and 2" (45 mm - 60 mm) | 1 part to 12 parts |

When lime is used instead of cement use **double** the amount. **Do not use the soil** if it has many cracks (not just three or four); if it has arched up out of the box; or if it has shrunk more than 2" (60 mm).

# MAKING ADOBE BLOCKS

To make adobe blocks, add water to the soil mix until it is plastic enough to mold. Water content should be between 16 and 20 percent of the soil by weight. The water and soil must be throughly mixed. Since all except the dryest soils will already contain some water, it is advisable to test the sample for water content first. Do this by weighing a soil sample, drying it, and then reweighing it to calculate water content.

Even the best adobe blocks may develop some cracks. To reduce the number of cracks, and also to make the blocks more weatherproof, stabilizing materials are often added to the mix. When stabilizers are used they must be thoroughly mixed with the soil or much of their benefits will be lost. The most widely used stabilizers are straw, rice husks, asphalt emulsion, Portland cement, and lime.

Asphalt emulsion can improve the waterproofing quality of the blocks, and also their elasticity and toughness, so that they are less likely to break during handling. Add asphalt emulsion between 5 and 15 percent by weight to the dry soil mix. For soil mixes with high sand content (55 to 75 percent sand) the asphalt emulsion should be nearer the 5 percent figure.

Portland cement stabilizers improve the bonding properties and add strength to the blocks. Only 5 to 6 percent cement by weight is needed for soil mixes with high sand content, but up to 20 percent by weight may be required for soils high in clay and silt. If the soil requires a large percentage of cement, it can be combined with an equal amount of lime, which costs less.

Equipment required for making adobe blocks is shown in Figure 3. The number of shovels, molds, etc, will depend on the size of the job. Using this equipment, and supplied with the mixed adobe, a team of two molders can produce about 1,000 blocks (10 x 4 x 14") per day.

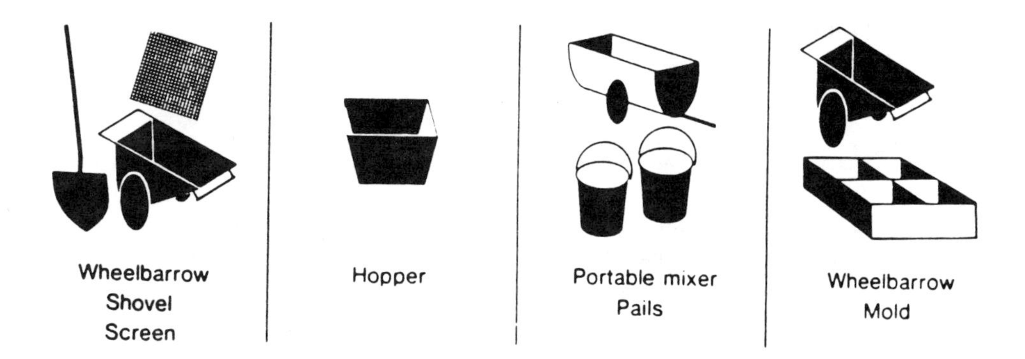


Figure 3. Equipment to Make Adobe Blocks

Select a large level area for mixing, molding, and curing the adobe. Mixing can be done in a hopper, or by making a shallow mixing pit in the ground. If possible, make the blocks near the construction site. If the mix is lumpy even after repeated working, let it soak overnight.

Block molds can be made in various sizes to fit the needs of the construction. But adobe blocks should not be larger than 81cm (32") around the outside. A gang form that will mold eight blocks of .009 cu. meter (one 1/3 cubic foot) can be operated by one worker. Before starting work, the mold should be thoroughly soaked with water to prevent the adobe mud from sticking to it.

Production steps are as follows:

- o Rake or drag a large ground area level.
- o Place mold on level area, on a piece of building paper if available, and dump the mud from a wheel barrow or hopper into the mold. Work the mud firmly into all corners of the mold.
- o Scrape off excess mud from top of mold to leave a smooth, flat surface.
- o Remove the mold by lifting it slowly and evenly up from the ground level. Move the mold to the next adjacent level area and repeat the process.

Blocks must be allowed to cure for about 14 days. After several days, the partially cured blocks may be carefully turned on edge so they dry more evenly. On very hot days, in direct sunlight the blocks may dry too rapidly and crack. To prevent this, cover the blocks with paper, leaves, or straw. Since rain will destroy unstabilized blocks, waterproof tarps may be needed.

To store the blocks after they are cured, stack them on edge. If left stacked flat, they will break of their own weight.

## MAKING COMPRESSED EARTH BLOCKS AND TILES

Compressed earth blocks can be made by ramming the earth in forms, or by using a block making machine, such as the CINVA-Ram Block Press. Blocks made by machine are less costly and have superior uniformity.

Some machine made blocks tested by the U.S. National Bureau of Standards had compressive strengths up to 800 pounds per square inch (56 Kg/cm), with 300 to 500 psi strength as the average. (This is three to eight times the compressive strength of adobe blocks). These test blocks contained 50 percent sand, and 50 percent clay and silt, mixed with 8 percent cement by weight.

Although one worker can make blocks with the CINVA-Ram, the process is best as a team effort with two to four workers each performing one task. (It is good to rotate tasks among the workers on an hourly or daily basis.) The CINVA-Ram is portable and can easily be moved about the work site to reduce carrying raw materials or finished blocks.

Floor tiles can also be made with the machine, using inserts to adjust for the thinner tiles. The mixture for floor tiles is two parts fine sand to one part cement. Mineral coloring can be added to produce colored tiles.

Average production rates and cement required are:

- o Average number cement blocks or tiles (made by two workers per day) 300-500

- o Average no. blocks for a two room house 2500

- o Typical block size: 9x14x29cm (3-1/2"x5-1/2"x11-1/2") which lay up to: 10x15x30cm (4x6x12 inches).

- o Average number blocks per 100 lbs cement: 150

Stacking the blocks for curing requires care. The blocks should be stacked on edge on clean planks. If planks are not available, stack on flat ground that has been covered with paper or leaves. The blocks should be covered with plastic or old cement bags that have been cut open. Stacks should not be greater than five blocks high, and some air space should be left between the blocks. For the first four days, sprinkle the blocks lightly with water to prevent them from drying too quickly. The total curing time is about 14 days, depending on the weather.

## BUILDING WITH STABILIZED EARTH BLOCKS

A firm, flat, water-resistant foundation should be built first using blocks with a higher percentage of cement and lime. Blocks should be joined by mortar about one half inch (1.25 cm) thick. The recommended mortar mix (by weight) is:

- o one part cement
- o two parts lime
- o nine parts soil (used to make the blocks)

Let the applied mortar dry for about a week; then paint the mortar joints with a thin, milk-like mix of cement and water. Stir this mixture often. After a day, the finished walls can be coated (3 coats recommended) with this same mixture, or with a coat of lime. Or, a waterproofing coat of silicone based wash may be added.

**Sources:**

Alfred Bush, Chris Ahrens, Balla Sidibe, VITA volunteers

**References:**

*Making Building Blocks with the CINVA-Ram Block Press.* Arlington, Virginia: VITA, 1977.

Bush, Alfred. *Understanding Stabilized Earth Construction.* Arlington, Virginia: Volunteers in Technical Assistance, 1984

"Building Materials and Structures Report BMS 78", Gaithersburg, Maryland: US National Bureau of Standards

Sidibe Balla. *Understanding Adobe.* Arlington, Virginia: Volunteers in Technical Assistance, 1985

U.S. Agency for International Development, "Handbook for Building Homes of Earth", Action Pamphlet No. 4200.36, Wolfkill, Dunlop, Callaway, Washington, DC, Peace Corps, 1979.

Ferm, Richard. *Stabilized Earth Construction: An Instructional Manual.* Washington, D.C.: The International Foundation for Earth Construction.

**Manufacturer:**

The CINVA-Ram Block Press is manufactured in Bogota, Colombia, by METALIBEC, S.A. The press may also be purchased in the USA for $400 (1987) from Schvader Bellows Inc., 200 West Exchange Street, Akvon, Ohio 44309-0631. Telephone: (216) 375-5202. Similar, locally manufactured presses can often be found in other developing countries.

# Construction Glues

## CASEIN GLUE

Strong, water-resistant casein glue, which produces joints as strong as or stronger than most of the common species of wood, is made from skim milk and common chemicals. Casein glue joints are water-resistant but not waterproof. They will withstand occasional soaking, but if soaked and dried, they will fail.

**Tools and Materials**

Mixer: paddle and bowl of wood, iron, or other material that won't be corroded by the alkali in the glue.
Containers
Scale or balance
Skim milk
Hydrated lime, $Ca(OH)_2$, also known as slaked lime. This should be a good quality lime: high in calcium and low in magnesia.
Silicate of soda, also called "waterglass" or sodium silicate. The preferred solution should have a density of about 40 degrees Baume (Density 1.38) with a ratio of silica to soda of approximately 3.25 to 1.
Cupric chloride, $CuCl_2$ (cupric sulfate, $CuSO_4$, also called "blue vitreol" can be substituted)
Wire screen or 20-mesh sieve with 0.033" (0.84mm) openings
Cloth for squeezing moisture out of curds

## Making Casein Powder

Casein powder is made from skim milk by the following steps:

- Let the milk sour naturally or sour it by slowly adding dilute hydrochloric or sulfuric acid until curds form. The milk will separate into curd and whey.
- Drain the whey off. Wash the curd by adding water and draining it off.
- Press the curd in a cloth to remove most of the moisture.
- Break the curd into small particles and spread it out to dry.
- Grind the dry curd to a powder and pass it through a 20-mesh screen.

## Mixing Casein Glue

**Proportions for Glue**

Formula 11 (not restricted by patent), U.S. Forest Products Laboratory

| | Parts by Weight |
|---|---|
| Casein (powder) | 100 |
| Water | 150 to 250 |
| Hydrated Lime (powder) | 20 to 30 |
| Water | 100 |
| Silicate of soda (solution) | 70 |
| Cupric chloride (powder) | 2 to 3 |
| Water | 30 to 50 |

If hydrated lime is not available, quicklime (CaO) can be used in the following ways:

A mixture of 15.1 parts CaO and 104.9 parts water by weight can be substituted for 20 hydrated lime and 100 water.

A mixture of 23.5 CaO and 106.5 water can substitute for 30 hydrated lime and 100 water.

When CaO is added to the water, it must be stirred for 15 minutes to get a uniform slurry.

The bowl and paddle for mixing casein glue should be made of wood, iron, or some other material that will not be corroded by the alkali in the glue and can be cleaned easily. All the ingredients should be weighed rather than measured by volume so that the proportions will be accurate. It is especially important not to use too much water.

- Put the **casein and water** in the mixing bowl and mix them well enough to distribute the water throughout the casein. If the casein used has been ground to pass through a 20-mesh screen, let it soak in the water for 15 to 30 minutes before going on to the next step. The soaking period can be reduced if the casein is ground more finely.

- Mix the **hydrated lime and water** in a separate container.

- Dissolve the **cupric chloride in water** in a separate container and add it, while stirring, to the moistened casein.

- o Immediately pour the **hydrated lime-water mixture** into the casein mixture. When casein and lime are mixed, large lumps form at first but they break up rapidly and finally disappear. The solution becomes somewhat thinner. **Thorough stirring is very important at this point.**

- o About a minute after the lime is mixed with the casein, the glue begins to thicken. Add the silicate of soda at this time.

- o The glue will thicken momentarily, but continue stirring the mixture until the glue is free of lumps. This should take no longer than 20 minutes.

  If the glue is a little too thick, a small amount of water can be added. If it is too thin, start the whole process over again, using a smaller proportion of water.

## Using Casein Glue

The working life of glue is the length of time it stays fluid enough to be workable. The silicate of soda extends this time. The glue produced by the formula used here will be useable for more than 7 hours at temperatures between 21C and 24C (70F and 75F). Working life will be shorter at higher temperatures.

Casein glue is fluid enough to be spread by a roll spreader or by hand with a brush or scraper. Very heavy spreads are wasteful because excess glue will be squeezed from the bond. Very light spreads can produce weak joints. A suggested minimum is 29.5 kilograms (65 pounds) of wet glue per 92.8 square meters (1,000 square feet) of glue-joint area.

To obtain good contact between wooden members of a joint, apply pressure while the glue is still wet. There is not much drying before 15 or 20 minutes. Under ordinary circumstances, a pressure of 105,450 to 140,600 kilograms per square meter (150 to 200 pounds per square inch) will give good results.

If casein glue joints are exposed for long periods to conditions that favor the growth of molds, they will eventually fail. The joints will be permanent only if the moisture content of the wood is not greater than 18 to 20 percent for long or repeated periods.

Dry casein can be kept for a long time in a cool, dry place.

**Sources:**

*Casein Glues: Their Manufacture, Preparation, and Application.* Madison, Wisconsin: Forest Products Laboratory, Forest Service, U.S. Department of Agriculture.

Dr. Louis Navias, VITA Volunteer, Schenectady, New York

# LIQUID FISH GLUE

Cold liquid glue can be made from the heads, skins, and skeletal wastes of cod, haddock, mackerel, hake, and pollack. A great advantage of liquid fish glue is that it remains in liquid form and consequently has an almost permanent working life. An advantage of using it to make wood joints is that it sets slowly and therefore penetrates further than other glues before hardening.

Since liquid fish glues are not very water-resistant, a casein or other glue should be used where water-resistance is needed. Thick fish glues produce stronger joints than thin solutions.

**Tools and Materials**

Fish heads, skins, and skeletal waste
Large pan for washing fish parts
Steam bath or double boiler
Paddle for stirring
Filter, such as cheese cloth

To make the glue:

- Wash the fish material thoroughly to remove blood, dirt and salt. If salted fish are used, wash them in running water for 12 hours.

- Once the material is washed and drained, put it into a large container, cover it with water, and cook it slowly at a low temperature, about 60°C (140°F). Cooking in an open pot helps to eliminate unpleasant odors in the glue. A steam bath or double boiler should be set up so that live steam surrounds the pot. Stir the contents occasionally. The length of the cooking period varies with the kind of fish material used.

- Let the cooked mixture settle. Skim off and discard the grease. Pour the remaining contents of the pot onto a filter.

- Concentrate the filtered fluid by slow heating to the desired thickness. This is the glue; it can be stored in convenient containers.

- Take the fish material remaining on the filter and cook it again to extract more glue, then repeat the filtering and concentrating.

**Sources:**

*Encyclopedia of Chemical Technology.*
Paul I. Smith. *Glue and Gelatine,* Chemical Publishing Co., Inc., 1943.
Thomas D. Perry. *Modern Wood Adhesives.* Pitman Publishing Co., 1944.

# Home Improvement

Margaret Crouch

# Simple Washing Machines

## PLUNGER TYPE CLOTHES WASHER

This hand-operated washer, which is simple for a tinsmith to build, makes washing clothes easier. It has been used successfully in Afghanistan.

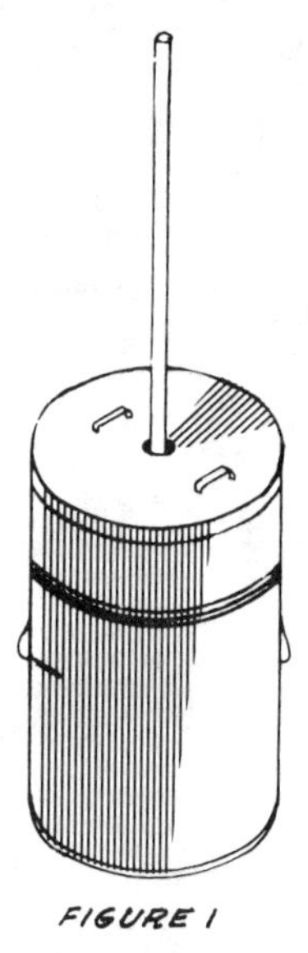

FIGURE 1

**Tools and Materials**

Tinsnips
Pliers
Hammer
Soldering equipment
Heavy galvanized sheet metal:
- 140cm x 70cm (55 1/8" x 27 9/16") for tub
- 100cm x 50cm (39 3/8" x 19 11/16") for lid and bottom
- 36cm x 18cm (14 3/16" x 7 1/16") for agitator

Wooden handle 140cm (55 1/8") long, about 4cm (1 1/2") diameter

## Making the Washer

Figures 1 to 4 show how this washing machine is made. The tub, lid, and agitator are made of heavy galvanized sheet metal.

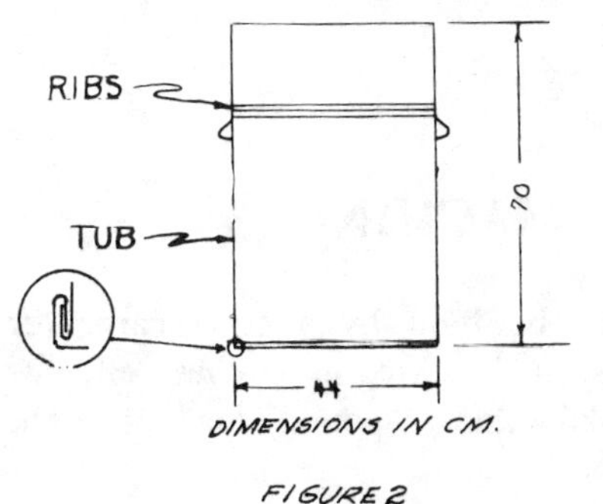


FIGURE 2

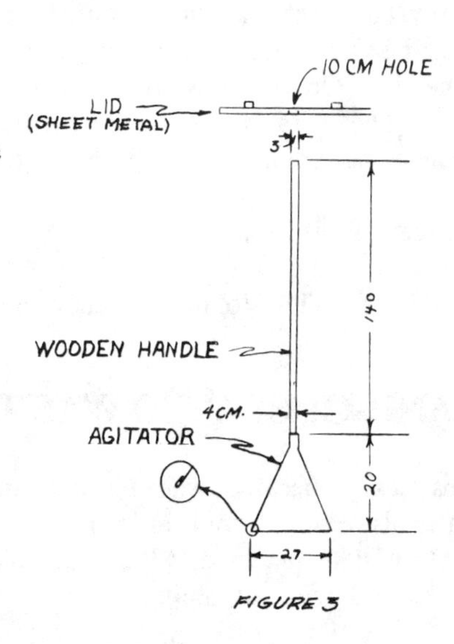


FIGURE 3

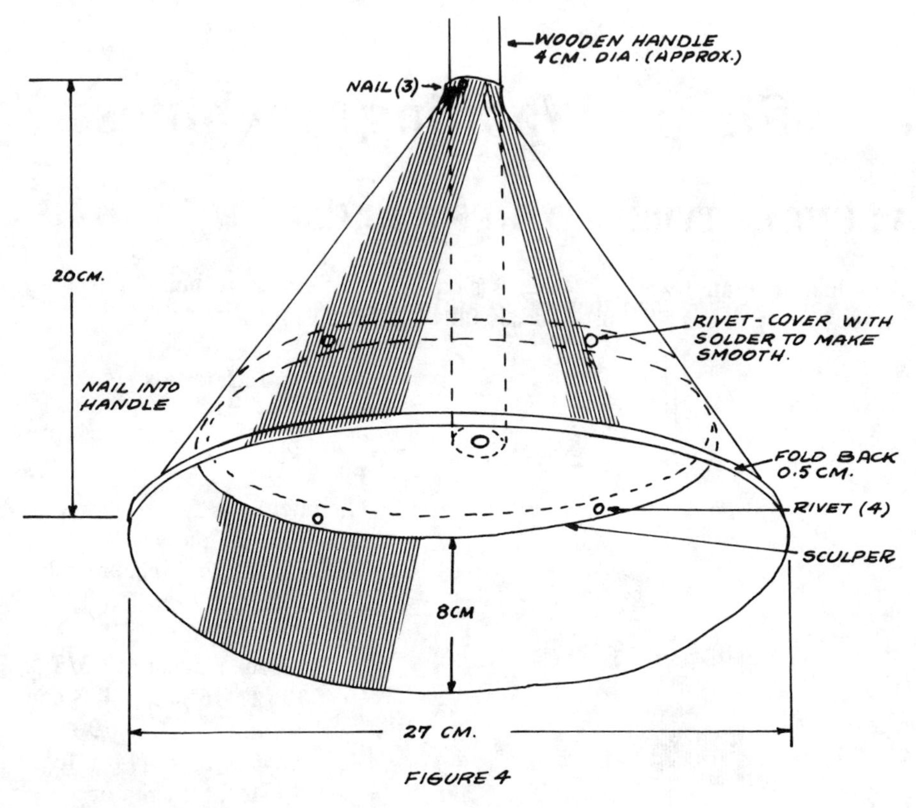


FIGURE 4

## Using the Washer

To operate the washing machine, work the agitator up and down with a quick motion but with a slight pause between strokes. The movement of the water caused by the agitator will continue for a few seconds before additional agitation is needed. On the upward stroke the agitator should come completely out of the water. The agitator should not hit the bottom of the tub on the downward stroke because this would damage both the tub and the clothes.

**Source:**

Dale Fritz, VITA Volunteer, Schenectady, New York

# HAND-OPERATED WASHING MACHINE

This easily-operated washing machine can be built by a good carpenter from materials easily found in most countries. It is easy on clothes, effective, and sanitary. The machine, which can take 3-kilogram (6-pound) load of clothes, can be shared by several families.

Clothes will last much longer if they are washed in this washing machine rather than beaten or scrubbed on rocks. Washing with the machine is also much less work. Under test conditions, a comparison with standard electric commercial washers was very favorable. If the cost of the machine is too much for one family, it can be used by several. However, if there are too many users, competition for times of use will become keen and the machine will wear faster.

The machine reverses the principle used in the usual commercial washer, in which the clothes are swished through the water for various degrees of a circle until the water is moving, and then reversed. In this machine, the clothes stay more or less stationary while water is forced back and forth through the clothes by the piston action of the plungers. One plunger creates suction as it rises and the other plunger creates pressure as it moves downward. The slopes at the ends of the tub bottom help the churning action of the water caused by the plungers (see Figure 1).

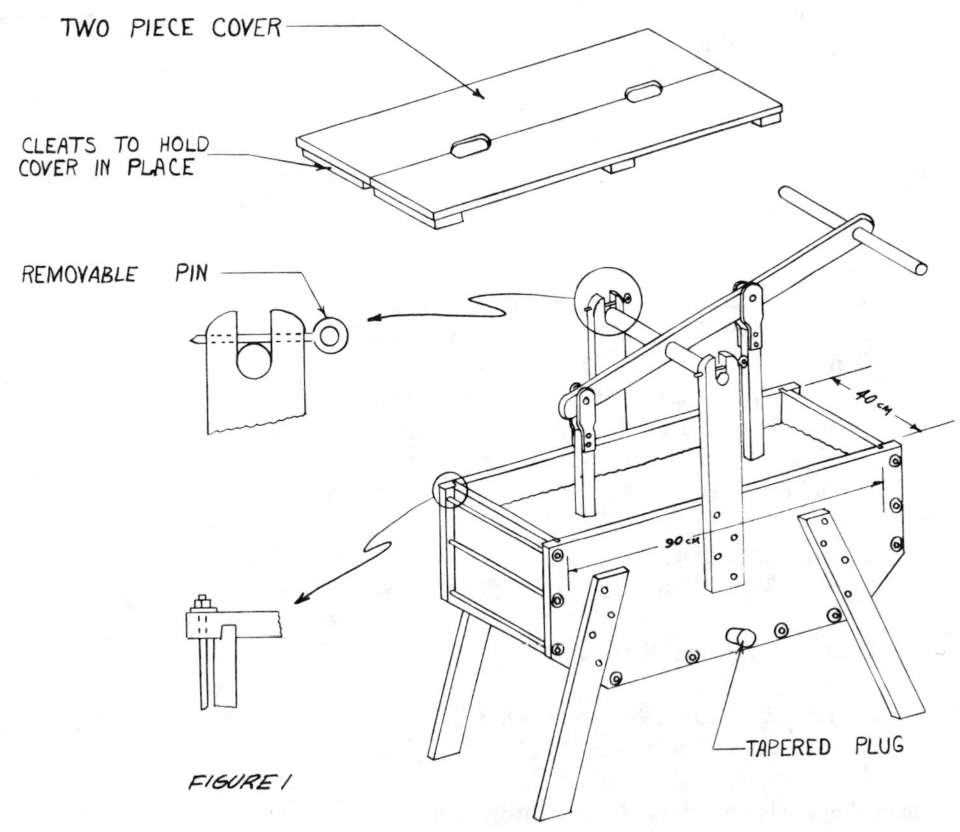


FIGURE 1

A rectangular tub is best for this method of operation. This is fortunate since the rectangular box is easy to build. In general, any moderately strong wood that will not warp excessively (such as cedro in Latin America) will be satisfactory. The sides should be grooved for the ends and bottom of the tub as indicated in Figure 1 and bolted with threaded rods extending through both sides with washers to draw them tight. The bolting is necessary to prevent leaks.

The size described in the drawings is large enough for an average family in the United States. The same principle may be used for a larger or smaller machine provided the basic proportions are maintained. The tub should be slightly less than half as wide as it is long to get a proper surge of water. The pistons should be wide enough to move within a couple of inches of each side of the tub. The lever pivot should be high enough to permit the plungers to move up and down several inches without the edge of the lever hitting the edge of the tub. Likewise, the length of the rods on the plungers must be such that the plungers go well into the water and the clothes, and then come completely out of the water at the highest position.

## Tools and Materials

**Tub Construction** - Moderately firm soft-wood free from large heartwood growth:

Tub

Sides–2 pieces, 2.5 x 45.7 x 96.5cm (1" x 18" x 38")
Ends–2 pieces, 2.5 x 30.5 x 40.6cm (1" x 12" x 16")
Bottom–2 pieces, 2.5 x 15.2 x 40.6cm(1" x 6" 16")
Bottom–1 piece, 2.5 x 40.6 x 66.0cm (1" x 16" x 26")
Legs–4 pieces, 2.5 x 10.2 x 76.2cm (1" x 4" x 30")

Round Plungers

2 pieces, 2.5 x 25.4cm diameter (1" x 10" diameter)
2 pieces, 3.8 x 12.7cm diameter (1.5" x 5" diameter)

Cover (may be omitted)

2 pieces, 2.5 x 20.3 x 91.4cm (1" x 8" x 36")
6 pieces, 2.5 x 7.6 x 20.3cm (1" x 3" x 8")

**Operating parts** - Moderately firm hardwood:

Lever–1 piece, 2.5 x 7.6 x 122cm long (1" x 3" x 48")
Plunger stems–2 pieces, 2.9cm square 38.1cm long (1 1/8" square 15" long)

Uprights

2 pieces–2.9 x 7.6 x 61.0cm long (1 1/8" x 3" x 24")

Pivot and Handle

2 pieces, 3.2cm diameter x 45.7cm long (1 1/4" diameter x 18")

**Metal Parts**

Plunger connections

4 pieces iron or brass plate, .64 x 3.8 x 15.2cm long (1/4" x 1 1/2" x 6")
10 rods, 3.6 or .79cm diameter (1.4" or 5/16") 45.7cm (18") long with threads and nuts on each end–iron or brass

20 washers about 2.5cm (1") diameter with hole to fit rods

1 rod, .64 x 15.2cm long (1/4" x 6") with loop end for retaining pivot

6 bolts, .64 x 5.1cm long (1/4" x 2" long)

24 screws, 4.4cm x #10 flat head (1 3/4" x #10)

50 nails, 6.35cm (2 1/2")

Strip sheet metal with turned edge, 6.4cm wide, 152.4cm long (2 1/2" wide, 72" long)

Small quantity of loose cotton or soft vegetable fiber for caulking seams

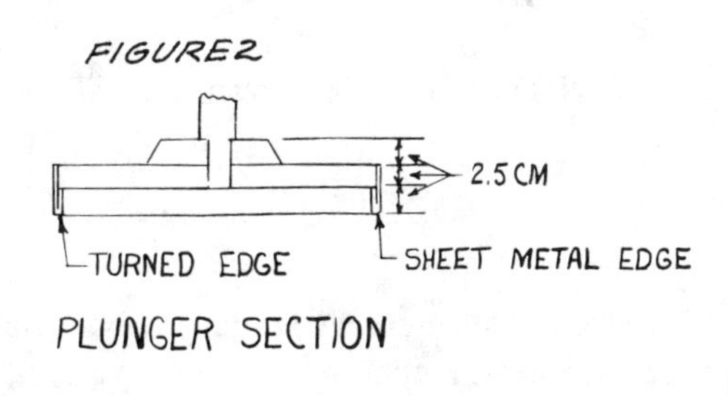

PLUNGER SECTION

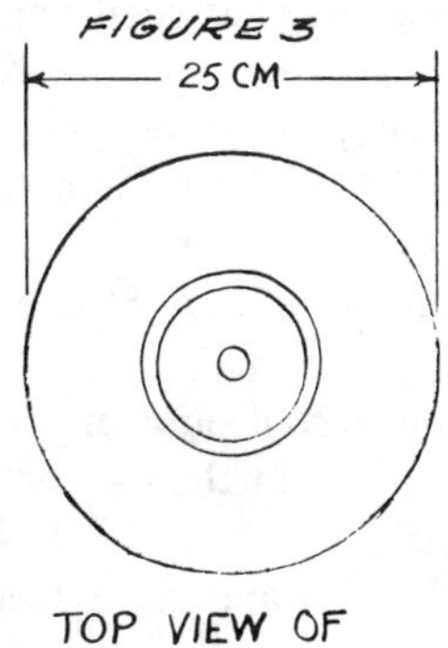

TOP VIEW OF PLUNGER

**Minimum Tools Needed**

Tape measure or ruler
Hammer
Saw
Wood chisel 1.3 or 1.9cm wide
(1/2" or 3/4")
Screwdriver
Pliers
Adjustable wrench
0.64cm (1/4") drill, gimlet or similar tool
Draw knife or plane and coping saw

## Making the Washing Machine

Mark and groove sides for end and bottom members (see Figures 1 and 4).

Drill holes for cross bolts.

Cut off corners and trim ends of side member to length.

Bevel ends and bottom pieces to fit into groove in side members.

Miter bottom and end members together.

Assemble and bolt.

Cut and install legs.

Caulk seams between ends and bottom members with loose cotton or other vegetable fiber to make seams water-tight. If joints to side members are carefully made, they may not need caulking.

Bore hole and make plug for draining tub. NOTE: This is shown on side in drawing but it is better in bottom of tub.

Make and install upright pivot members.

Make and install plunger lever. NOTE: The cross pivot member (round) should be shouldered or notched at each pivot to prevent side movement.

Make plungers and install (see Figures 2, 3 and 4).

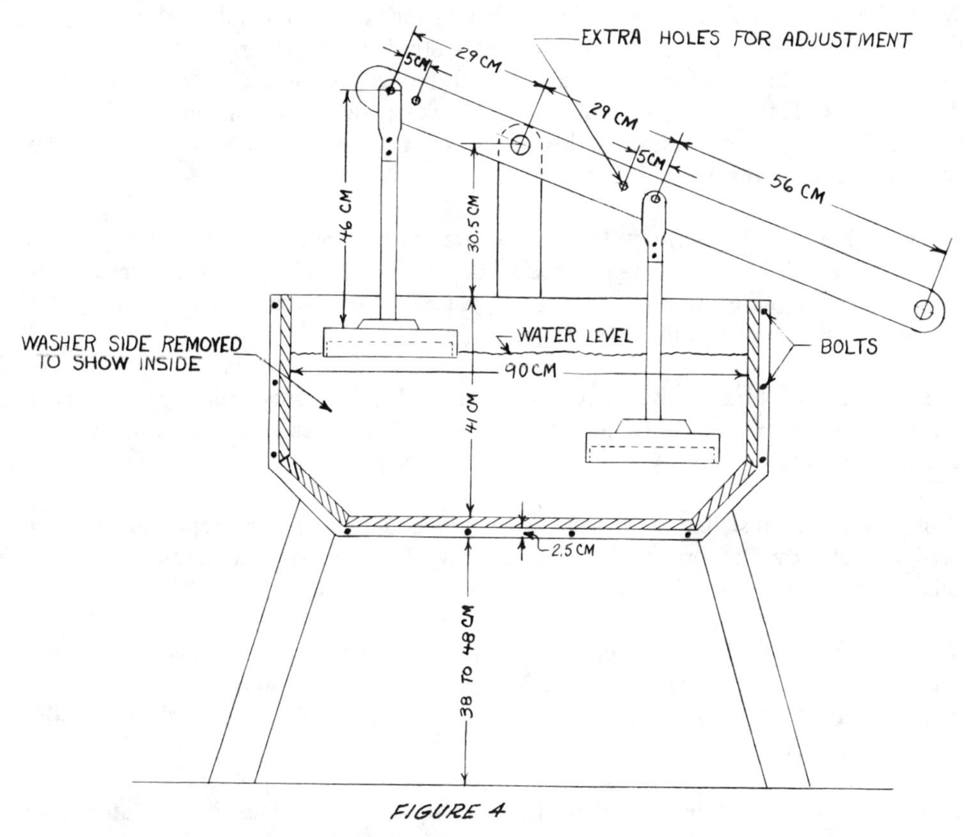


FIGURE 4

# Using the Washing Machine

Here are several suggestions for using this washing machine: Fill the washer with approximately 55 liters (15 gallons) of warm or hot water depending on what is available. Try to remove stains in clothing before putting it in the wash water. Rub soap into the areas of garments like cuffs and collars that come in close contact with the body. Soak very dirty clothes before putting them in the washer. Soap can be dissolved by shaving it into strips and then heating it in a small quantity of water before adding it to the wash water. A 3kg load of clothes is the right size load for best cleaning. Wash at a moderate speed, about 50 strokes a minute, for ten minutes–longer if it seems necessary.

If more than one load of clothes is to be washed, some basic procedures will help to simplify the job and conserve water. (Water used for washing and rinsing can help irrigate a garden plot.)

First divide the clothes so that whites and light colors are separate from dark clothes. Try to keep small items together so they won't get lost. Heavily soiled or greasy clothes should be washed alone.

Wash the white or light-colored things first in the hottest possible water (remember that you will have to handle the wet clothes–don't get the water too hot!), then move on through darker clothes. The water will become discolored. Much of the color is dirt, of course, but some is excess dye. The lightest clothes are washed in the cleanest water; dark clothes won't be as noticeably affected by the coloring matter in the water.

After each load, the wash water can be warmed, if necessary, by adding some boiling water. A bit more soap may also be needed. Probably at least three loads of clothes–depending on how dirty they are–can be washed before the water becomes too murky to be used again.

The clothes, of course, will have to be rinsed thoroughly. Soap or detergent residues can damage fabrics and may cause allergic reactions. Two rinses are usually necessary.

Probably the easiest, but most expensive, procedure is to have separate tubs for rinsing. Tubs can be of either wood or galvanized metal, and may be used for other purposes provided they are cleaned thoroughly on wash day.

When clothes are clean, squeeze out as much excess water as possible and put them into the rinse water. The next load of wash can be soaking while the first is rinsed and put to dry. Then the clothes in the machine are washed and the process repeated.

If no separate rinse tubs are available, wash up to three loads (if the water stays clean enough that long) and set each aside. Be sure to keep loads separate, as dyes from wet clothes may stain lighter colored fabrics. Then drain and rinse the washing machine and refill it with clean water. Rinse the clothes, again starting with the lightest colored load, and put out to dry. Repeat the whole wash-rinse process as often as necessary.

Another method is to wash the first load of clothes and squeeze out excess water. Drain the wash water and refill the machine with clean warm water. Rinse the clothes, squeeze out excess water, and put to dry. Warm the rinse water with boiling water and and some soap. Then wash the next load. Repeat the procedure as often as necessary.

After washing and rinsing the clothes, rinse the washer clean and then replace the stopper. To keep the wood from drying out and causing the tub to leak, put about 3cm (1") of water in the washer when it is not in use.

**Source:**

Petit, V.C. and Holtzclaw, Dr. K. *How to Make a Washing Machine.* Washington, D.C.: U.S. Agency for International Development.

# Cookers and Stoves

## FIRELESS COOKER

Where fuel is scarce, this easy-to-build fireless cooker can be a contribution to better cooking. It keeps food cooking with a small amount of heat stored in hot stones; loss of heat is prevented by a thick layer of insulating material around the pot.

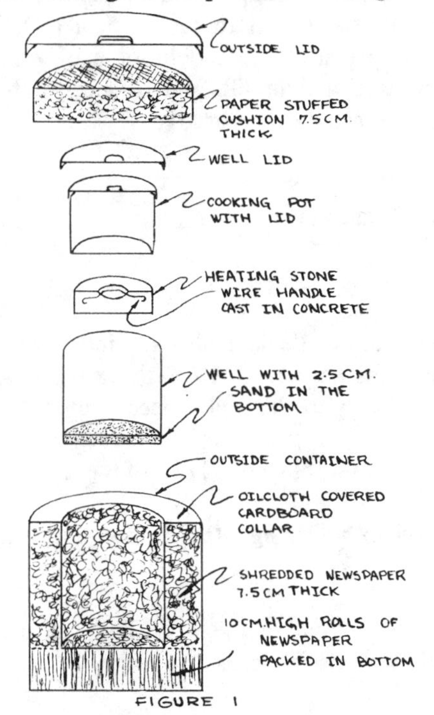

FIGURE 1

Fireless cookers have been successfully used in many countries. Once the principle of operation, heat retention through insulation, is understood, the reader may develop plans that are better suited to local resources than those described here. In some countries, fireless cookers are built into the ground. In others, they are built from surplus tin cans, one can fitted into another tin can or box but separated by paper, sawdust, or other layers of insulation.

### Materials

Outside container with lid, 37.5cm to 60cm (15" to 24") in diameter
Inside container or well, at least 15cm (6") smaller in diameter and 15cm (6") shorter than outside container
Cooking pot with lid
Cloth for cushion, 1.2 square meters (1 1/2 square yards)
50 sheets newspaper or other insulation
Cardboard
Sand, .95 liter (4 cups)
Cement, .95 liter (4 cups)
Oilcloth for collar (optional), 0.4 square meters (1/2 square yard)

The outside container can be a wooden bucket, kerosene can, garbage can, packing crate, or even a hole in dry ground. The inside container or well can be a pail or can with a lid. It must allow for 7.5cm (3") of insulation between it and the outside container and should hold the stone and cooking pot without much vacant space.

Insulation can be made of shredded newspapers, wool, cotton, sawdust, straw, rockwool, fiberglass, or other material. The insulation should be at least 7.5cm (3") thick on all sides, top and bottom. Be sure that it is very dry. The bottom layer of insulation must be strong enough to support the weight of the well, stone, and cooking pot. A natural stone carved to shape or a piece of concrete may be used for the heating stone. The cushion is a cloth sack, 7.5cm (3") thick, filled with shredded newspapers or other insulation. It should fit snugly in the outside container. The cooking pot must have a tight lid, and fit nicely into the well when the stone is in place. Be sure it can be removed easily when full of hot food.

## Making the Fireless Cooker (See Figure 1)

Wash and dry the containers and lids.

Cut 10cm-wide strips of newspaper several layers thick. Roll each into a cylinder with a center hole no greater in diameter than a pencil. Pack these on end into the bottom of the outside container. They will support the well, stone, and pot.

Put the well in place. Pack insulation around it to within 1cm (1/2") of the top.

Make a cardboard collar covered with oilcloth. Though this is not necessary, it improves appearance and cleanliness.

Place about 2.5cm (1") of clean sand in the bottom of the well. This will prevent the hot stone from scorching the paper rolls and possibly causing a fire.

To make a concrete heating stone, place a 5cm-wide cardboard band or collar on heavy paper or board to form a circle the size of the stone desired. Mix .95 liter (4 cups) each of cement and sand (the sand should first be washed free of silt); then mix in enough water (about .35 liter or 1 1/2 cups) to form a stiff mush. Fill the collar, casting in a wire handle for lifting the hot stone. Let the stone stand for 48 hours, then remove the collar, place it in cold water, and boil for 30 minutes. Cool it slowly.

## Using the Fireless Cooker

It is important to keep the cooking pot and well carefully washed and open, in the sunshine if possible, when not in use. The cooker's lid should be left partly open and the stone kept clean and dry.

It is not necessary to use much water when cooking in a fireless cooker for there is little loss by evaporation. Most foods should be brought to a boil and cooked for 4 to 5 minutes on another stove. The heating stone is heated and placed in the cooker. Then the covered cooking pot is set on the hot stone in the cooker and the lid is placed on the well. Cereal may be left in the cooker all night. Rice and cracked or whole wheat are especially good. Beans should be soaked over night, boiled for 5 minutes and then placed in the cooker for 4 to 5 hours. Dried fruit should be washed, soaked for an hour in 2 parts water to 1 part fruit, boiled for 5 minutes, then placed in the cooker for 4 hours.

**Source:**

*Home Making Around the World,* Washington, D.C.: U.S. Agency for International Development, 1963.

# CHARCOAL OVEN

This simple charcoal-fired oven is made from two 5-gallon oil tin cans. With practice, all types of baking and roasting can be done effectively.

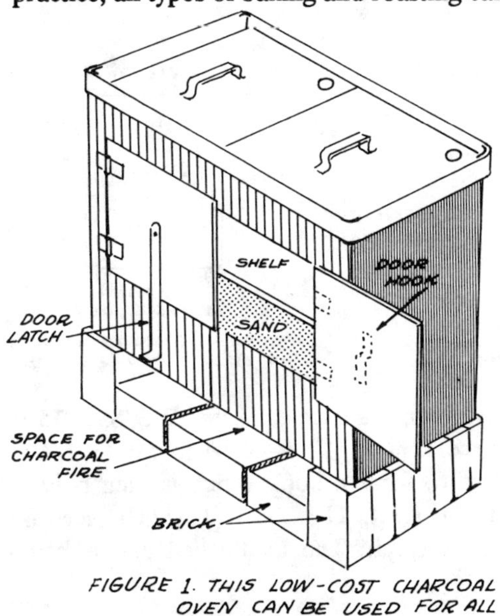


FIGURE 1. THIS LOW-COST CHARCOAL OVEN CAN BE USED FOR ALL KINDS OF BAKING.

**Tools and Materials**

Tin snips
Heavy knife
Nail for scriber and punch
Hammer
Screwdriver
Pliers
Bricks and sand
Metal bar, 20cm (7 5/8") long with square edge for bending tin
5-gallon cans (2)
Tin (for shelf, top strip, and latch)
Light rod, 50cm (19 5/8") long
Light hinges with bolts (2 pairs)
Stove bolts, 5mm x 13mm (3/16" x 1/2") (15)

## How To Build the Charcoal Oven

Mark the two 5-gallon cans for cutting (see Figure 2), making sure that the second can is marked the reverse of the first. Do not cut the corner that has a vertical seam: Besides being hard to cut, the seam will strengthen the oven. The material removed will be easier to make into doors if it is seamless.

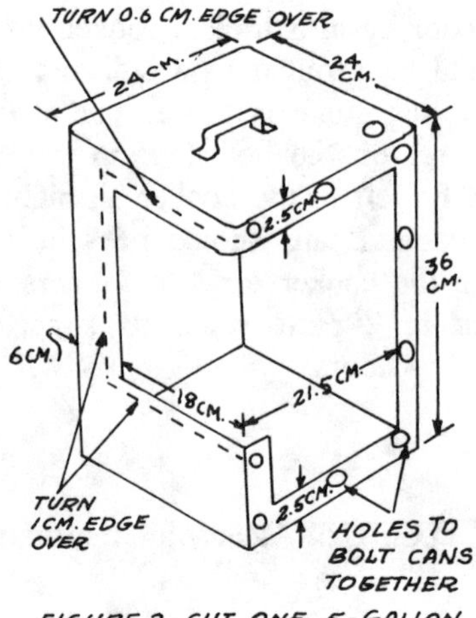


FIGURE 2. CUT ONE 5-GALLON CAN AS SHOWN ABOVE. THE OTHER CAN IS CUT THE REVERSE OF THE ONE HERE.

Cut along the marks with a heavy knife, keeping the cut-out sections as undamaged as possible. Fold the edges of the oven-door openings back 1cm (3/8") (see Figure 2).

With the nail, punch 5mm (3/16") holes around the opening in the side of the can to be used for the left hand section of the oven (see Figure 2). Place the second can against the one just punched and mark the holes with the nail. Punch holes in the second can. Bolt the cans together, using 10 stove bolts.

Flatten sections cut from cans and mark for doors (see Figure 3). Using the tin snips, cut doors to size and fold back the 1cm (3/8") edge (see Figure 4). Position doors as shown in Figure 1, butting the edge of each door against the edge of the opening to which it will be attached. Install hinges.

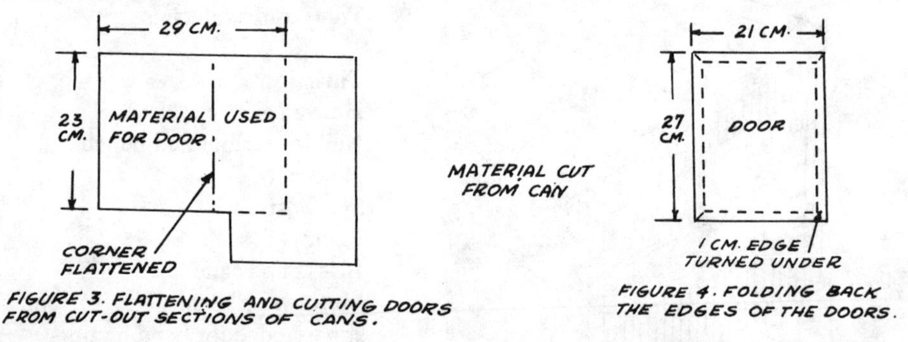


FIGURE 3. FLATTENING AND CUTTING DOORS FROM CUT-OUT SECTIONS OF CANS.

FIGURE 4. FOLDING BACK THE EDGES OF THE DOORS.

The door latch (see Figure 1) is made by folding a 6cm x 38cm (2 3/8" x 15") strip three times lengthwise, forming a piece 2cm (3/4") wide. An 8cm (3 1/4") piece is cut from the end of the folded strip to form a hook–which is then bolted (use 2 bolts) to the center of the door on the right. The 30cm (11 3/4") piece is bolted loosely to the center of the door on the left. The unattached end is bent up to form a handle.

A triangular-shaped hole at the top of the doors where the two cans come together must be plugged to keep heat from escaping. This can be done by shaping a small piece of tin to fit the opening, with a tail that is inserted between the joined cans to hold it in place.

Construct shelf as shown in Figure 5 and install (see Figure 1). The shelf should be bolted in place 15cm (5 7/8") from the floor of the oven (see Figure 5).

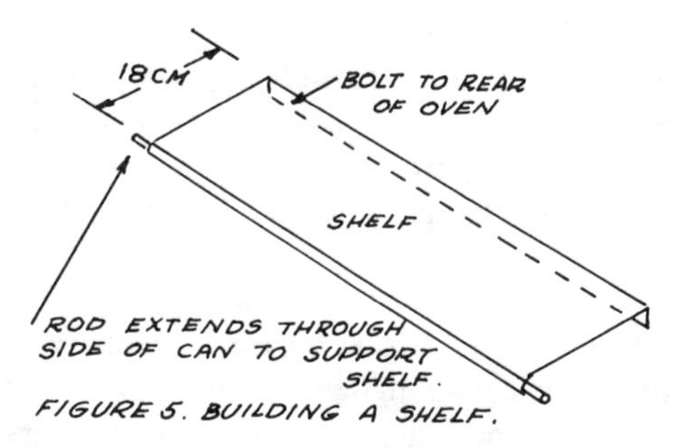

FIGURE 5. BUILDING A SHELF.

The oven should be cleaned thoroughly and heated at least once before use to burn out any remaining oil in the cans.

## How to Use the Charcoal Oven

Place 1cm (3/8") of sand in the bottom of the oven and place the oven on bricks as in Figure 1. The oven can be removed until the charcoal starts to burn, then put in place.

A little time is required before the proper temperature is reached because the sand must first absorb and dissipate the heat. For very high baking temperatures, or to brown the top surface of baked goods, additional pieces of charcoal can be placed on top of the oven. An extra rim can be added to the top edge of cans for this purpose (see Figure 1).

**Sources:**

V.C. Pettit, United States Agency for International Development
Dale Fritz, VITA Volunteer, Schenectady, New York

# PORTABLE METAL COOKSTOVES

Loss of forest cover is a serious problem around the world, particularly in developing countries. In some of these countries, forest cover has decreased from over 60 percent to under 20 percent in just a few decades. One consequence of this loss of wood supply is that it is becoming increasingly difficult for people in these areas to obtain fuel to cook their food.

Improving the fuel efficiency of cookstoves is one way to reduce the drain on forests and the wood supply. Improved stoves can also reduce the cost of cooking fuel--an expense that consumes up to one third of the income for some families.

## Principles of Energy-Efficient Stoves

Traditional stoves are generally of three types. The simplest is the three-stone design, where the cooking pot rests on stones over an open fire. The second type is the massive stove, made of clay and sand, that may hold several cook pots, but which takes a long time and much fuel to heat up. The third type is the light-weight portable stove made of sheet metal or ceramic.

The traditional portable stove has been studied intensively and modified to achieve a very high level of efficiency–40 to 50 percent, or more than twice the efficiency of traditional stoves. In addition, the portable stoves are easily mass produced by local artisans and find a ready market alongside more traditional goods.

In a stove, heat is transferred from the fire to the pot by the convective heat process. To get the most convective heat transfer–and hence fuel efficiency–it is necessary to pass the hot gases from the fire over as much of the surface of the cook pot as possible, and through as narrow a channel as possible (see Figure 1).

*Figure 1. Narrow channel for cookstove efficiency.*

Narrower channels give higher convective heat transfer efficiencies, and thus reduce the overall size of the fire needed for cooking. But if the channel is too narrow, the fire may be choked off, and either smoke or die. Experimental work has shown that a channel between 4mm and 8mm wide (about 1/4 inch) is best.

If families already have their cook pots, then the stove(s) must be designed and built to fit the cook pots in order to obtain the narrow channels for the hot gases from the fire. This means that one should not design and build the cookstoves until the sizes of the cook pots have been measured. An alternative is to design a cookstove that can be efficient with a variety of pot sizes, using a selection of inserts provided with the stove so that the channel can be just right for a variety of pot diameters. It is recommended that a survey be made of the pot diameters in common use in the local area before the cookstove design is made final.

In order to sell fuel-efficient cookstoves, local artisans must not only be able to produce them, but people must want to buy them and must have the means to do so. In addition to determining the usual cook pot size in the market area, it is useful to ask potential customers what they want in a cookstove and how much they think they would be willing to pay. Market surveys in some countries show most people want a stove that can cook food quickly and use less fuel. The selling price of the stove described below is approximately US$3.00 (1987) in one West African country, a price people were willing to pay.

# Cookstove Design

If you plan to make more than one or two cookstoves, it is best to make templates (patterns) for the stove parts first. The templates shown in Figures 2, 3, and 4 will produce stoves suitable for spherical or cylindrical cook pots. Templates may be made of cardboard, plywood, or, better yet, sheet metal.

The stove presented here requires some welding and the use of concrete reinforcing rod (re-rod) as the pot support. Other designs, equally efficient, may use rivets or hammered seams and pot supports made of the same material as the stove.

1. The length of the template is given by

$$\mathbf{L = C + G + S + T}$$

C is determined by the measure of the pot around its widest circumference. G is determined by the desired pot-to-wall gap, G = 2pi. For a gap of 4 mm, G = 2.5 cm; for 6 mm, G = 3.8; for 8 mm, G = 5.0 cm. A gap of four to six mm (3/16-1/4") is preferred. Increase it only if excessive smoke comes out the door or the heating rate is too slow. S is determined by the amount of overlap in the seam. It is preferable to weld the stove together end to end (thus S = 0) to prevent the creation of a small vertical channel by which the heat can by-pass the pot. If the seam is crosswelded or folded, typical values for S will be 1 cm. T is determined by the thickness of the metal used. One typically uses 1 mm (T = 0.3 cm) or 1.5 mm (T = 0.47 cm) thick metal. Thus, for a 90 cm circumference pot, a 4 mm gap, an end-to-end welded seam, and 1 mm thick metal we find:

$$\mathbf{L = 90 + 2.5 + 0.3 = 92.8\ cm}$$

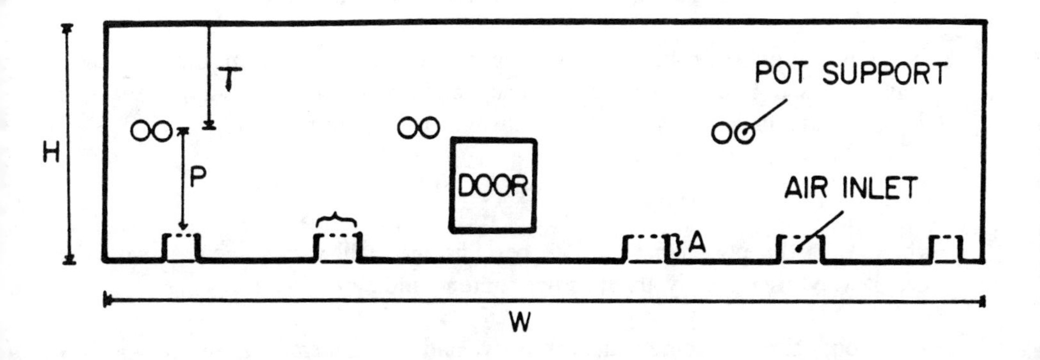


Figure 2. Template for a cylindrical metal stove shell

2. For spherical pots, template height **H** is determined by the sum of the airhole height (A), the grate-to-pot height (P), and the amount necessary to extend a few centimeters above the pot's maximum circumference when in place on the stove (T).

$$H = A + P + T$$

Typical values for A are 3 cm (1 13/16") and for P 0.4 of the pot diameter. For cylindrical pots the height **T** is typically 5 to 10 cm (2 to 4"). The best height **T** is determined more precisely by comparing the increased efficiency and reduced fuel use caused by the additional height versus the increased cost of the extra metal. Additional height can also be provided at the top and bottom of the template, typically 1 cm (3/8") each, to allow the edge to be folded over to protect against sharp edges and increase the stove's rigidity and strength.

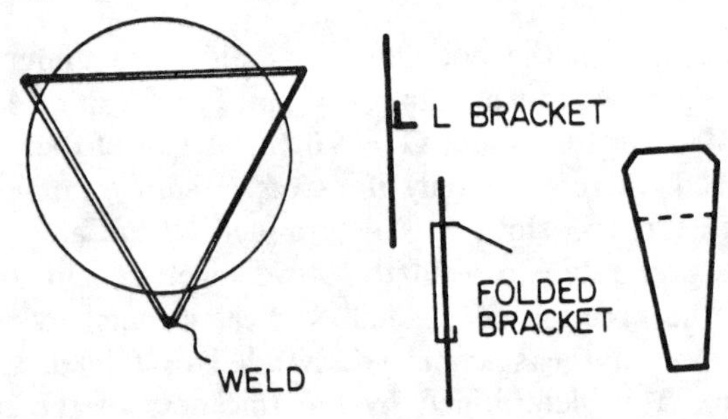


*Figure 3. Templates for a folded pot support, a welded L-bracket, and a complete triangular support of re-rod. Supports should be kept small so they don't keep heat away from the pot.*

3. Stoves usually have four air holes, about 3 cm by 3 cm (1 13/16" by 1 13/16") each (A = 3 cm). Space them symmetrically, but far enough away from the door and the seams to avoid weakening the stove. Cut the airholes on two sides only so that when bent upward and inward they can act as supports for the grate. for larger pots or soft soil where the stove will sink in, larger airholes may be necessary. Alternatively, for soft soil conditions a ring-shaped platform can be cut and attached to the stove.

4. Space pot supports evenly around the stove, but offset from the door and edges so as not to weaken them. The height **P** for the pot supports above the top of the air holes (where the rate will rest) is given roughly by

$$P = 0.4C/pi \text{ or } 0.4D$$

where **D** is the pot diameter. The best distance will vary somewhat with the size of wood used locally, its moisture content, and other factors.

5. The door size is somewhat arbitrary and is determined by the locally available wood size. Typical sizes for a 90 cm (35") circumference pot are 12 cm wide by 9 cm high (4 3/4" x 3 1/2"). Place the bottom of the door at the grate position-the top of the air holes. Make the top of the door several

centimeters below the bottom of the pot so that the hot gases are guided up around the pot rather than out the door. If necessary, decrease the door height to ensure that it is below the bottom of the pot.

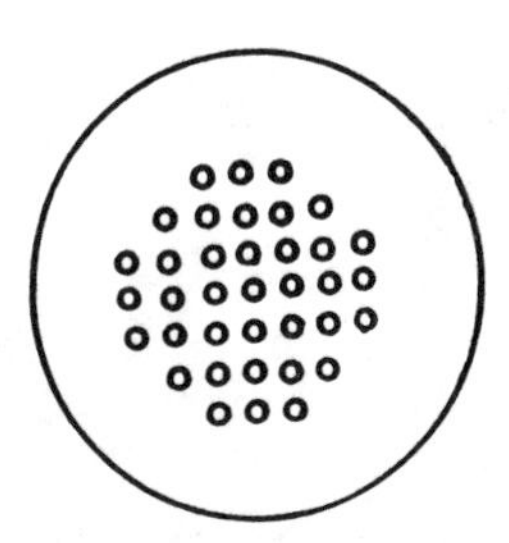

6. The grate is a circle of sheet metal cut to fit snugly into the finished cylinder. Punch the center diameter with a 30 percent hole density of 1 cm (3/8") holes.

*Figure 4. Template for a grate. Grate holes are not to scale.*

## Producing the Cookstoves

The stoves can be produced in villages in nearly all countries by metal working artisans with modest skills.

**Tools and Materials**

Sheet metal shears
Ball-peen hammer
Hole punch
Anvil
Welder
Sheet metal, approx. 1 mm (.04") thick (2 to 3 stoves per sq. meter)
Heavy wire (for handle)
Heat resistant paint (optional)

To produce stoves in quantity:

- Trace the template on a sheet of metal as many times as desired or as space permits.
- Cut each form out in outline. Cut the door, pot support holes, and strips for the airholes.
- Roll the metal into a cylinder. The cylinder should be as straight and smooth as possible.
- Cut out other components such as pot supports and stabilizers and put them into place.

- Cut the grate and punch the holes in it.
- Weld the stove together. Weld pot supports into place. Alternatively, fold all seams together. Hammer smooth.
- Place the grate in the stove, fold the tabs from the airholes inward and upward.
- Paint it with heat resistant paint where available.
- Add wire loop if desired to lift stove.

The finished stoves are shown in Figures 5 and 6.

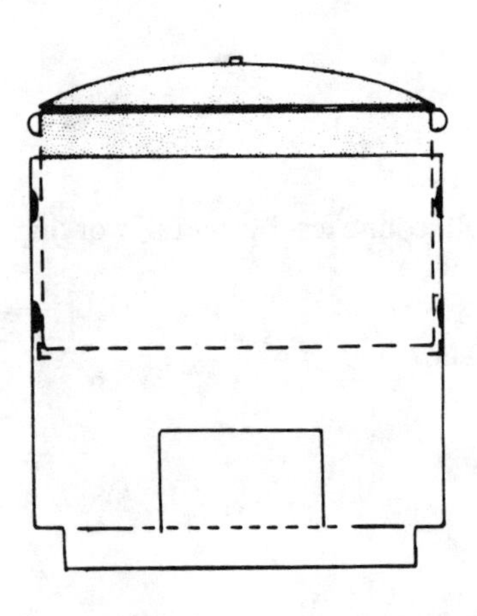

*Figure 5. Cross-section of the metal stove showing how the pot fits down inside.*

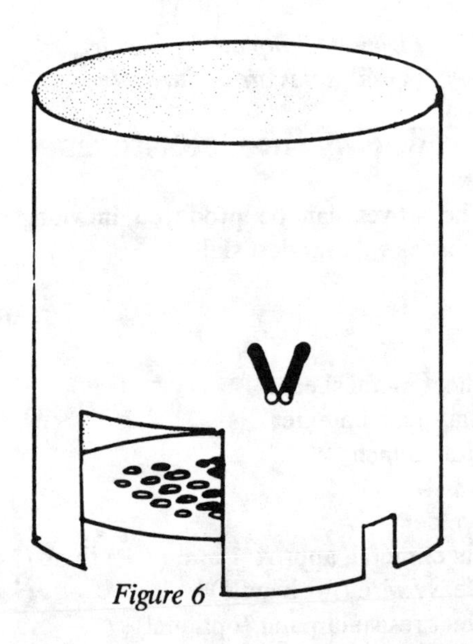

*Figure 6*

**Source:** Sam Baldwin, VITA Volunteer, Princeton, New Jersey.

# OUTDOOR OVEN

An outdoor oven is easy to build and good for baking bread, potatoes, beans, cereals, cakes, and other foods.

### Tools and Materials

Adobe blocks or brick: 35cm x 25cm x 10cm (14" x 10" x 4")
Wood or metal for door and smoke hole covers
Clay or cement for plastering

Lay bricks on the ground to make a base, 120cm x 120cm square and 30cm high (4' x 4' x 1'), on which to build the oven. After making the base, build the oven walls in an oval shape as shown in Figures 1 and 2. Lay the bricks flat and lengthwise starting from each side of the door opening, using the center of the square base as a guide. To form the dome shape and oval door opening, cut the corners of the bricks as you lay them. The inside space should be about 75cm (30") in diameter and 90cm (3') high. Leave a front opening for the oven door and a small opening at the top to let the smoke escape (see Figure 2).

Now make wooden or metal covers to fit tightly over the door opening and the smoke-hole (see Figure 3). These should be tight-fitting so that hot air will not leave the oven when the openings are closed.

FIGURE 1. THE OUTDOOR OVEN BEFORE BEING PLASTERED.

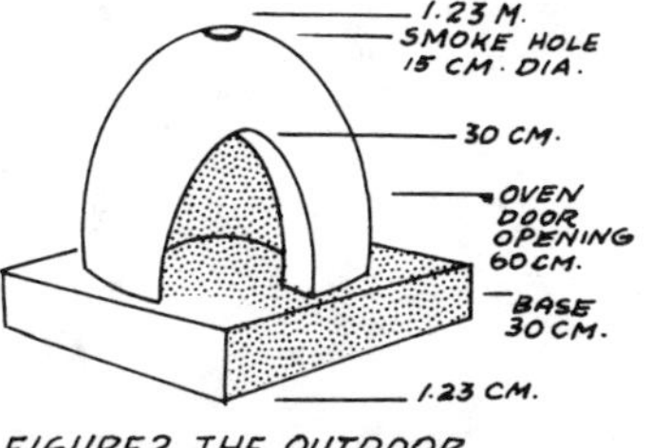


FIGURE 2. THE OUTDOOR OVEN AFTER BEING PLASTERED.

FIGURE 3. TIGHT FITTING WOODEN OR METAL COVERS ARE USED TO COVER DOOR OPENING AND SMOKE HOLE.

Plaster the inside and outside with a clay mixture or cement. The stove should be re-plastered at least once a year.

With the door and smoke-hole open, build a fire in the oven.

When the fire has burned to ashes, sweep out the ashes.

Put the food to be cooked inside the oven. Use trays or be sure the oven floor is very clean.

Cover the door opening and smokehole tightly.

Experience will teach how long food should be cooked. Bread, for example, can be expected to take an hour to an hour and a half.

This type of oven was used traditionally in many areas of Europe, the south-western United States, and in villages throughout South Asia.

**Source:**

*Home Making Around the World.* Washington, D.C.: U.S. Agency for International Development.

# Home Soap Making

Soap is an essential cleaning agent, helping people to keep themselves and their surroundings clean. When soap is mixed with water, it forms a lather that washes out dirt and grease far better than water alone.

Soap can be made on a small scale in the home or village cheaply and easily. The main ingredients are fats and lye, both of which can be made from materials found throughout the world. Making soap at home is practical when there is waste fat or oil and when there is no cheap source of soap.

## TWO BASIC METHODS

The two basic methods for small-scale soap making are:

Method 1. With commercial lye: This method is used when commercially-prepared lye or caustic soda (sodium hydroxide crystals) is available.

Method 2. With lye leached from ashes (potash): This method is patterned after a process used by early settlers of North America.

The first method, soap-making with commercial lye, is recommended because it is simpler and more reliable.

## INGREDIENTS FOR SOAP

### Fats and Oils

Soap can be made from either animal fat or vegetable oil. Mineral oil cannot be used. Animal fats commonly used are tallow, mutton, and lard. Vegetable oils used include coconut, palm nut, maize, olive, cottonseed, soybean, groundnut, safflower, and castor. Chicken fat, which is not a hard fat, is considered an oil. The best soap is made from a mixture of fat and oil.

o If you want a hard soap for use in hot water, use only tallow, made from melting rendered sheep, cattle, or horse fat.

o If you want a good laundry soap, use 1 part tallow to 1 part lard or cooking grease from melted hog fat, skin, and bones.

o If you want a fine toilet soap, use 1 part tallow to 1 part vegetable oil.

The best vegetable oils are made from crushing dried coconut meat, palm nut kernels, or the outer pulp of the palm nut. The last makes a harder soap than the coconut meat or kernels.

## Lye

Either commercially-prepared lye, also called caustic soda or sodium hydroxide (NaOH) crystals, or lye leached from ashes, called potash, can be used. Caustic soda is cheap and is sold in the markets of most countries.

*** CAUTION ***

Lye is a corrosive poison. It can cause serious burns. Do not let the lye crystals, dust, or solution touch your skin. Wear rubber gloves and protective eye wear. If any crystals do touch your skin, wash with water and then with a vinegar solution. If any of them are swallowed, take as much vinegar, citrus juice, or rhubarb as possible and call a doctor.

*** KEEP LYE AWAY FROM CHILDREN ***

## Borax

Borax is not necessary for making soap, but it improves the soap's appearance and increases the amount of suds produced.

## Perfume

Artificial perfumes or essential oils are not necessary ingredients but they can be used to make a more pleasant soap, particularly if rancid fat is used. If soap is made from tallow, citrus oil or juice will improve its smell and help preserve it.

## Water

The best water to use is soft water. Water that is not too hard can be used, but if it is very hard it is best to soften it. Hard water contains mineral salts that hinder the cleansing action of soap. To soften hard water: Add 15ml (1 tablespoon) of lye to 3.8 liters (1 gallon) of hard water, stirring the water as it is added. Let the mixture stand for several days. Pour off the water from the top. This is the soft water for soap making. The water and particle mixture at the bottom of the container can be thrown away. Soft water can also be obtained by collecting rain water.

# SOAP MAKING WITH COMMERCIAL LYE

The directions given here will make 4.1kg (9 pounds) of good quality soap. But the amount can be changed as long as the techniques and proportions are followed.

**Equipment and Materials**

Bowls, buckets, pots, or tubs made of enamel, iron, or clay. Never use aluminum; lye destroys it.

Measuring cups of glass or enamel.

Wood or enamel spoons, paddles, or smooth sticks for stirring.

Wood, cardboard, or waxed containers for molding soap. The molds can be of any size but those that are 5cm to 7.5cm (2" to 3") deep are best. Gourds or coconut shells can also be used for molds.

Cotton cloth or waxed paper for lining the molds. Cut the cloth or paper into two strips: one should be a little wider than the mold and the other should be a little longer. This lining will make it easier to remove the soap from the molds.

A thermometer that ranges from -18$^{o}$ to 65$^{o}$C (0$^{o}$ to 150$^{o}$F) is helpful, but not necessary.

## Recipes

For 4.1kg (9 pounds) of soap:

Oil or clean, hard fat: 13 cups (3 liters) or 2.75kg (6 pounds)
Borax (optional): 57ml (1/4 cup)
Lye (sodium hydroxide crystals): 370g (13 ounces)
Water: 1.2 liters (5 cups)*
Perfume (optional), use one of the following;
- Oil of sassafras: 20ml (4 teaspoons)
- Oil of wintergreen: 10ml (2 teaspoons)
- Oil of citronella: 10ml (2 teaspoons)
- Oil of lavender: 10ml (2 teaspoons)
- Oil of cloves: 5ml (1 teaspoon)
- Oil of lemon: 5ml (1 teaspoon)

* Note: Some experienced soap makers prefer to use twice this amount of water (i.e., 10 cups) and to boil the solution for three hours. Your own experience and the amount of water and fuel you have available are your best guide.

For one bar of soap:

Oil or clean, hard fat: 230ml (1 cup)
Borax (optional): 5ml (1 teaspoon)
Lye (sodium hydroxide crystals): 23.5g (5 teaspoons)
Water: 115ml (1/2 cup)
Perfume (optional): a few drops

## How to Make the Soap

The fat used in making the soap should be clarified. To do this: put the fat in a kettle with an equal amount of water; boil this mixture. Remove the kettle from the fire and strain the mixture through a sieve or a piece of cheesecloth. Add 1 part cold water to 4 parts of hot liquid. Do not stir the mixture; let it stand until it cools. The clarified fat can then be removed from the top. To help in cleaning the fat, a sliced unpared potato can be added before the mixture is boiled.

Measure carefully the amount of fat required and melt it down in the kettle to be used for soap making.

Measure the amount of water required.

Measure the lye required.

To the water previously measured slowly add the measured lye. **For safety always add the lye to the water; never add water to lye.** The resulting solution will become very hot and may spatter. Cool the lye mixture down to a body temperature. To test when the solution has reached body temperature, place your hand under the vessel holding the lye solution: there should be no noticeable difference between the temperature of your hand and that of the vessel. **DO NOT PUT YOUR FINGER IN THE SOLUTION.**

Cool the melted fat to body temperature. If borax is used, add it to the fat when it has cooled.

Then add the lye mixture to the melted fat. The lye mixture should be poured into the fat very slowly in a small stream. As this is being done the whole mixture is stirred slowly and evenly in one direction. After the lye solution is added, the mixture is stirred until the spoon makes a track. This usually takes about 30 minutes. After this let the mixture stand, stirring it once or twice every 15 or 20 minutes for several hours. When the mixture is very thick and honey-like in consistency, pour it into the molds lined with cloth or waxed paper (see Figure 1).

Cover the mold and let it set for 48 hours. Keep it dry and at room temperature. If it is moved or struck while it is setting, the ingredients may separate.

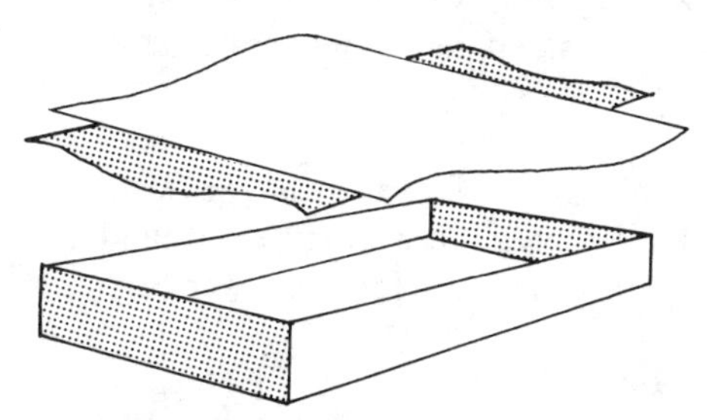

FIGURE 1. LINE THE MOLD BOX WITH TWO STRIPS OF COTTON CLOTH OR WAXED PAPER TO MAKE IT EASY TO REMOVE THE SOAP.

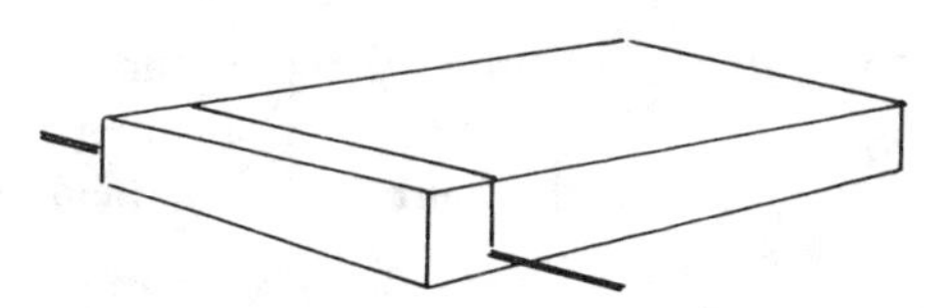

FIGURE 2. WHEN THE SOAP IS FIRM, REMOVE IT FROM THE MOLD AND, USING A THIN WIRE OR KNIFE, CUT IT INTO BARS.

At the end of this period, the soap should be firm and can be removed from the mold. If it is not firm, let it set longer until it is.

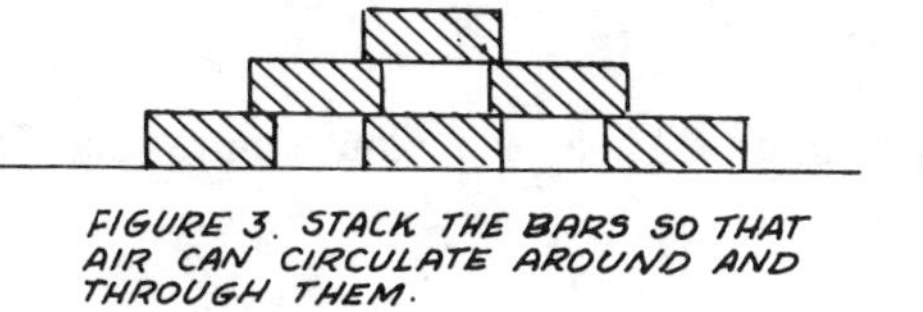

FIGURE 3. STACK THE BARS SO THAT AIR CAN CIRCULATE AROUND AND THROUGH THEM.

If grease is visible on the top of the soap at the end of the 48-hour curing period, the soap should stand a while longer. If there is liquid at the bottom of the box, cut the soap into bars and let them stand a day or two to see if the liquid will be absorbed.

## How to Know Good Soap

The soap should be hard, white, clean smelling, and almost tasteless. It should shave from the bar in a curl (see Figure 4). It should not be greasy or taste harsh when touched by the tongue.

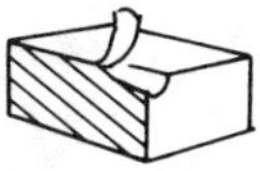

FIGURE 4. WHEN THE SOAP IS COMPLETELY CURED, IT SHOULD SHAVE FROM THE BAR IN CURLS.

## Reclaiming Unsatisfactory Soap

If some of the ingredients are still separated after this curing period, if the soap is curdled or grainy, or if you want a finer, smoother soap, do this:

Cut the soap into small pieces and put it into a pot with 2.8 liters (12 cups) of water and any liquid left in the molding box. Avoid touching the soap with your hands by wearing rubber gloves if possible, as there may be some free lye on the surfaces of the pieces of soap.

Bring it slowly to a boil and boil for 10 minutes, stirring occasionally. If you wish, you can add 10ml (2 teaspoons) of wintergreen, lemon, or other oil at this stage for perfume. Pour into a mold box, let stand 48 hours, and follow the procedure below.

Empty the soap from the box and cut it into bars with a string or wire (see Figure 2). Place the bars in an open stack so that air can circulate around and through them (see Figure 3). Leave them in a warm, dry place for 2 to 4 weeks.

**References:**

Bramson, Ann, *Soap.* New York.: Workman Publishing Co., 1975

Donkor, Peter, *Small-Scale Soapmaking.* London: Intermediate Technology Development Group, 1986

Francioni, J.B. and Collings, M.L. *Soap Making.* Extension circular 246. Baton Rouge, Louisiana: Louisiana State University, 1943

*Making Soaps and Candles.* Pownal, Vermont: P.H. Storey Communications Inc., 1973.

# SOFT SOAP WITH LYE LEACHED FROM ASHES

This method, patterned after one used by the early settlers of North America, produces soft soap by combining fat and potash (lye obtained by leaching wood or plant ashes.) The recipe has been tried successfully with waste cooking grease, olive oil, peanut oil, and cocoa butter.

## Leaching the Lye

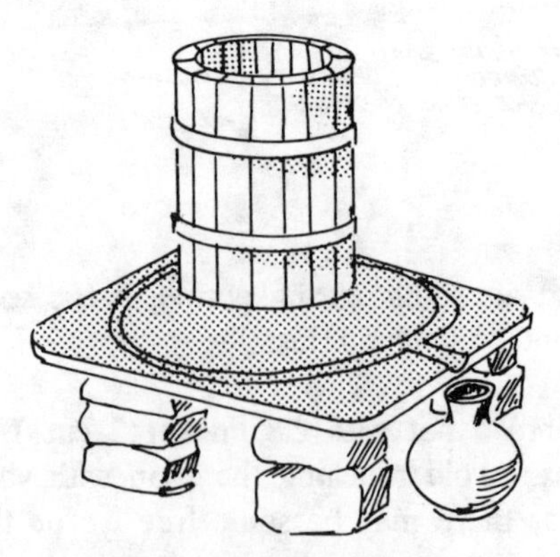

FIGURE 5. ROCKS ARE PILED TO FORM A BASE FOR THE FLAT, GROOVED STONE ON WHICH THE BUCKET IS PLACED.

**Tools and Ingredients**

Several medium sized rocks
A flat stone with a groove and a run-off lip chipped into it.
19-liter (5-gallon) wooden bucket with several small holes in the bottom. A hollowed log with the same capacity can be used.
Collection vessels for the lye. These should be made of iron, steel, enamel, or clay. An aluminum vessel should not be used, since lye would corrode it.
Small twigs, straw

19 liters (5 gallons) of wood ashes. The ashes may be from all types of woods. Ashes from hardwoods yield the best lye, but ashes from the burning of plants and leaves of trees may be used (see Table 1). Ashes of burnt seaweed are particularly useful as these produce a sodium-based lye from which hard soap can be made. Lye leached from the ashes of plant life (excepting seaweed) is potash or potassium carbonate ($K_2CO_3$), an alkali. This alkali reacts with fat to form soft soap. Ashes from other materials such as paper, cloth, or garbage cannot be used.

7.6 liters (2 gallons) of soft or medium-hard water.

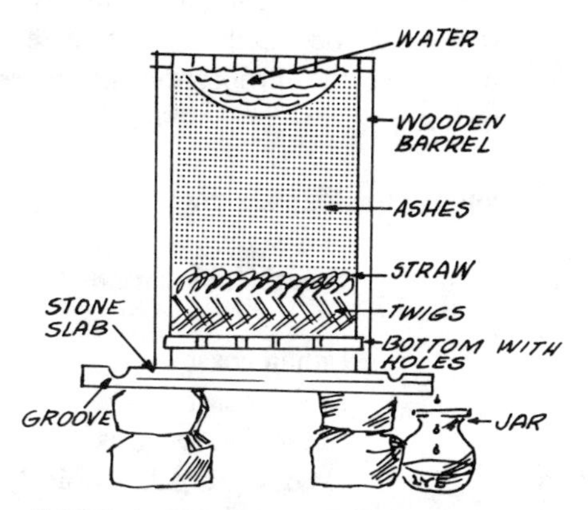


FIGURE 6. TWO LAYERS OF SMALL TWIGS ARE CRISS-CROSSED TO FORM A FILTER IN THE BOTTOM OF THE BUCKET. WHEN THE BUCKET IS FILLED WITH ASHES, WATER IS POURED IN AND THE LYE SOLUTION, A BROWN LIQUID, SLOWLY DRIPS INTO A CONTAINER.

Pile the rocks so that the flat, grooved stone rests evenly on top (see Figure 5). Set the wooden bucket on this stone.

In the bottom of the bucket, make a filter to trap the ashes by criss-crossing two layers of small twigs and placing a layer of straw on top (see Figure 6).

Fill the bucket with dry ashes. To keep the lye from being leached accidentally, the ashes must be kept dry before they are used.

Pour warm water into the bucket, making the ashes moist and sticky. To make sure that the water passes through the ashes at the correct rate for leaching the lye, move the ashes up at the sides of the bucket to form a depression in the center.

Add all the remaining water in small amounts in the following manner: Fill the center depression with water; let the water be absorbed; fill the depression again. When about two-thirds of the water has been added, the lye or potash, a brown liquid, will start to flow from the bottom of the bucket. Use more water, if necessary, to start this flow. The lye flows over the flat stone into the groove and then into the collection vessel below the run-off lip. It takes about an hour to start the flow of lye.

The yield from the amounts given here is about 1.8 liter (7 3/4 cups) lye. The results vary according to the amount of water loss from evaporation and the kind of ashes used.

If the lye is of the correct strength, an egg or potato should float in it. A chicken feather dipped in the solution should be coated, but not eaten away. If

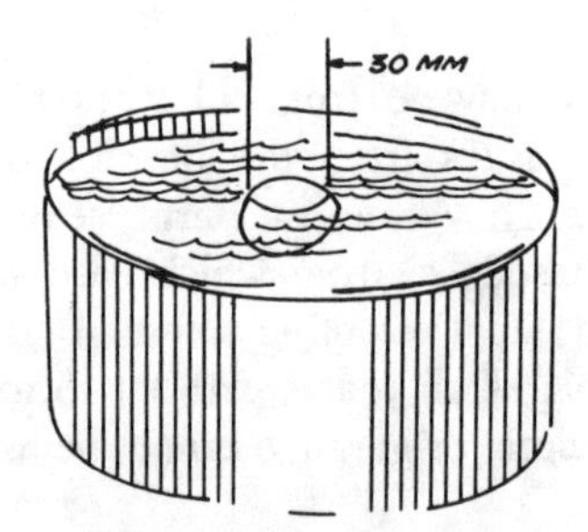


FIGURE 7. THE STRENGTH OF A LYE SOLUTION CAN BE TESTED BY FLOATING AN EGG IN IT. A SOLUTION OF THE CORRECT STRENGTH WILL RAISE PART OF THE EGG OUT OF THE WATER, FORMING AN OVAL ABOUT 30 MM LONG AT THE WATER'S SURFACE.

the solution is weak, pour it through the barrel again, or through a new barrel of ashes, or concentrate it by boiling. Thirty-five liters of ashes is about the right amount for 2 kilograms of fat (a bushel of ashes for 4 pounds of fat). This proportion is cited in soap-making recipes of the colonial period in the United States, but many of the recipes of that era differ on the proportion of ashes to fat.

Here is a list of tropical plants whose leaf ashes yield lye for soap making:

| Scientific Name | Common Name | Prominent location |
|---|---|---|
| *Arthrocnemum indicum* | mangrove | Indian coast |
| *Atriplex repers* | salt bush | Indian coast |
| *Avicennia nitida* | mangrove | Philippino swamps |
| *Cocos nucifera* | coconut palm | Coasts of all tropical regions |
| *Halocharis violocea* | | Indian coast |
| *Haloxylon recurm* | camel food | Indian coast |
| *Haloxylon multiflorum* | | Indian coast |
| *Haloxylon salicornicum* | | Indian coast |
| *Kochia indica* | | Indian coast |
| *Salicornia brachiata* | | Indian coast |
| *Salsola foetida* | Aden balsam | Indian coast |
| *Suaeda fruticosa* | | Indian coast |
| *Suaeda monoica* | | Indian coast |
| *Suaeda maritima* | | Indian coast |
| *Suaeda nudiflora* | | Indian coast |

# Making the Soap

### Equipment and Materials

Iron kettle
Wooden spoon or stick for stirring
Measuring vessels
Wooden, steel, iron, glass, or clay vessels for storing the soap

Clarified fat (see the entry on Soap Making with Commercial Lye for cleaning process)

Lye that floats an egg or potato (see Figure 7)

Put 115ml (1/2 cup) of lye in the kettle for every 230ml (1 cup) of fats or oils.

Add the measured amount of fat.

Boil the lye and fat together until the mixture becomes thick, rubbery, and foamy.

Remove the kettle from the fire and let it cool.

The soap is a thick jelly substance that ranges in color from tan to dark brown depending on the fats or oils used and the length of boiling time.

Upon strong mixing in water, the soap will lather up into white suds and serve as an effective cleaning agent. This soap greatly improves with age. Store it in a container for at least a month before using it.

230ml (1 cup) of fat yields 230ml (1 cup) of soft soap.

**Sources:**

Marietta Ellis, VITA Volunteer, Bedford, Massachusetts
Dr. S. K. Barat, VITA Volunteer, Adyar, Madras, India

Earl, Alice Morse. *Home Life in Colonial Days.* New York: MacMillan Company.

*Make Your Own Soap.* Washington, D.C: Federal Extension Service, U.S. Department of Agriculture.

## LARGER-SCALE SOAP PRODUCTION

In many areas in developing countries soap-making can be an important small business, providing a needed product and earning income with minimal investment. The Intermediate Technology Development Group, for example, has worked with the University of Science and Technology in Ghana to develop equipment for small manufacturing operations. One such set up uses specially made tanks heated by wood fires. The diagrams below show the parts for the tank. Soap-making processes are the same as those described above. Recipe quantities change according to the amount of soap produced. For example, one small manufacturer in Brazil supplied the following recipe for 45 kgs (100 lbs):

10 kgs tallow
2 kgs lye
2 kgs rosin
36 liters water

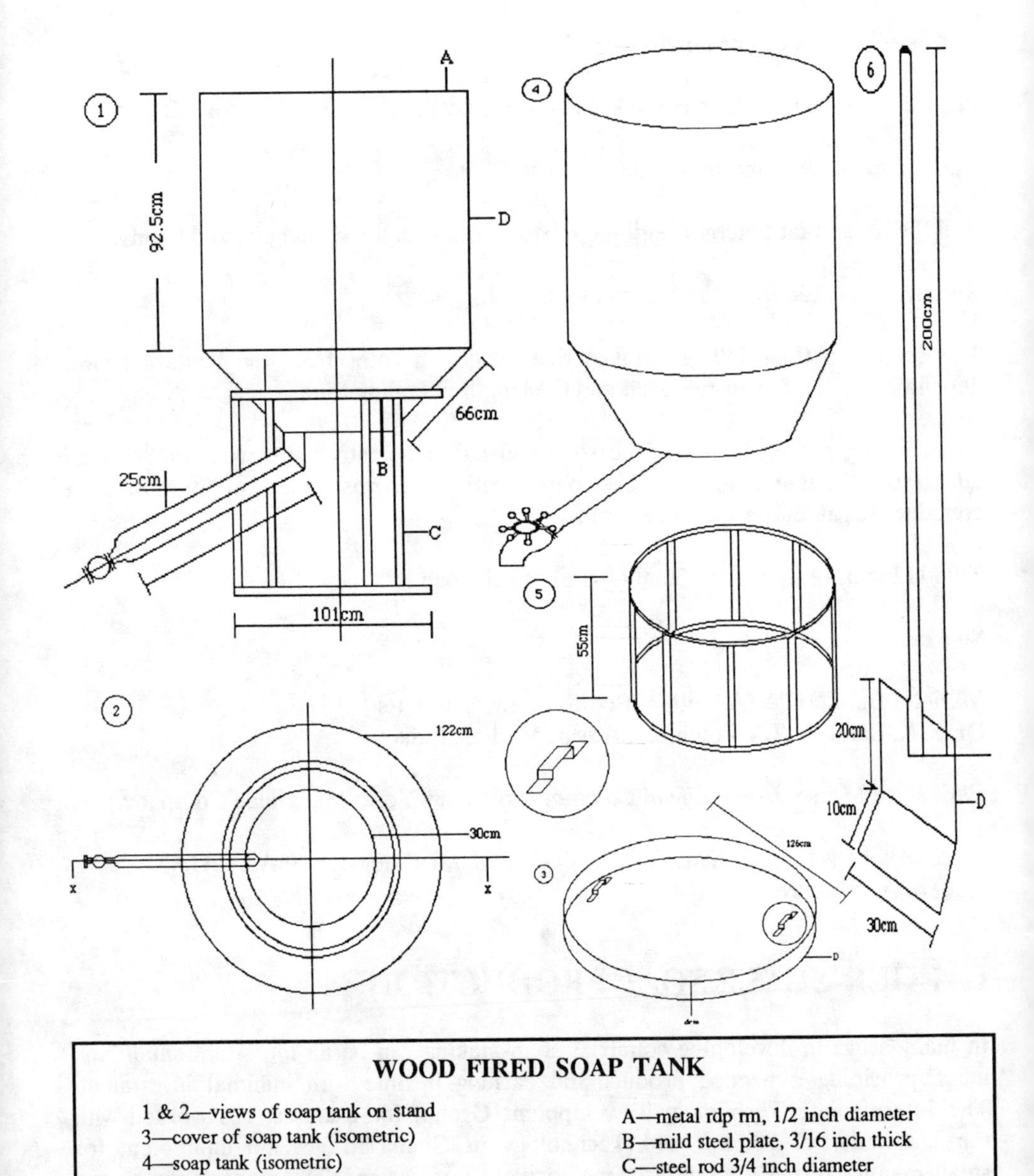


## WOOD FIRED SOAP TANK

1 & 2—views of soap tank on stand
3—cover of soap tank (isometric)
4—soap tank (isometric)
5—soap tank stand (isometric)
6—soap tank stirrer (isometric)

A—metal rdp rm, 1/2 inch diameter
B—mild steel plate, 3/16 inch thick
C—steel rod 3/4 inch diameter
D—galvanized steel gauge 16

**Sources:**

Donkor, Peter, *Small-Scale Soapmaking.* London: Intermediate Technology Development Group, 1986

Rezende Iriner, VITA correspondent, Recife, Brazil

# Bedding

## A NEST OF LOW-COST BEDS

This nest of three beds will save space in a small room during the daytime because it takes up only the space needed for one bed. The beds are low in cost and easy to make from local materials. Dimensions suggested here are approximate. The exact dimensions depend on the kind of wood used.

**Tools and Materials**

Carpenter's tools
Wooden boards 2.5cm x 7.5cm (1" x 3"), of varying lengths
Wooden posts 5cm x 5cm (2" x 2"), of varying lengths
Nails
Paint
Baling wire, burlap strips, rope, or wood for the "spring" of the beds.

All of the beds are the same width but the length and height of each bed varies so they fit under each other. The nest of beds can then be used as a sofa in the daytime (Figure 1).

FIGURE 1. THE BEDS CAN SERVE AS A SOFA IN THE DAYTIME.

The wood used in the largest bed is:

- o 2 boards, 2.5cm x 7.5cm x 183cm (1" x 3" x 72")
- o 2 boards, 2.5cm x 7.5cm x 91.5cm (1" x 3" x 36")
- o 4 legs, 5cm x 5cm x 51cm (2" x 2" x 20")

Nail the legs to the ends of each of the 91.5cm (36") boards. Then join these boards by nailing the 183cm (72") boards to them as in Figure 1. This completes the framework, which is now ready for the spring to be attached.

The spring can be made by nailing baling wire, burlap strips used as webbing, or wood to the frame. Another method is to bore holes in the framework and pass rope through the holes as shown in the middle-size bed in Figure 2.

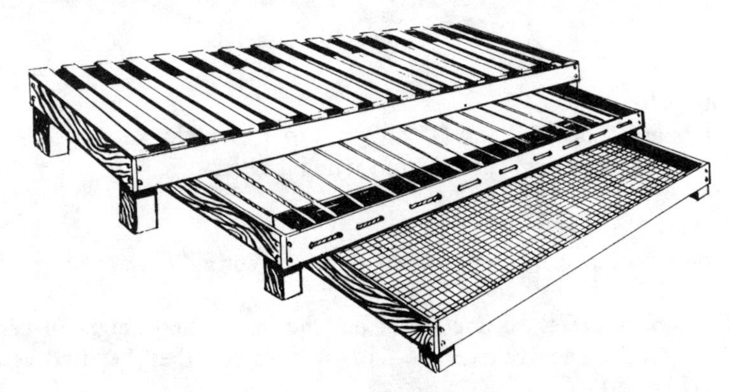

FIGURE 2. THE THREE BEDS ARE MADE TO FIT ONE UNDER THE OTHER, SAVING ON SPACE WHEN NOT IN USE. THREE DIFFERENT KINDS OF SPRING ARE SHOWN: WOOD ON THE LARGEST BED, ROPE ON THE MIDDLE-SIZE BED AND CHICKEN WIRE ON THE SMALLEST BED.

The other two beds are made the same way. They use the following materials:

Middle-size bed:

- o 2 boards, 2.5cm x 7.5cm x 168cm (1" x 3" x 66")
- o 2 boards, 2.5cm x 7.5cm x 91.5cm (1" x 3" x 36")
- o 4 legs, 5cm x 5cm x 38cm (2" x 2" x 15")

Smallest bed:

- o 2 boards, 2.5cm x 7.5cm x 152cm (1" x 3" x 60")
- o 2 boards, 2.5cm x 7.5cm x 91.5cm (1" x 3" x 36")
- o 4 legs, 5cm x 5cm x 25cm (2" x 2" x 10")

# HOW TO MAKE A MATTRESS

This low-cost mattress is made from materials available in most areas. It can be used as a bed at night and as a sofa by day. The mattresses are widely used.

**Tools and Materials**

Cotton
Corn shucks, rice or wheat straw, hay, banana or palm leaves
Smooth, heavy cloth (ticking)
Strong needles
Waxed cord
Oil felt or double-thickness ticking cut in a round shape, for tufts
Hand paddle with small nails
Sharp knife

## Making the Mattress

The first step is to dip the corn shucks in boiling water and, while they are still moist, shred them into small strips with a hand paddle that has small nails in it. The tough top part of the shuck is then cut off with a sharp knife. When dry, the shredded corn shucks are ready for use.

FIGURE 3. CORN SHUCKS OR SIMILAR FILLING MATERIALS ARE STUFFED INTO THE COVER, WHICH IS MADE FROM SIX PIECES OF CLOTH SEWN TO FORM A BOX WITH SQUARE CORNERS.

Cut six pieces of cloth as follows:

- o two pieces the size of the bed, to make the top and bottom of the mattress.
- o two pieces 15cm (6") wide and the length of the bed for the mattress sides.
- o two pieces 15cm (6") wide and slightly longer than the width of the bed, for the ends of the mattress.

Sew the pieces together to form a box with rounded corners. Attach the bottom piece on just one side, leaving the bottom open for filling the mattress. Twelve feed sacks full of tightly-packed corn shucks are enough for a double-bed mattress. A single bed mattress needs less.

Pack the filling material into the cloth cover in even layers. Otherwise the mattress will be lumpy. After each layer, pull the bottom piece over the filling material and beat the mattress gently to distribute the material evenly. Then pull the bottom piece back and continue filling the mattress. When the mattress is filled, sew the bottom piece in place. If there are still high and low spots in the mattress, beat it gently again: hitting the high spots to drive the filler into the low spots. Only a few strokes should be needed.

## Making a Rolled Edge

A rolled edge will keep the cotton in place and help the mattress to hold its shape. Mark a faint line 6cm (2 1/4") in from the edge seam all around the mattress top. Mark another faint line 1.5cm (1/2") below the seam. Sew the two lines together with stitches about 1.5cm (1/2") apart, working enough filler into the roll with each stitch to make the roll firm. Fill the roll evenly. In rounding the corners, make the stitches closer and take shorter stitches on top of the roll than on the bottom.

Turn the mattress over and make a roll edge on the other side.

Use a strong needle and waxed cord to sew the round pieces of oil felt or doubled ticking for simple tufts that will hold the filling in place.

More detailed instructions are given in: *Making a Cotton Mattress,* Federal Extension Service, U.S. Department of Agriculture.

# Crafts and Village Industry

*Len Doak/Cover potter by Sam Baldwin*

# Pottery

## WASTE OIL FIRED KILN

Ceramic kilns that burn waste oil from automobiles and other industries have been operating in Tanzania, Haiti, and several other developing countries for several years. These kilns offer the advantages of good operational control that is easily achieved with fuel oil, but at lower fuel cost because waste oil is used.

The waste oil fired system presented here (Figure 1) was designed by Ali Sheriff and his assistant, Bashir Lalji, in Tanzania for Mr. Sherriff's pottery plant. Mr. Sherriff also helped entrepreneurs in Djibouti build and operate kilns for use in their brick making and pottery businesses.

## Cost Advantage of Waste Oil

Originally, waste oil was collected free from auto service stations and industries, but by 1983 $0.35 (US) per liter was charged. At these rates, it cost Mr. Sheriff US$105 for each firing of his six-burner kiln, compared with $165 for fresh oil.

Some alternative fuels such as electricity are too expensive in developing countries to be economically feasible for kilns. One alternative, wood, may be less costly than waste oil in some countries, but wood supplies are being reduced rapidly and costs are rising.

## Design of Kiln and Fire Box

The kiln shown in Figure 1 is a down draft type with three fireboxes on each side. The height of the chimney is determined by the intensity of the heat

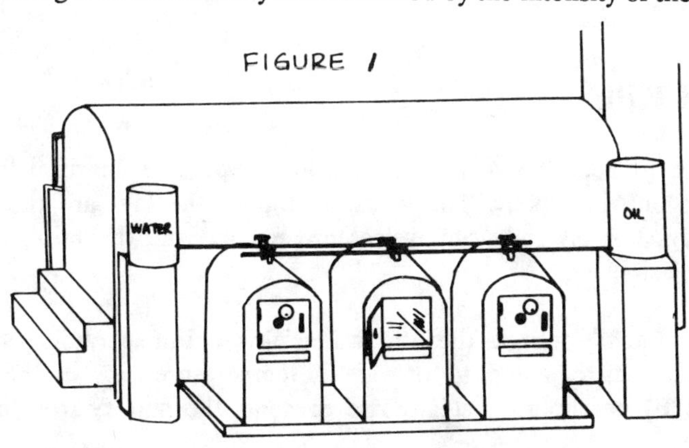

required. The hotter the fire, the higher the chimney. Other kiln designs may have more or fewer fireboxes depending on the size.

Fuel and water are metered by gate valves connected to the distribution pipes from their respective tanks. A ratio of about 75 percent waste oil and 25 percent water is about optimum.

*** CAUTION ***

Use only crankcase oil.

Under no circumstances should lubricating oils from electrical transformers be used for any purpose. They may contain PCB compounds.

Preheated splash plates serve as a grate to ignite the oil-water mixture. The grates, made of pieces of sheet steel (Figure 2), slope down so that any fuel not burned on the upper grate will spill off onto the lower grates for combustion.

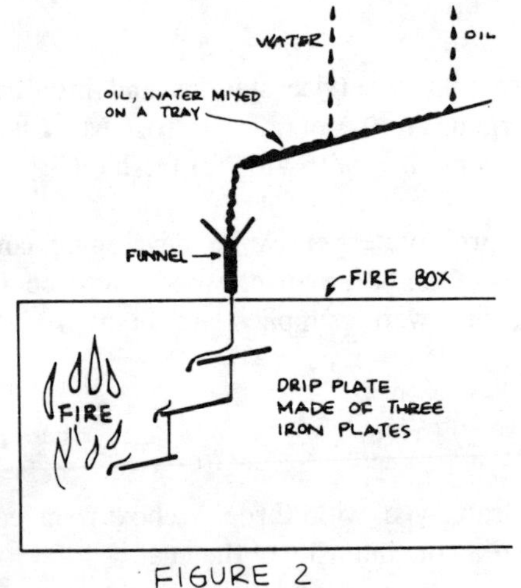

FIGURE 2

The waste oil must be treated before it can be used as fuel. The oil is first filtered through a screen of 60 mesh or finer to remove solid particles. It is then allowed to stand in a drum for a few minutes to let the water settle to the bottom. A tap at the bottom of the drum allows water to be removed.

# Operating the Kiln

The first step is to preheat the splash plates using a wood or charcoal fire. This should take about one half hour. The vents on top of the kiln are then closed with brick and clay. The oil and water valves are opened and the mixture should ignite on the hot splash plates.

Workers must continually check the fuel-water flows. For ceramics, the fuel burning rate should be regulated to provide a temperature rise in the kiln of about 100°C (212°F) per hour. A steady rise prevents the pottery from cracking.

In Mr. Sheriff's kiln, peak temperature is reached in about 18 hours. At this time, all entrances to the kiln, including the chimney, are closed and the kiln is allowed to cool slowly.

**Source:**

Sheriff, A. and Lalji, B. *Waste Oil Fired Kiln.* VITA Technical Bulletin. Arlington, Virginia: Volunteers in Technical Assistance, 1983

"Ceramic Kiln Burns Waste Oil," *VITA News,* April 1983, pp. 3-6.

# SMALL RECTANGULAR KILN

The small rectangular kiln was designed for both bisque and glaze firing of small pottery pieces. In bisque firing, pottery is cured but not glazed. It can be glazed either in the first firing or in subsequent firings. The kiln can be larger or smaller than the dimensions given here. Its capacity depends on the size of the base.

**Materials**

Common (pressed) brick
Firebrick (Note: Sandstone blocks were used before the invention of firebrick)
Clay or mortar
The dimensions shown in Figures 4 to 8 are based on the 23cm x 11.5cm x 6.5cm (9" x 4 1/2" x 2 1/2") straight brick commonly found in the United States. The dimensions can be changed to suit the size of locally available brick.

## Construction

The joints in the kiln, except for those in the loading area, should be mortared. The preferable mortar is a refractory cement; that is, one that is highly resistant to the action of heat. If there is a brick plant in the area, find out what material is used there. If refractory cement is not available, make it by mixing crushed firebrick with your purest clay, which will be white or light in color. As a last resort, use clay alone. In any case, have the mortar fill as much of the joint as possible. Each time the temporary door for loading is rebuilt it should be mortared with the purest clay available.

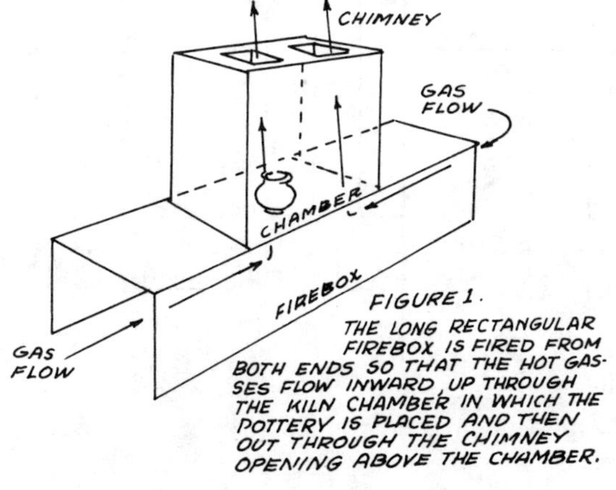


FIGURE 1. THE LONG RECTANGULAR FIREBOX IS FIRED FROM BOTH ENDS SO THAT THE HOT GASSES FLOW INWARD, UP THROUGH THE KILN CHAMBER IN WHICH THE POTTERY IS PLACED AND THEN OUT THROUGH THE CHIMNEY OPENING ABOVE THE CHAMBER.

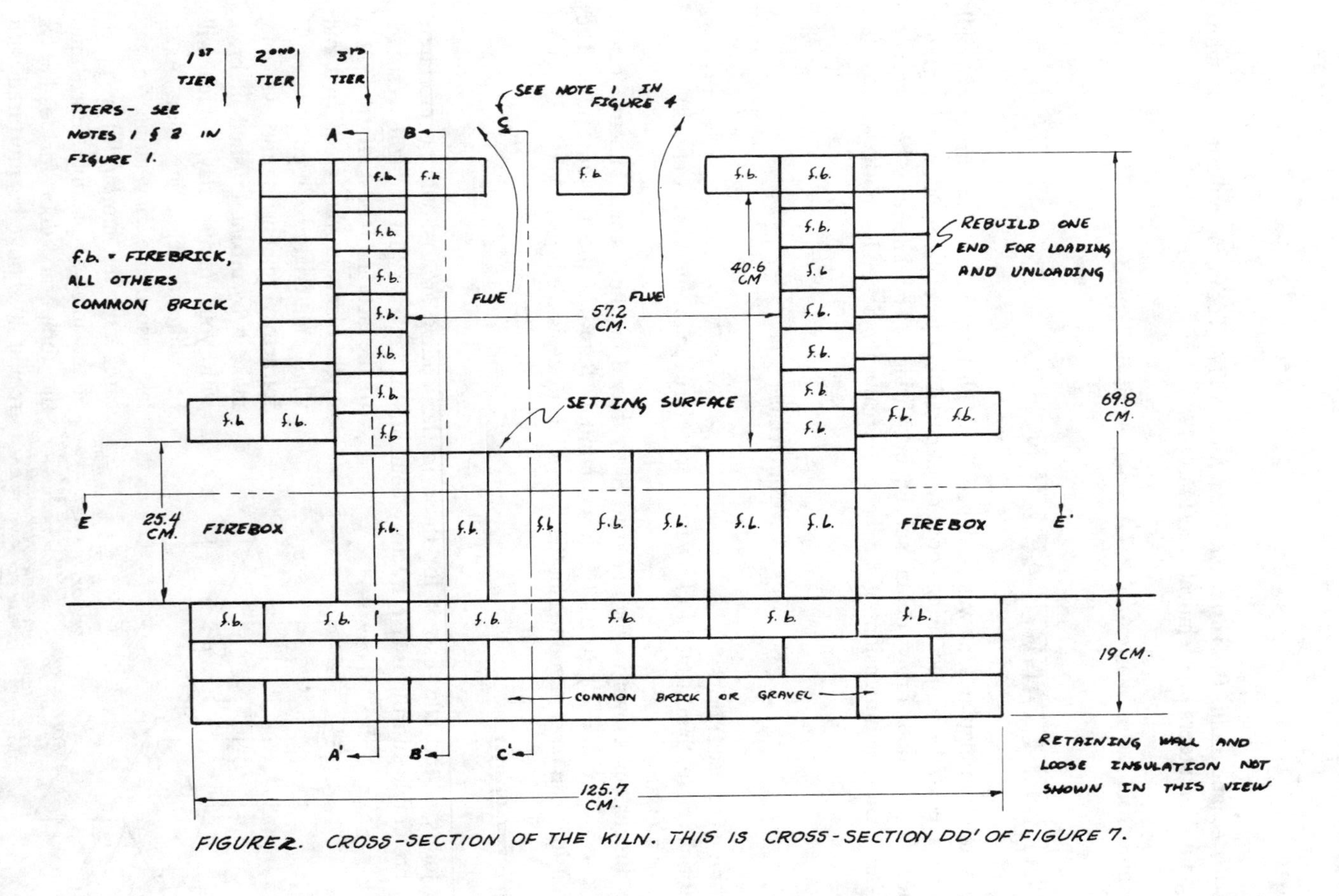


FIGURE 2. CROSS-SECTION OF THE KILN. THIS IS CROSS-SECTION DD' OF FIGURE 7.

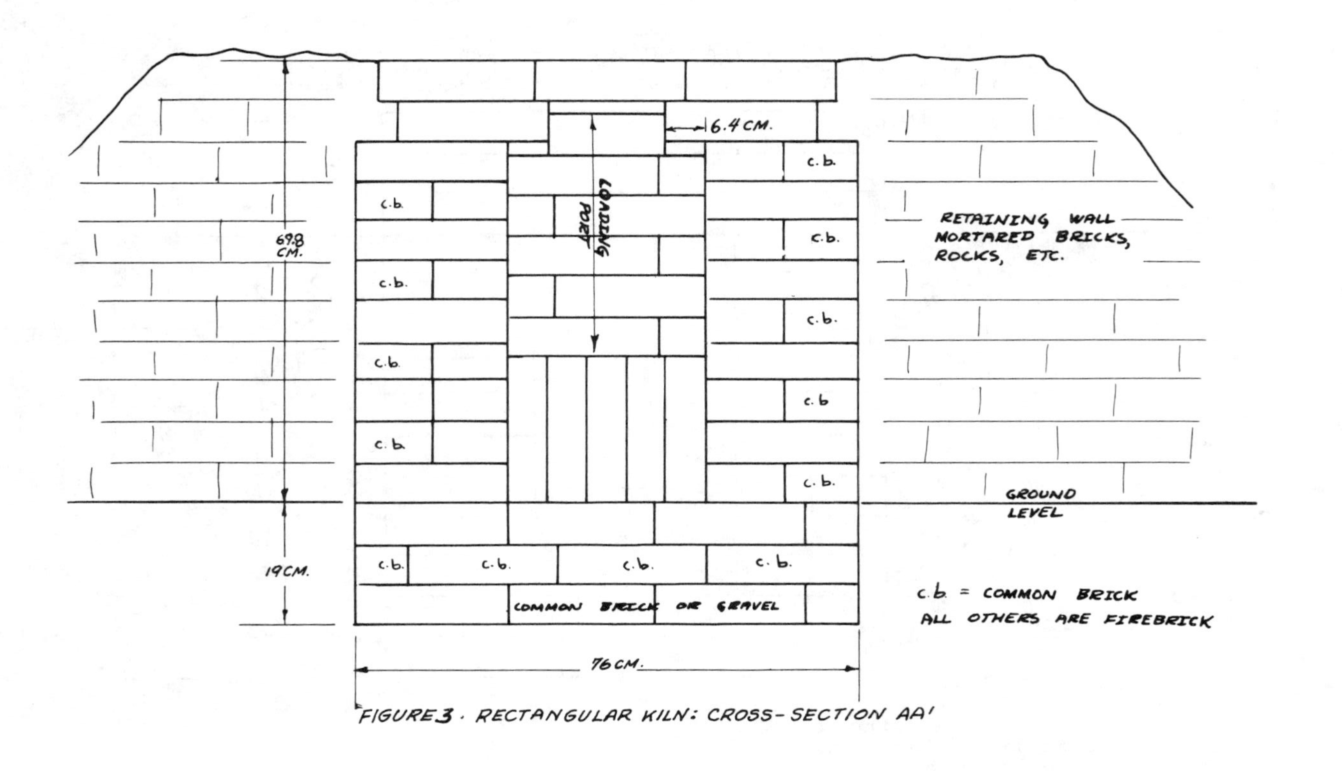


FIGURE 3. RECTANGULAR KILN: CROSS-SECTION AA'

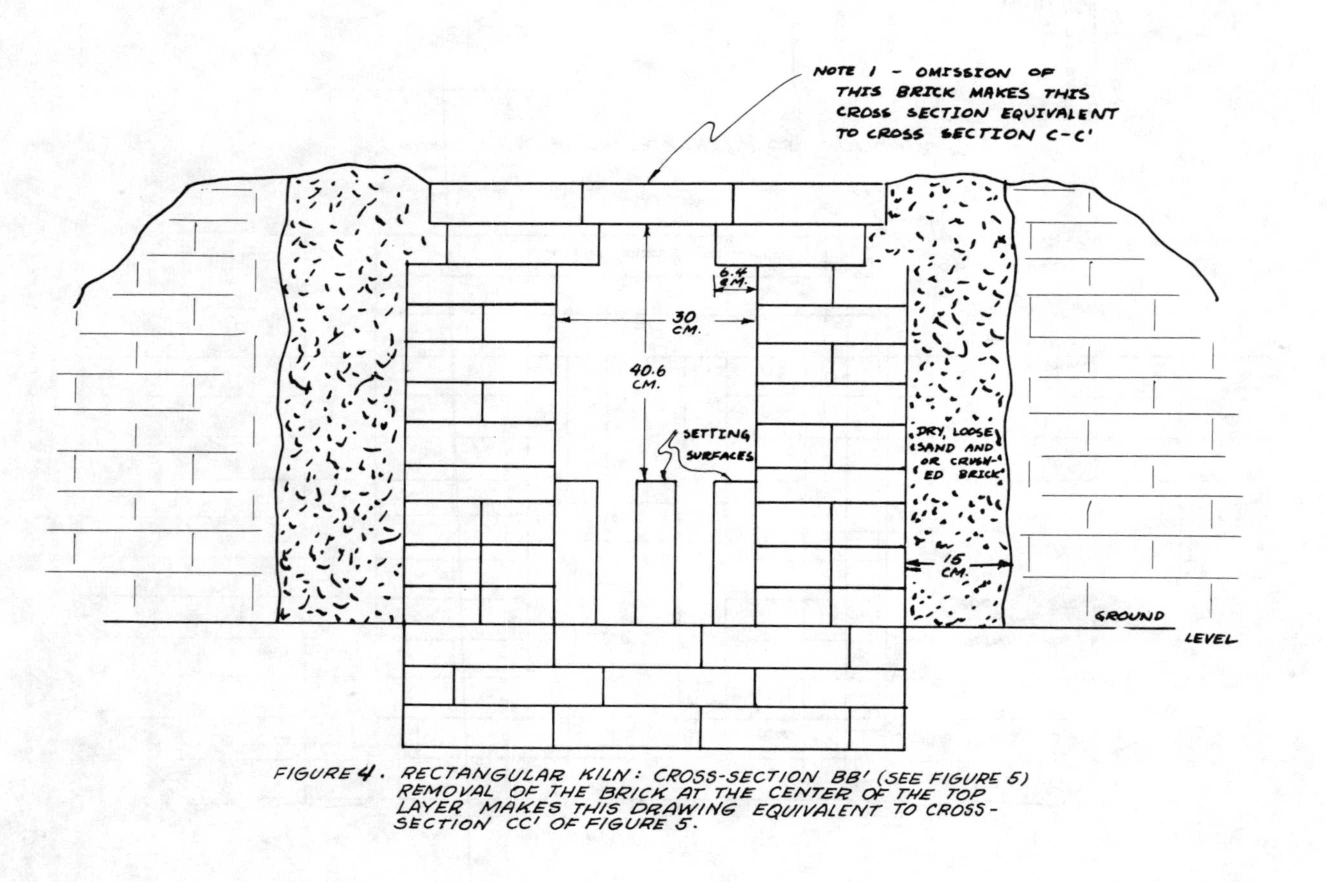


FIGURE 4. RECTANGULAR KILN: CROSS-SECTION BB' (SEE FIGURE 5) REMOVAL OF THE BRICK AT THE CENTER OF THE TOP LAYER MAKES THIS DRAWING EQUIVALENT TO CROSS-SECTION CC' OF FIGURE 5.

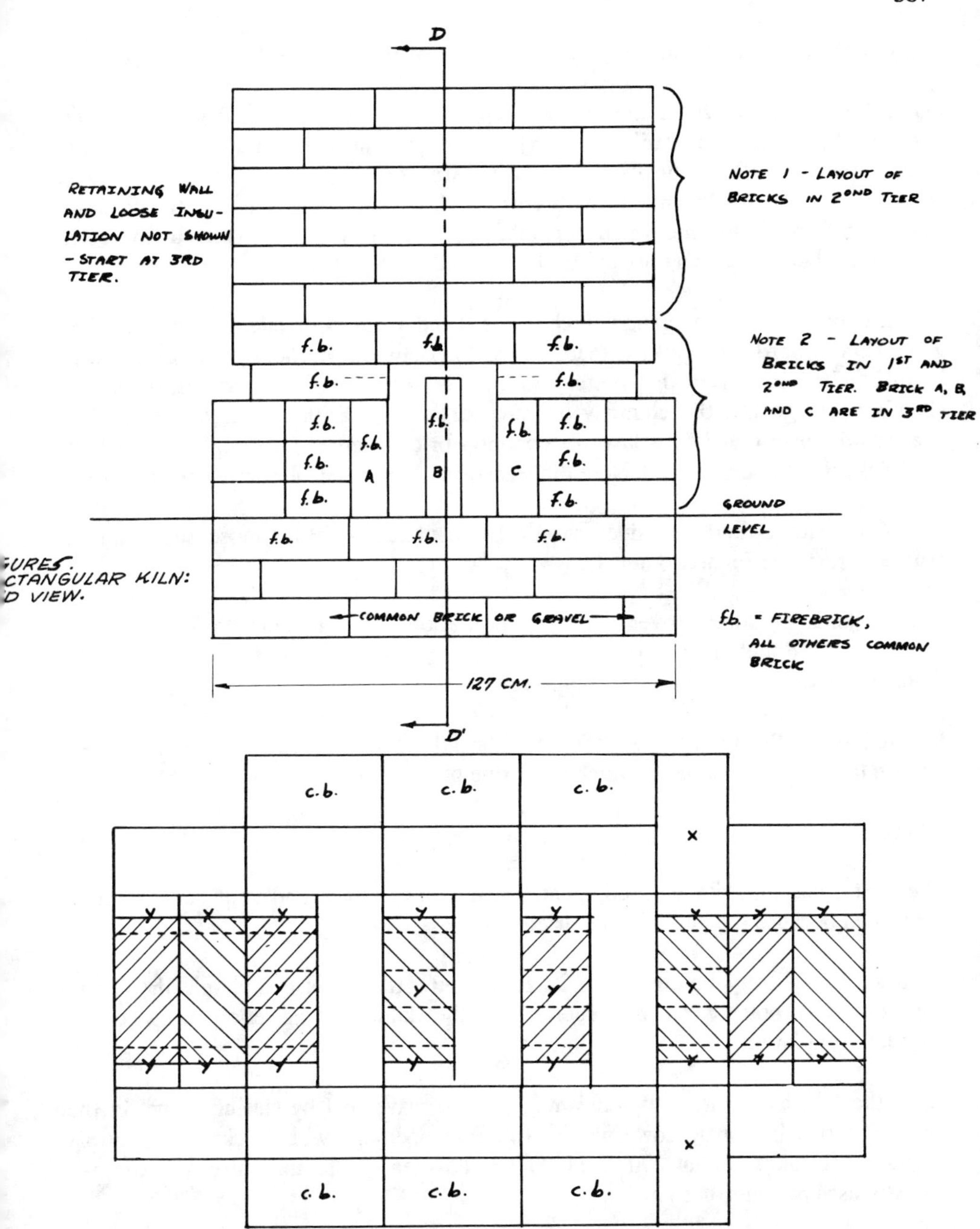


FIGURE 6. RECTANGULAR KILN: CROSS-SECTION EE' TOP VIEW (SEE FIGURE 3)

C.B. = COMMON BRICK, ALL OTHERS ARE FIREBRICK.
SHADED BRICKS ARE PLACED ON TOP TO FORM SHELVES AND TOP OF FIREBOX.
ALTERNATE LAYERS FORM MIRROR IMAGES - i.e. BRICKS "X" ARE ON LEFT.
BRICKS "Y" ARE STOOD ON END TO SUPPORT SHELVES

In laying the brickwork, stagger the joints in each layer to cut heat loss.

Dig a hole 76cm x 126cm and 19cm deep (30" x 49 1/2" x 7 1/2")–or whatever base size is needed for available brick–in level ground. Note in Figures 4, 5, and 7 that the first three horizontal courses are: first, gravel or common brick; second, common brick; and third, firebrick. This foundation is under the firebox. The firebox, with its end open for loading, is built with firebrick. If charcoal, coke, or coal are used as a fuel, the firebox should have grates.

The firebox is a long rectangular chamber, fired from both ends so that the hot gases flow inward and upward (see Figure 1). Between and above the fires is the kiln chamber in which the pottery is placed. The hot gases rise through the chamber and go out the chimney opening at the top. Both the chamber and the firebox are surrounded by a layer of common brick. Figures 4 through 8 show how the bricks should be arranged. Note the staggering of joints in alternate courses.

When the kiln is built, its sides should be insulated with dry loose sand and/or crushed brick (see Figures 3 and 4).

If the kiln is outdoors, cover the loose insulation and brickwork to keep it from getting wet. Sheet metal is suitable. If large pieces are not available, use flattened tin cans to build a shingle-type cover.

In building up the temporary door after the kiln is loaded (see Figures 3 and 4), be sure to leave a peephole to watch the inside of the kiln.

## Firing

The first time the kiln is fired, heat-up will take longer and require more fuel than usual because the kiln must be dried out.

Sunbake the pottery before firing it, to be sure that it is completely dry. Load the sun-dried pottery on the shelves of the kiln, leaving enough space for adequate ventilation.

After the kiln has heated up somewhat, you can save fuel by cutting down on the draft. Do this by partly covering the top flue openings with bricks. The pottery begins to shrink at about 870°C (1600°F). To measure temperature, the ceramic industry uses pyrometric cones.

If no temperature-measuring devices are available, the color of the glow in the inner mass of the kiln can indicate the approximate temperature of the kiln. See Table 1.

The kiln should be heated slowly to 870°C (1600°F). This process should take about eight hours. Chemical and physical changes caused during the heating of the

**TABLE 1**
**COLOR/TEMPERATURE GUIDE**

| °C | Color | °F |
|---|---|---|
| 475 C. | Lowest visible red | 885° F. |
| 475 - 650 C. | Lowest visible red to dark red | 885 - 1200° F. |
| 650 - 750 C. | Dark red to cherry red | 1200 - 1380° F. |
| 750 - 815 C. | Cherry red to bright cherry red | 1380 - 1500° F. |
| 815 - 900 C. | Bright cherry red to orange | 1500 - 1650° F. |
| 900 - 1095 C. | Orange to yellow | 1650 - 2000° F. |
| 1095 - 1315 C. | Yellow to light yellow | 2000 - 2400° F. |

The glow of the inner mass of the kiln gives a rough indication of temperature

kiln can destroy the pottery if they take place too quickly. For example, dehydration of clay and other minerals takes place throughout the whole temperature range, but particularly between 480°C (900°F) and 815°C (1500°F); organics and sulfides are oxidized between 595°C (1100°F) and 980° (1800°F).

Several hours at 870°C (1600°F) and higher are needed to complete the firing.

When the firing is completed and the fire is out, block the flue and firebox openings so that the kiln will cool slowly. Let the kiln stand this way overnight. When the temperature of the kiln has dropped, open the flue and firebox openings. This slow cooling keeps the pottery from being cracked by thermal stresses. Slow cooling through the dark red heat range is most critical.

The time and temperature required to fire an unknown clay can be learned only be experimenting. Heating and firing times may vary from those given here.

**Source:**

Irwin M. Lachman, VITA Volunteer, Corning, New York

Suppliers of temperature cones are:

The Edward Orton (Jr.) Ceramic Foundation
144 Summit Street, Columbus, Ohio USA

Bell Research, Inc.
Box 757, East Liverpool, Ohio USA

Bell Clay Co.
Gleason, Tennessee USA

# SALT GLAZE FOR POTTERY

This method can be used for applying a very thin, transparent glaze to pottery such as clayware and stoneware. Examples are: brick, sewer-pipe, stoneware shapes, and containers.

Open pieces, such as bowls, will become glazed inside and out. Narrow-necked pieces must be glazed inside by a slip-glaze method in which the pottery is dipped into the glaze.

## Considerations

Some ceramic articles will take a salt glaze. Others, under certain conditions, will not. Experimentation is the best way to discover how to glaze an unknown clay.

Common salt (NaCl) may be used alone, and this is common practice. Boric acid or borax may be added to the salt to improve the glaze and lower the firing temperature.

Salt glazing can be done in a wide range of temperatures, 670°C to 1360°C (1230°F to 2470°F); the more usual range is 1200°C to 1300°C (2185°F to 2375°F).

## How to Fire the Pottery

Place the pottery on the shelves of the kiln. The pieces should not touch so that there is plenty of room for ventilation.

Mix 9 parts salt with 1 part borax or boric acid. This mixture can be dampened with water: 5 to 10 percent by weight of the mixture. For ordinary fire-clay pottery, about 285 to 570gm (10 to 20 ounces) of salt is needed for 0.028 cubic meter (1 cubic foot) of kiln capacity.

When the kiln is as hot as it will get, throw the mixture into the fire heating the kiln.

This step may be repeated several times when the temperature gets back up to the hottest point. The kiln is then gradually cooled.

The sodium (Na) separates from the heated salt and combines with the clay body to form a very thin, uniform glaze that shows the colors of the ceramic body.

**Sources:**

Dr. Louis Navias, VITA Volunteer, Schenectady, New York

Parmalee, Cullen W. *Ceramic Glazes.* Chicago: Cahners Publishing Company.

# Hand Papermaking

In many areas of developing countries paper is scarce. Rural schools may not have enough paper for their students and market goods may be wrapped in old newspapers if at all. Often this is because resources are not available to invest in modern papermaking factories, which require large amounts of energy and raw materials if they are to be economical.

But paper can be made in small shops in small quantities. Access to electricity makes some of the steps easier, but is not absolutely necessary. (Indeed, paper was made this way for many years before electricity was discovered.) In a situation where paper is scarce and expensive, it may be worthwhile to consider small-scale papermaking as a source of school supplies or as a small business. Such a business might produce heavy coarse paper for packaging or even thick paper egg cartons, plant pots, and so on.

## PAPERMAKING PROCESSES

Whether paper is made in a home or school workshop or a small factory, the production processes for making paper by hand are quite similar. The scale of the equipment changes with the volume of production and the raw materials vary with what is available and the quality of paper to be produced.

### Pre-processing

Cotton or other rags and waste paper to be recycled are sorted thoroughly to remove all non-fibrous materials such as staples, paper clips, cellophane, nails, buttons, zippers, etc. Both rags and paper are cut or shredded into small pieces.

### Pulping

The cleaned and shredded raw materials are brought to the boiling point and cooked for two to six hours. They are rinsed thoroughly to remove impurities that might have separated out during the cooking process.

The beater–this can range from a kitchen blender to a specially made tank–is filled with the required quantity of water, and the cooked, chopped rags or paper are added gradually with high speed agitation. Bleaching powder or liquid bleach (1 percent) is then added. The pulp is washed thoroughly, a process that may take another six to eight hours. Additives used may include titanium dioxide or other fillers, dyes (for colored paper), or optical bleaching agents (for white paper). Rosin soap and alum are added later.

## Lifting, Couching, and Stacking

When the pulp process is complete, the pulp is transferred to storage containiers or vats. Depending on the scale of the operation, the pulp is then mixed with a sufficient quantity of water to dilute it to form a uniform suspension, free of lumps. In the home workshop, the pulp is mixed in quantities to make one sheet at a time. In the small factory, a larger quantity may be mixed at one time. The diluted pulp is then lifted from the water on wire screens, and the resulting sheets are covered by felt or other absorbent cloth. With the cloth is place, the still wet pulp layer is carefully lifted from the screen. This process is called couching (pronounced cooching). The couching cloth, paper side down, is placed on a felt covered board and smoothed to remove wrinkles or air bubbles. Each succeeding sheet is placed in a stack over the first.

## Pressing and Drying

When a sufficient number of sheets have been formed, they are put under a press to remove the water. The sheets are then separated and, to avoid shrinkage, placed under absorbent boards and pressed again. The sheets are hung to dry in bunches of three to six, according to thickness, or dried in a warm oven.

## Sizing

Sizing gives paper a harder finish so that water based paints and inks will not bleed or run. Paper may be sized internally, by adding the sizing agents to the pulp, or externally, by painting or dipping the dried sheets. For internal sizing, alum, rosin, gelatin, cornstarch, or linseed oil may be added in very small quantities at the end of the pulping stage. For external sizing, the dried sheets are dipped in a dilute glue or starch solution, pressed to remove the excess, and hung up to dry again. In the home workshop, the individual sheets may be painted with the dilute solution.

Blotting paper, filter paper, toilet tissue, grey board, and some art papers may be require very little, if any, sizing.

## Calendering

The dried sheets are placed alternately between metal plates into a stack or "post." The stack is passed between calender rollers to obtain the desired smoothness. This can be done in the home workshop by pressing the paper sheets between sheets of aluminum foil with a hot iron.

## Sorting and Cutting

After calendering, the sheets are carefully sorted and cut to size for packing, storage, and/or shipment.

# MAKING PAPER IN THE SMALL WORKSHOP

Papermaking at this scale can be done as hobby, for gifts, or to supply schools. The necessary equipment may already be available in some kitchens, but the markets should be considered carefully before any investment is made.

This process assumes that waste paper or cotton cloth will be used to make the paper. Approximately 50 sheets of 21.5cm x 28cm (8 1/2" x 11") paper can be made from a half kilo (about a pound) of waste paper. Household bleach, alum, gelatin, cornstarch, and animal glue may also be needed. And ordinary fabric dyes can be used to produce tinted or colored papers. As described here, the availability of adequate water and electrical or other power supplies is also assumed.

### Equipment and Materials

The following equipment is needed:

Deckle box and mold, made of oiled wood (figure 1)
Power food mixer or blender
Stainless steel or enamel pot (not almuinum)
Steam iron
Stove with oven
Sink, tub or wash basin
Couching cloth (e.g., cotton sheeting), cut to size
Felt or absorbent terry cloth, cut to size
Thin metal sheet
Flat "receiving" board, 1cm (1/4") plywood or other board

## Pulping

Choose paper with minimal printing. Old envelopes are good for this reason; the glue on the flap won't matter. Colored paper is acceptable; the dye usually comes out when it is boiled. Avoid paper that has "wet strength" such as paper towels. Be careful how many brown paper bags you use. Unbleached kraft paper lowers the brightness or whiteness of the pulp, but it is strong and will give your paper toughness.

Newsprint alone makes a weak pulp, grey in color. It adds little but bulk. Cotton or other cloth or yarns may also be used. They must be cut or shredded into very small pieces to avoid jamming the mixer.

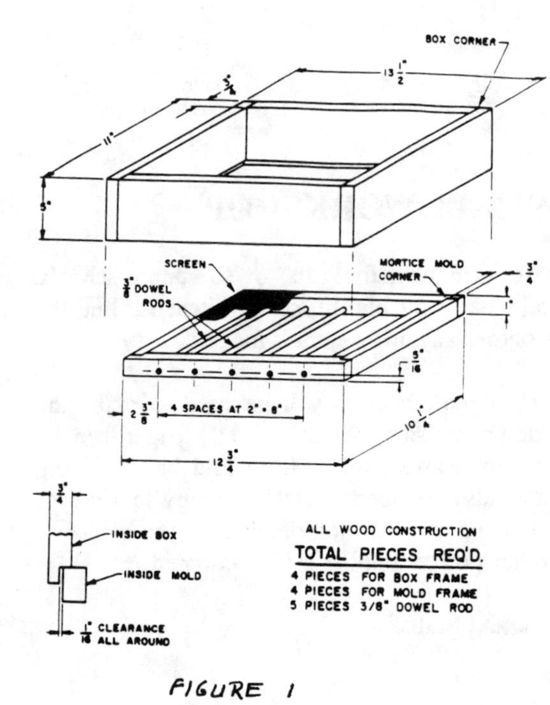


FIGURE 1

Cut or tear the paper into small pieces, about 5cm x 5cm (2" x 2"). Shred any cloth that may be used. Put the pieces in the pot, cover with water, and add a few tablespoonfuls of household bleach. Turn on the heat, cover the pot, and bring to a gentle boil. Stir occasionally for a couple of hours to ensure that the bleach is mixed and all the paper is wetted down well, then cool.

After the batch has cooled, try to break up the lumps and any remaining pieces of paper still holding together. The smaller the pieces in the beginning, the easier this step is now. (The pulp can then be drained and stored in plastic bags in a refrigerator, if you have one, until you are ready to make the sheets. It will keep for weeks without any change.)

## Making the Sheets

Take a lump of the semi-moist pulp you have prepared. Press as much moisture out of it as possible to leave a ball about the size of a pigeon egg (7g--1/4 oz--dry weight). This is enough pulp to make one 21.5cm x 28cm sheet. Make the sheets, one at a time, as follows:

1. Blend and mix pulp in blender 3/4 full of water. Add additives.

2. Put mold in box, screen side up and immerse in sink.
   Rap box to get rid of air bubbles.

3. Pour pulp into box.

4. Holding box down, agitate the water in the box with fingers so that the pulp spreads evenly over the mesh.

5. Grasp box and mold firmly and lift quickly and evenly to surface (feel suction).

6. Hold for 10 seconds or so to drain.

7. Lift up out of water and hold vertically to drain. If sheet looks okay, proceed; if flawed, put box and mold back into sink. Repeat steps 4 to 7.

8. Set box on flat surface and carefully remove box. Note: Water drops on wet web will make marks!

9. Carefully lay cotton couching cloth over web and smooth gently.

10. Place absorbent felt over couching cloth. Smooth and press down from center out.

11. Remove felt and wring out water.

12. Repeat 10 and 11 until no more water comes out.

13. Couch off sheet, starting at corner and peeling back quickly.

14. Place couched sheet, paper side up or down on flat absorbent surface. Smooth and press down to remove trapped air.

15. Repeat for each sheet until a neat stack is built up.

## Pressing and Drying

The sheets can be dried quickly by pressing them with a hot iron and an aluminum sheet or slowly (2-3 hours) by placing them in a 120°C (280°F) oven, with the couching sheets tacked down to the receiving board all along the edges of the paper sheets. The first method gives a smooth surface on one side, embossing with cloth marks on the other; the second gives embossing on both sides.

A very slick surface can be obtained by smoothing the couching cloth, paper side down, against an aluminum or oiled galvanized sheet. A squeegee can be used to get rid of all the air. Dry in air or in a 120°C (280°F) degree oven.

## Sizing and Coating

A simple method of internal sizing uses a combination of pure gelatin and cornstarch (either laundry or cooking type). The gelatin is dissolved in boiling water and cornstarch is added to make a clear, thick mixture to add to the pulp. Use about one teaspoon of this per 21.5cm x 28cm sheet.

Another simple internal sizing procedure is to add about 1/4 teaspoon of linseed and/or a teaspoon of cornstarch solution while the pulp is being mixed at step 1. The oil is dispersed in the water and precipitates on the fiber. The starch will be caught on the fibers and during the drying stage will set to give a stiffer sheet. External sizing is done when the sheet is coated with a water based solution after the paper has been dried. With an ordinary 4cm (1 1/2") paint brush, coat each sheet with a 7 percent straight corn starch solution. One tablespoonful of cornstarch added to a cup of water will be enough for 20 to 25 sheets (both sides). Animal glue can be added to the starch to improve the water resistance. Modern glues can be added also.

When the coated sheets are nearly dry to the touch, place them in a neat stack. They should be somewhat limp but not wet. Put a metal sheet or smooth board on top. Allow the stack to dry overnight. The sheets can then be trimmed if necessary and packaged for sale.

## MAKING PAPER IN THE MICRO FACTORY

On a somewhat larger scale, but still in an essentially hand process, paper can be made in a micro factory capable of producing about 240kg (1/4 ton) of paper per day. Such small factories are fairly common in India, and VITA has assisted at least one such operation in Tanzania. This process uses wastepaper or rags to make pulp, or pulp purchased from a pulp mill. It can produce good quality bond or drawing paper, card stock, school tablets, filter paper, toilet tissue, grey board, and album or blotting paper. It can also turn out such articles as egg cartons, flower pots, seed flats, hospital trays, and so on.

In addition to an identified, reliable market, the small factory requires a steady, reliable supply of raw materials, water, and power. Suggested facilities include a building of about 300 square meters for operations and a shed of about 185 square meters for collecting and sorting the materials. Six administrative staff and as many as 100 laborers working in two or three shifts are needed.

The U.N. Industrial Development Organization (UNIDO) estimates an investment of approximately US$26,000 (1984) for the total cost of installation. Production may be increased by installing one or two more beaters and operating the vats in three shifts. (Beyond this capacity, however, economies of scale decline, and production moves up to small-scale mechanized plants.)

**Sources:**

Vogler, Jon, and Sarjeant, Peter. *Understanding Small-Scale Papermaking.* Arlington, Virginia: Volunteers in Technical Assistance, 1986.

*Appropriate Industrial Technology for Paper Products and Small Pulp Mills.* Vienna Austria: United Nations Industrial Development Organization (UNIDO), 1979

Sheriff Dewji and Sons, Arusha, Tanzania

American Paper Institute, 260 Madison Avenue, New York, New York

# Candle Making

In areas without electricity, lanterns, candles, and cooking hearths often provide the only source of light at night. Candles are easy to make at home for home use. With attention to quality control, they can be made in a small workshop for sale in the shops and markets.

The directions given here are for dipped candles, which are made by repeatedly dipping a length of wick into melted wax until the candle is the desired size. Dipped candles often cost more in the shops than other kinds, but they usually burn longer and with less smoke. This system, developed by the Environmental and Development Agency in South Africa, uses a special jig that holds up to four candles at a time.

**Tools and Materials**

Paraffin wax (you may wish to experiment with bee's wax if it is available)
Stearic acid
Candle wicking (the string inside the candle)
Container to melt the wax (this has to be as deep as the candles are tall)
Wire for the jig
Thermometer, in a brass case
Rod or rope to hang the candles on while they cool
A gas or kerosene stove

It is suggested that a small business or candle making cooperative would likely need to make an initial investment in 40kgs (88 lbs.) of wax, stearic acid in quantity to make a ratio of 10 parts wax to 1 part stearic acid, and 20 wire jigs.

## MAKING THE JIGS

A jig is the hanger that holds the wicking while you dip it into the melted wax. Make 20 or so jigs for your business. Even working at home it is convenient to have a half dozen.

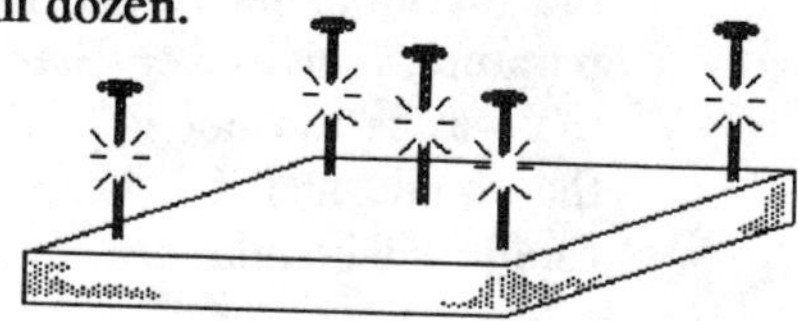

To make the jig, hammer 5 nails into a piece of wood as shown and cut off the heads.

Cut one piece of wire 60cm long and one piece 50cm long.

Take the shorter piece of wire and wrap it around the nails as shown in Figure 1. Start at nail 1, bend the wire around nail 2 and then up around nail 3. Then bend it back to nail 4 and up around nail 5. Take the wire off the frame. This is the bottom of the jig.

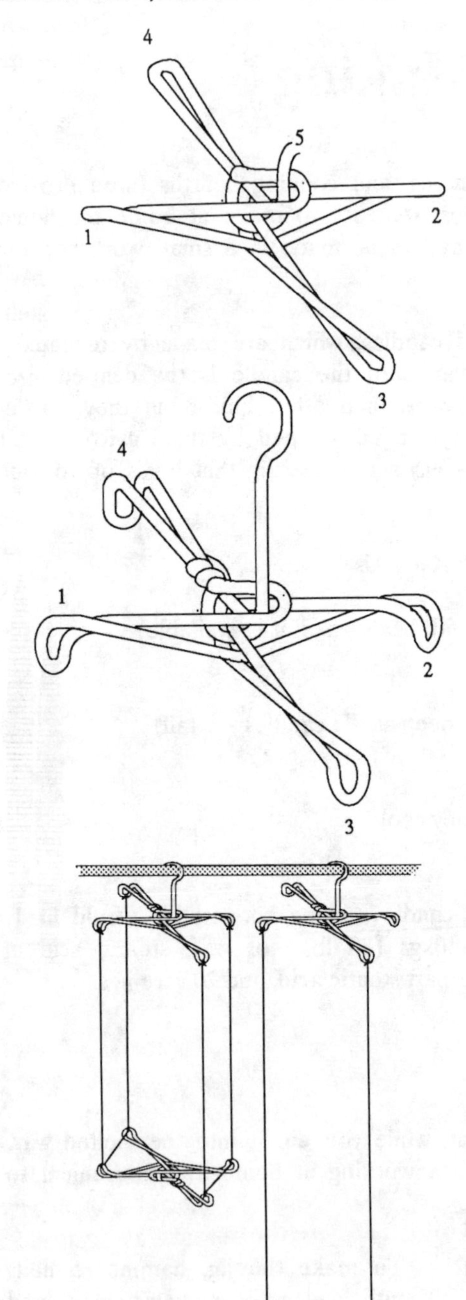


Make the top of the jig with the longer piece of wire. Bend the wire around the nails as described above. You will have some left over. Bend this part into a hook to hang up the jig with. Take the wire off the frame and bend down the corners as Shown in Figure 3.

Take 4 pieces of wicking, as long as you want your candles plus a little bit. Tic one end of each piece to the top part of the jig and the other end to the bottom part (Figure 4). Fix as many jigs as you think you will need at one time.

## PREPARING THE WAX

Cut the wax into small pieces. Make sure no dirt gets mixed up with it. Melt enough wax and stearic acid to fill the container almost full. Use 1 part stearic acid to 10 parts wax.

Heat the wax to 70°C (158°F). Use the thermometer to check the temperature. This is very important. If the wax is too hot it won't stay on the candle and if it is too cool the candle will be lumpy.

The safest way to melt the wax is to set the container with the wax into a pot of water so that the wax is not directly over the flame. It is very dangerous to let the wax get too hot. Wax catches fire easily, and a wax fire is difficult to put out. In case of fire, cover the container and turn off the stove as quickly as possible. Be careful not to splash the hot wax. It will catch fire if it falls into the flame and it will burn your skin if it touches you.

## DIPPING THE CANDLES

Take one of the jigs you have put the wicking on and dip it into the melted wax. Hang the jig on the rod to cool. Dip another jig with wicking into the melted wax and hang it on the rod. When you have dipped all the jigs you have prepared, start with the first one and dip again. Each time you dip the jig a little more wax will stick to the wick and the candle will get thicker. Continue dipping until candles are the size you want.

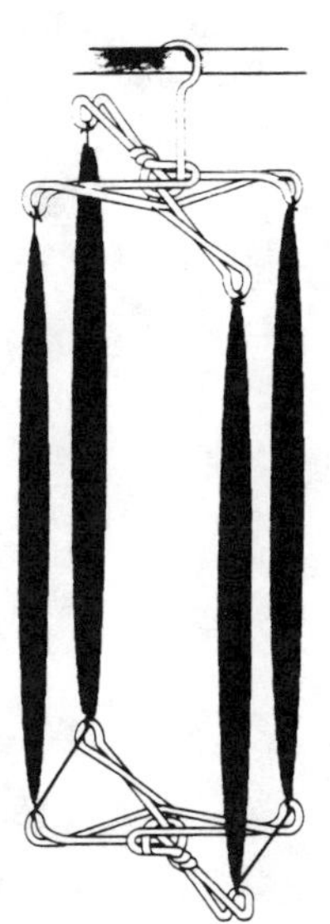

Don't handle the candles until they are cool and hard. Then, cut them off the jigs. Trim the wicks to an even length. Store candles out of the sun and away from heat.

Put a wide board or plastic sheet under the rod where you are hanging the jigs. Any excess wax will drip onto them and you can scrape it off and melt it down again. Be sure to keep this area clean; any dirt that gets in the wax will get into your candles. The wax that sticks to the metal jigs can also be scraped off and used again.

Do not dispose of excess melted wax by pouring it down a drain. When it cools and hardens it will clog the drain. Besides, any extra wax can be melted down and used again. If you find that you have to get rid of a batch of wax, let it harden and then throw it away.

If the market is good and you can get the materials, you may want to try scenting your candles with essential oils like vanilla or sandalwood. Or you might try making colored candles. These oils and pigments must be specially made for use in candles, however, and are not always available.

**Source:**

Berold, Robert, and Caine, Collette (eds.). *People's Workbook.* Johannesburg, South Africa: Environmental and Development Agency, 1981.

*Simple Methods of Candle Manufacture.* London: Intermediate Technology Publications, Inc., 1985.

# Communications

VITA Archives

# Bamboo or Reed Writing Pens

This low-cost, easy-to-make pen has been in use in Jordan since 3000 B.C. Pens of different sizes can be made for work ranging from fine writing to large block letters. Similar pens have also been used in Thailand.

FIGURE 1. PENS CAN BE MADE FROM BAMBOO FOR WORK RANGING FROM FINE WRITING TO BLOCK LETTERS.

**Tools and Materials**

Dry bamboo, 15cm x 1cm x 0.5cm (6" x 3/8" x 3/16")
Small rubber band or fine wire
Sharp knife
Fine sandpaper

Whittle one end of the bamboo to the desired width, and then shave it down to make it flexible (see Figure 2). Be sure that the writing tip is made from the more durable material near the outside of the bamboo.

Cut the writing end straight across with a sharp knife. Use sandpaper to make the end smooth. The point of the pen can be shaped to the proper writing angle for your hand by gently writing on the sandpaper with the dry pen.

To make a retaining hole for ink, place the tip of the knife on the pen, at least 3mm (1/8") up from the point of the pen, and then rotate the knife to drill a hole about 2mm (3/32") in diameter.

The pen can now be used for writing, but it will need to be reinked frequently. To make a reservoir pen, attach a thin bamboo cover plate to the pen as shown in Figure 3. Attach the cover plate by wrapping a small rubber band or a piece of fine wire around the notches provided for this.

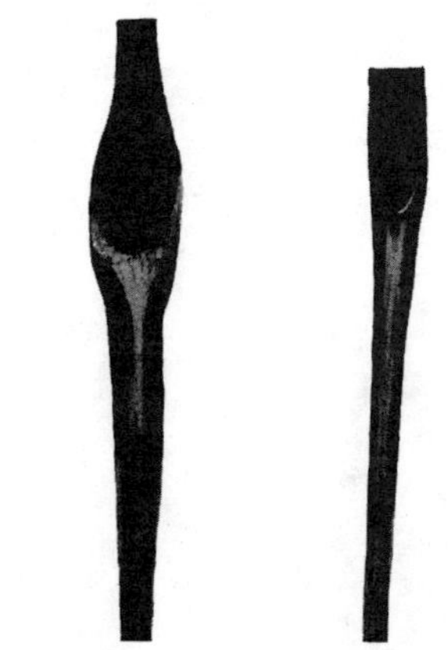

FIGURE 2. ONE END OF THE BAMBOO IS WHITTLED TO THE DESIRED WIDTH. THEN IT IS SHAVED DOWN TO MAKE IT FLEXIBLE.

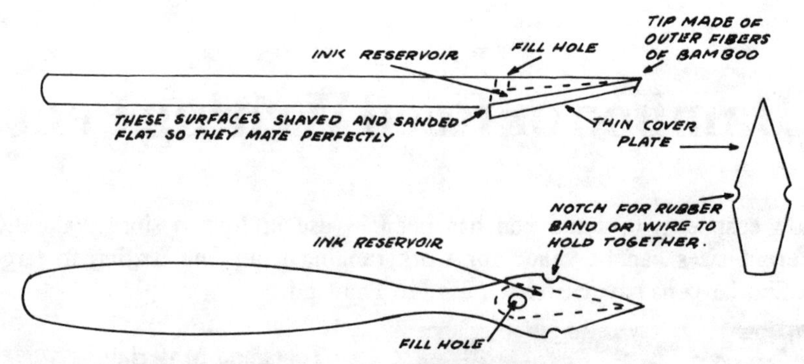


FIGURE 3. IF A THIN, NOTCHED COVER PLATE IS ADDED, THE PEN NEEDS REFILLING LESS OFTEN.

**Source:**

*The Multiplier,* Vol. 3, No. 10. Washington, D.C.: U.S. Department of State, Agency for International Development, 1960.

# Silk Screen Printing

Silk screen printing is a simple, inexpensive method of producing multiple copies of attractive visual aids, posters, and other materials, including typewritten pages. A squeegee forces very thick paint through those parts of the silk screen that are exposed by the stencil onto paper placed underneath the screen. The silk-screen process presented here is used for educators and trainers who must prepare their own training materials. It would require considerable upgrading of equipment and materials to be appropriate for commercial painting operations.

## BUILDING THE SILK SCREEN PRINTER

### Tools and Materials

Hinges, about 2.5cm x 7.5cm (1" x 3")
Wing or regular nuts
Squeegee
Trigger support
Wood for frame
Baseboard or smooth table top
Silk or other sheer cloth
Thumbtacks
Silk screen paint
Paper for copies
Water-soluble paint, e.g., finger paint
(Oil-soluble paint also works well,
but a solvent is needed to clean it off the screen.)

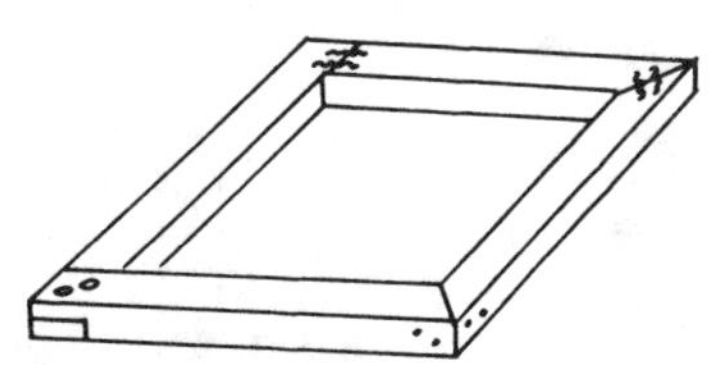

FIGURE 1. MAKING THE FRAME OF THE SILK SCREEN. DIFFERENT JOINT CONSTRUCTIONS ARE SHOWN AT EACH CORNER; ANY ONE OF THESE MAY BE USED FOR THE JOINTS OF THE FRAME.

1. Build a frame (see Figures 1 and 2), using 1.9cm x 5cm (3/4" x 2") plywood or other wood. The frame should be big enough for the largest prints to be made. Average inside frame dimensions would be 38.1cm x 50.8cm (18" x 24"). Make sure that the corners are square and the frame lies flat against a flat baseboard or table top. The baseboard can also be made of 1.9cm (3/4") plywood. A few coats of shellac on the wooden frame will make it longer lasting and less apt to warp.

2. Stretch the silk very tightly over the underside of the frame, using tacks every 2.5cm (1"). Make sure that the threads of the silk run parallel with the edges of the frame, pull the silk over the outside bottom edges and tack the silk around the outside of the frame (see Figure 2).

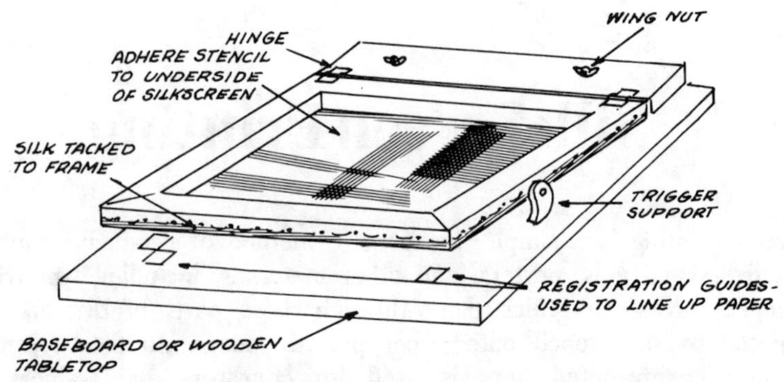


FIGURE 2. BOLT THE HINGED END OF THE FRAME TO A FLAT BASEBOARD OR WOODEN TABLE TOP.

3. Make a squeegee (see Figure 3).

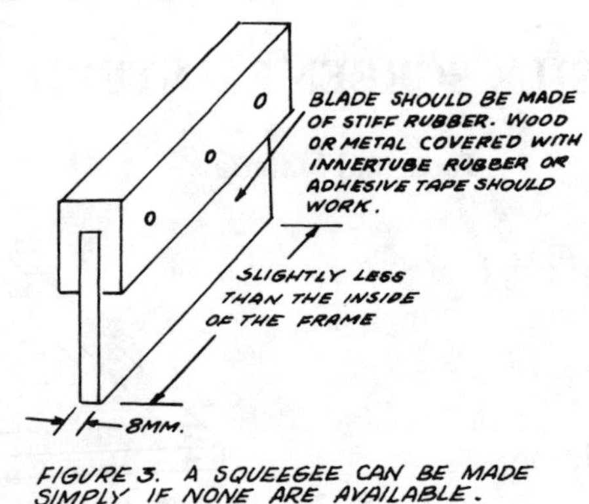


FIGURE 3. A SQUEEGEE CAN BE MADE SIMPLY IF NONE ARE AVAILABLE.

# PRINTING

1. Cut the stencil and attach it to the screen (see "Preparing a Paper Stencil").

2. Place the paper or cardboard to be printed under the screen and stencil. Draw about 10ml (2 teaspoons) of water-soluble paint (for example, finger paint) in a line along the edge of the silk just inside one end of the frame. The paint should be thick, about like auto transmission grease, so that it will not just fall through the screen without being pushed by the squeegee.

3. Using an edge of the squeegee, pull the paint across the surface of the silk. This squeezes the paint through all the open areas of the paper stencil. Lift the screen and remove the print, replacing it with the next piece to be printed. Pull the paint back in the opposite direction for this print. The correct technique is to put an amount of paint on the screen that will, combined with the right pressure on the squeegee, produce a good print with one stroke of the squeegee.

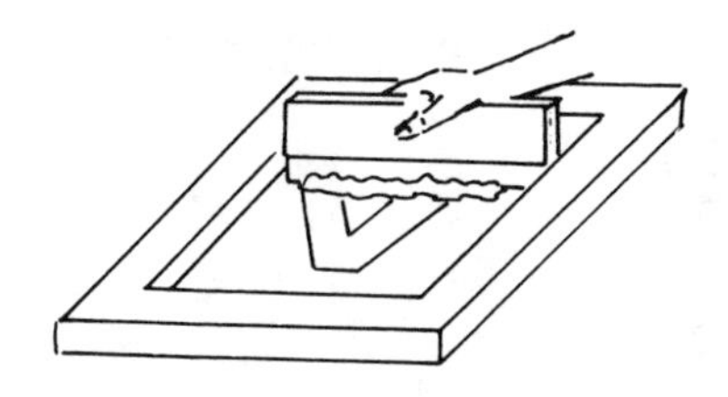

FIGURE 4. THE SQUEEGEE IS USED TO DRAW THE THICK PAINT ACROSS THE SILK SCREEN.

Make sure that the paint contains no dried paint particles. They could damage the screen.

4. When a printing is completed, pull the stencil off the screen. Remove the wing nuts and wash the frame under running water.

5. The pieces to be printed can be **registered** (lined up so that the printed image appears in exactly the same place on each piece). Registration guides can be made of thin cardboard or several layers of tape (see Figure 2).

   Thicker guides could break the silk when the squeegee presses the screen against them. The guides should be taped on the baseboard at the edges of three sides of the sheets to be printed.

6. If more than one color is to be printed, registration becomes very important. The procedure to follow is this:

   - Print the first color, **using registration guides.**
   - Wash the screen as in Step 4 above, and attach the next stencil.
   - Place a piece of waxed paper or thin translucent paper under the second screen to be printed, and tape this paper on one edge.
   - Print an image of the second screen on this paper.
   - Raise the screen.
   - Slide a sample of the first printing into position beneath the taped paper. Adjust the sample so that the second image will appear in the right place on the pieces already printed.
   - When the sample is lined up, **carefully** hold the first printing sample in position and remove the wax paper.
   - Tape new registration guides on three sides of the sample.
   - More colors can be printed by returning to Step 6.

7. Several colors can be printed over one another if transparent paints are used.

8. A drying rack (see Figure 5) is helpful when many prints are to be dried.

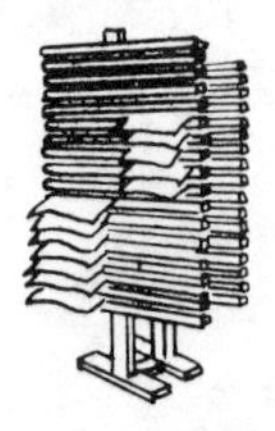

FIGURE 5. A SIMPLE DRYING RACK CAN BE MADE FROM 5CM. X 5CM. UPRIGHTS WITH 2.5CM. X 2.5CM. CROSSBARS ABOUT 2.5CM. APART.

**Source:**

John Tomlinson, VITA Volunteer, Rochester, New York

# PREPARING A PAPER STENCIL

This method of preparing a stencil for silk screen printing is more versatile for some effects than the usual stencil technique: for example, the letter "O" can be formed without connecting lines to hold the center in place. But the method has these limitations: Images must be bold and simple designs. The stencil will last for only a few hundred impressions; will not hold up with water-base paint; and cannot be stored.

### Tools and Materials

Stencil paper–Somewhat-transparent white bond paper works well. Commercial stencil paper can be used, but the edge of the printing may be fuzzy. Thick paper leaves a thick layer of paint when the squeegee draws the paint across the screen.
Mimeograph stencils can be used to reproduce typing.
Stencil knife
A small-blade knife with a handle about as thick as a pencil.

To prepare and use the paper stencil, follow these steps:

- Place the stencil paper over the image to be reproduced and fasten both to a hard level surface, like the baseboard of the silk screen.

- Trace the design and then cut around the areas where one color is to be printed. Press just hard enough to cut through the stencil paper without cutting the original. Do not strip the cut-out parts away yet; leave the stencil intact.

- o Put a pad of newspaper on the baseboard of the silk screen so that when the screen is lowered it will hit the stencil firmly.
- o Place the stencil on this pad in the position desired. Slip several pieces of tape, sticky side up, under the edges of the stencil; this will tape the stencil to the screen when the screen is lowered. Mask the open areas of the screen beyond the edges of the stencil.
- o To make the stencil stick to the screen, draw paint across the screen with the squeegee.
- o Remove the cut-out parts of the stencil.
- o At the end of the printing run, peel the paper stencil and masking from the screen. Clean the screen.

A mimeograph stencil is prepared as it would be for a mimeograph machine. Attach it to the screen the same way a paper stencil is attached.

**Source:**

Mrs. Benjamin P. Coe, VITA Volunteer, Schenectady, New York

# MAKING SILK SCREEN PAINT

The paints described here for silk screen printing should have a shelf life of several months when they are stored in jars with tight-fitting lids. The recipes have been tried successfully in a temperate climate. Paints colored with powdered tempera are more brilliant than those colored with food colors or ink. Other water-soluble dyes can probably be used also.

**Materials**

Starch or cornstarch
Soap Flakes
Gelatin (optional)
Coloring matter (food color, tempera powder, ink, or a dye of some sort that is water soluble)

## Recipe #1

Linit starch (not instant) 115 ml (1/2 cup)
Boiling water 345ml (1 1/2 cup)
Soap flakes 115ml (1/2 cup)

Mix starch with enough cold water to make a smooth paste. Add boiling water and cool until glossy. Stir in soap flakes while mixture is warm. When cool, add coloring.

## Recipe #2

Cornstarch 57.5ml (1/4 cup)
Water 460ml (2 cups)
Soap flakes 29ml (1/8 cup)

Bring water to a boil. Mix cornstarch with a small amount of cold water and stir the two together. Bring to a boil and stir until thickened. Add soap flakes while warm. Color.

This recipe produces paint that seems quite lumpy but this does not affect the printing quality.

## Recipe #3

Dissolve 115ml (1/2 cup) cornstarch in 172.5ml (3/4 cup) cold water

Dissolve 1 envelope gelatin (15ml or 1 tablespoon, unflavored) in 57.5ml (1/4 cup) cold water

Heat 460ml (2 cups) of water, pour in cornstarch. Add dissolved gelatin. Boil, and stir until thickened. Cool and add 115ml (1/2 cup) soap flakes. Color.

**NOTE:** Adding 5 to 10ml (1 to 2 teaspoons) of glycerine to any of these recipes will make the paint easier to use.

Never let dried particles of paint get mixed into the paint or fall onto the screen because they may puncture the silk during the printing. A small hole in the silk can be repaired with a small drop of **shellac.**

**Source:**

Mrs. Benjamin P. Coe, VITA Volunteer, Schenectady, New York

# Inexpensive Rubber Cement

Inexpensive rubber cement can be made easily with ordinary gasoline and raw sheet rubber.

Imported pastes are often expensive. Many of these are not good for mounting pictures and similar materials; they soak through the paper and wrinkle both the picture and the mount.

Rubber cement does not wrinkle the pieces to be joined. It has another advantage: if it smears, it can be rubbed off with the fingers when it is dry.

**Tools and Materials**

Ordinary gasoline: 250cc (16 ounces)
Raw sheet rubber in one piece:
5gm (1/5 ounce)
Jar with lid
Stirring rod
Brown bottle
*Tin can
*Charcoal
*Small pieces of cloth

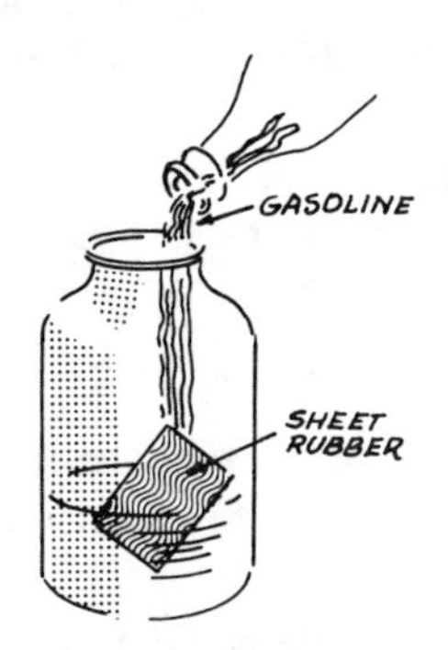


FIGURE 1. RUBBER CEMENT IS MADE BY MIXING RAW SHEET RUBBER WITH ORDINARY GASOLINE. GASOLINE IS EXPLOSIVE. BE CAREFUL WHEN MIXING OR USING CEMENT.

---

* Needed only if gasoline is colored.

*** CAUTION ***

**Gasoline will burn and explode, and the vapors can be a health hazard. Be careful when mixing or applying the cement. Do not inhale the gasoline vapors. Make the rubber cement in a well-ventilated place.**

The rubber to be used should be a translucent, light-brown sheet. Any brand of gasoline can be used. Some gasolines are highly colored. This coloring should be removed so that the rubber cement will not stain when it is used. To remove the coloring, pour the gasoline over common charcoal several times (see Figure 2).

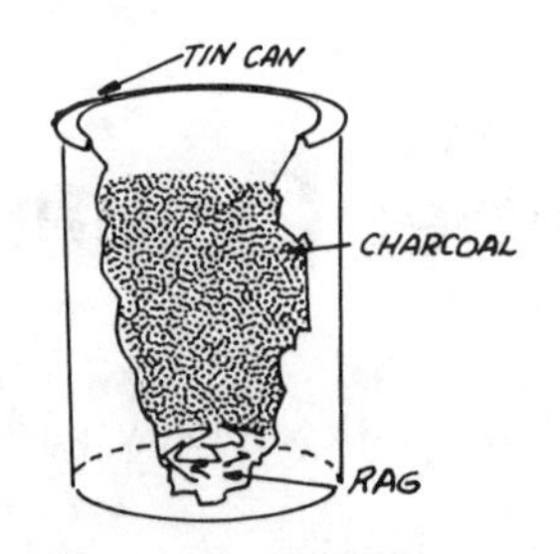


FIGURE 2. IF THE GASOLINE IS COLORED IT CAN BE FILTERED THROUGH CHARCOAL TO REMOVE THE COLOR. THIS WILL KEEP THE RUBBER CEMENT FROM STAINING.

Use a clean tin can with a hole in the bottom. Put a small piece of cloth in the bottom of the can to keep the charcoal from falling into the filtered gasoline. You may have to change the charcoal several times before the gasoline is clear.

Put the 5 grams (1/5 ounce) of raw sheet rubber in a jar and pour in the 250cc (16 ounces) of ordinary gasoline (see Figure 1). Cover the jar.

It takes about three days for the rubber to dissolve completely in the gasoline. Stir the mixture several times during this period, especially when the mixture becomes thick. If some of the rubber does not dissolve, more stirring will break it up. When the rubber is dissolved, you will have a smooth, milky-colored cement.

To store the rubber cement, it is best to use a brown bottle because the cement will become thin if it is exposed to sunlight for a long time.

Mark the bottle:

**DANGER: EXTREMELY FLAMMABLE,**
**HARMFUL OR FATAL IF SWALLOWED**

The cement should be kept in a ventilated cupboard when it is not being used.

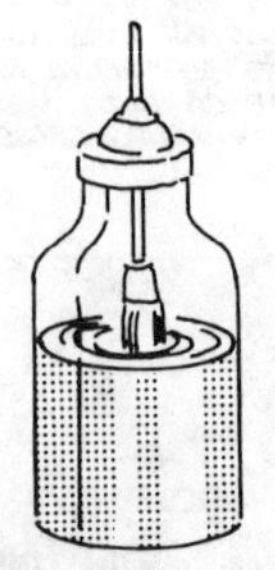

FIGURE 3. A DISPENSER CAN BE MADE BY CUTTING A HOLE IN THE COVER LARGE ENOUGH FOR THE HANDLE OF A 2.5 CM. BRUSH. THE DISPENSER MUST BE AIRTIGHT SO THAT THE CEMENT WILL NOT DRY OUT.

To make a handy dispenser for the cement: Cut a hole in the cover of the jar, large enough for the handle of a 2.5cm (1") brush (see Figure 3). Push the handle through the hole and leave the brush in the jar. This should be airtight because the cement hardens quickly when exposed to air.

**Source:**

Bunyard, Robert J. "Rubber Cement in a Tropical Climate," *The Multiplier,* Vol. 2, No. 6, July 1956.

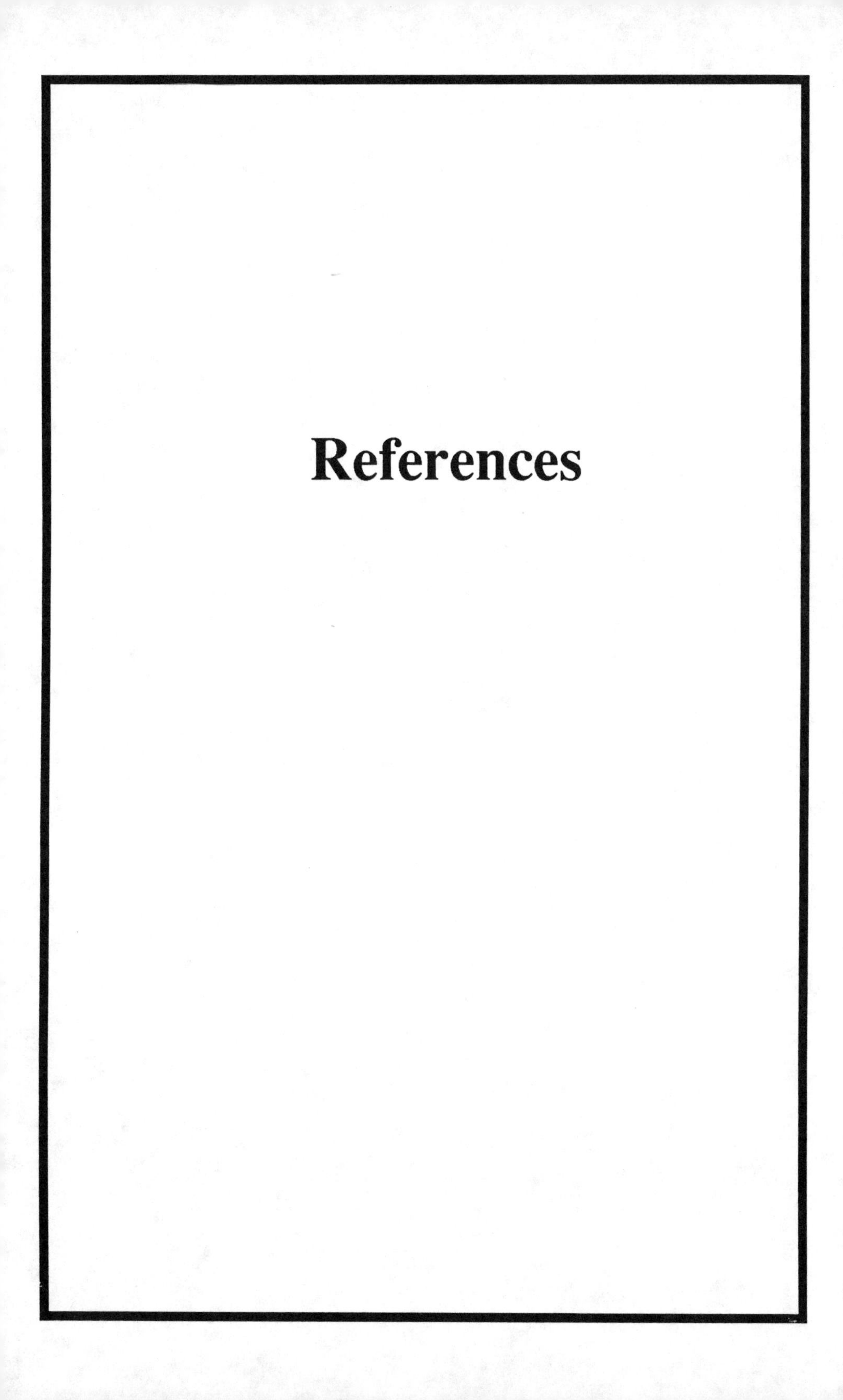

# References

# REFERENCES

## WATER RESOURCES

American Water Works Association. "AWWA Standard D-100-79 for Welded Steel Water Storage Tanks." Denver, Colorado: American Water Works Association, 1979.

American Water Works Association. "AWWA Standard D-105-80 for Disinfection of Water Storage Facilities." Denver, Colorado: American Water Works Association, 1980.

American Water Works Association. *Water Distribution Operator Training Handbook.* Denver, Colorado: American Water Works Association, 1976.

Anchor, R.D. *Design of Liquid-Retaining Concrete Structures*. New York: Wiley and Sons, 1982.

Blackwell, F.O., Farding, P.S., and Hilbert, M.S. *Understanding Water Supply and Treatment for Individual and Small Community Systems.* Arlington, Virginia: Volunteers in Technical Assistance, 1985.

Brown, J.H. "Flexible Membrane: An Economical Reservoir Liner and Cover." *Journal of the American Water Works Association*. Vol. 71, No. 6, June 1979.

Cairncross, S., and Feachem, R. *Small Water Supplies*. London: Ross Institute, 1978.

Crouch, Margaret (ed.). *Six Simple Pumps*. Arlington, Virginia: Volunteers in Technical Assistance, 1983.

Helweg, O.J., and Smith, G. "Appropriate Technology for Artificial Aquifers," *Ground Water.* Vol. 18, No. 3, May-June 1978.

Maddocks, D. *Methods of Creating Low Cost Waterproof Membranes for Use in the Construction of Rainwater Catchment and Storage Systems.* London: Intermediate Technology Publications, Ltd., 1975

Mazariegos, J. F., and de Zeissig, Julia A. A. *Water Purification Using Small Artisan Filters.* Guatemala: Central American Research Institute for Industry, 1981.

McJunkin, F. and Pineo, C. U.S. Agency for International Development. *Water Supply and Sanitation in Developing Countries.* Washington, D.C.: USAID, 1976.

Pacey, Arnold, and Cullis, Adrian. *Rainwater Harvesting: The Collection of Rainfall and Runoff in Rural Areas.* London: Intermediate Technology Publications, Ltd., 1986.

Remmers, J. *Understanding Water Supply: General Considerations.* Arlington, Virginia: Volunteers in Technical Assistance. 1985.

Ritter, C.M. *Understanding Potable Water Storage.* Arlington, Virginia: Volunteers in Technical Assistance (VITA), 1985.

Ryden, D.E. "Evaluating the Safety and Seismic Stability of Embankment Reservoirs." *Journal of the American Water Works Association.* Vol. 76, No. 1. Denver, Colorado: American Water Works Association, January 1984.

Salvato, J.A., Jr. *Environmental Engineering and Sanitation.* New York: Wiley-Interscience, 1972.

Schiller, E.J., and Droste, R.L., eds. *Water Supply and Sanitation in Developing Countries.* Ann Arbor, Michigan: Ann Arbor Science Publishers, 1982

Sharma, P.N., and Helweg, O.J. "Optimum Design of Small Reservoir Systems." *Journal of Irrigation and Drainage Division--American Society of Civil Engineers.* Vol. 108, IR4, December 1982.

Sherer, K. "Technical Training of Peace Corps Volunteers in Rural Water Supply systems in Morocco." Water and Sanitation for Health Project (WASH) Field Report No. 43. Washington, D.C.: U.S. Agency for International Development, May 1982.

Silverman, G.S.; Nagy, L.A.; and Olson, B.H. "Variations in Particulate Matter, Algae, and Bacteria in An Uncovered, Finished Drinking-Water Reservoir." *Journal of the American Water Works Association*. Vol. 75, No. 4. Denver, Colorado: American Water Works Association, April 1983.

Spangler, C.D. United Nations and World Bank. *Low-Cost Water Distribution: A Field Manual.* Washington, D.C.: World Bank, December 1980.

Swiss Association for Technical Assistance, ed. *Manual for Rural Water Supply.* Zurich, Switzerland: Swiss Center for Appropriate Technology, 1980.

Sylvestre, Emilio. "Water, Water Everywhere: Island Communities Install Water Systems." *VITA News,* October 1986, pp. 8-10.

United Nations. World Health Organization. "WHO Guidelines for Drinking Water Quality," by H.G. Gorchev and G. Ozolins. Geneva, Switzerland: World Health Organization, 1982.

United Nations. World Health Organization. "The Purification of Water on a Small Scale. WHO Technical paper No. 3. The Hague, The Netherlands: WHO International Reference Centre for Community Water Supply, March 1973.

United Nations. World Health Organization. "Preliminary List of References on Slow Sand Filtration and Related Simple pretreatment Methods." The Hague, The Netherlands: WHO International Reference Centre for Community Water Supply, July 1976.

Upmeyer, D.W. "Estimating Water Storage Requirements." *Public Works.* Vol. 109, No. 7, July 1978.

U.S. Environmental Protection Agency. *Manual of Individual Water Supply Systems.* Washington, D.C.: EPA, 1975.

"Wind Power for Roatan Island: Pumping Water in Honduras." *VITA News,* October 1982, pp. 3-7.

## HEALTH AND SANITATION

American Concrete Institute. "Concrete Sanitary Engineering Structures." Report No. ACI 350R-83. Detriot, Michigan: American Concrete Institute, 1983.

Baumann, Werner, and Karpe, Hans Jurgen. *Wastewater Treatment and Excreta Disposal in Developing Countries.* West Germany: German Appropriate Technology Report, 1980.

Bull, David. *A Growing Problem: Pesticides and the Third World Poor.* Oxford: OXFAM, 1982.

Canter, L.W. and Malina, J.F. *Sewege Treatment in Developing Countries.* Norman, Oklahoma: The University of Oklahoma (under contract to USAID), December 1976.

Cointreau, Sandra J. *Environmental Management or Urban Solid Wastes in Developing Countries (A Project Guide).* Washington, D.C.: World Bank, June 1982.

Davis, B.P. *Understanding Sanitation at the Community Level.* Arlington, Virginia: Volunteers in Technical Assistance, 1985.

Feachem, Richard G.; Bradley, David; Garelick, Hemda; and Mara, D. Duncan. "Health Aspects of Excreta and Sullage Management: A State-of-the-Art Review." *(Appropriate Technology for Water Supply and Sanitation,* vol. 3). Washington, D.C.: World Bank, 1980.

Feachem, Richard, et al. *Water, Health and Development: An Inter-disciplinary Evaluation.* London: Tri-Med Books, Ltd., 1977.

Feachem, Richard, McGarry, Michael, and Mara, D. Duncan (eds.). *Water, Wastes and Health in Hot Climates.* New York: John Wiley and Sons, 1980.

Goldstein, Steven N., and Moberg, Walter J., Jr. *Wastewater Treatment Systems for Rural Communities.* Washington, D.C.: Commission on Rural Water, 1973.

Golveke, C.G. *Biological Reclamation of Solid Wastes.* Emmaus, Pennsylvania: Rodale Press, 1977.

Grover, Brian. *Water Supply and Sanitation Project Preparation Handbook* (vol. 1, Guidelines). Washington, D.C.: World Bank, 1982.

Herrington, J.E. *Understanding Primary Health Care for a Rural Population.* Arlington, Virginia: Volunteers in Technical Assistance, 1985.

Kalbermatten, John M., et al. "A Planner's Guide." *(Appropriate Technology for Water Supply and Sanitation,* vol. 2). Washington, D.C.: World Bank. 1981.

Kalbermatten, John M.; Julius, DeAnne S.; and Gunnerson, Charles G. *Appropriate Sanitation Alternatives: A Technical and Economic Appraisal.* Baltimore, Maryland: Johns Hopkins University Press (for the World Bank), 1982.

Mann, H.T., and Williamson, D. *Water Treatment and Sanitation: Simple Methods for Rural Areas.* London. Intermediate Technology Publications, 1982.

Patel, Ishwarbhai. *Safai-Marg Darshika (A Guide Book on Sanitation).* New Delhi: Udyogshala Press, 1970.

Reid, George and Coffey, Kay. (eds.). *Appropriate Methods of Treating Water and Wastewater in Developing Countries.* Norman, Oklahoma: Bureau of Water and Environmental Resources Research (University of Oklahoma), 1978.

Rybczynski, Witold, Polprasert, Changrak, and McGarry, Michael. *Low-Cost Technology Options for Sanitation* (A State-of-the-Art Review and Annotated Bibliography). Ottawa: International Development Research Centre, 1978.

Salvato, J.A., Jr. *Environmental Engineering and Sanitation.* New York: Wiley-Interscience, 1972.

*Sanitation in Developing Countries* (Proceedings of a workshop on training held in Lobatse, Botswana, 14-20 August 1980). Ottawa: International Development Research Centre, 1981.

Stonerook, H. *Understanding Sewage Treatment and Disposal.* Arlington, Virginia: Volunteers in Technical Assistance, 1984.

Strauss, Martin. *Sanitation Handbook* (Community Water Supply and Sanitation, Nepal). Pokhara, Nepal: Pokhara Centre Press, June 1982.

van Wijk-Sijbesma, Christine. *Participation and Education in Community Water Supply and Sanitation Programmes - A Literature Review.* The Hague: WHO International Reference Centre for Community Water Supply, 1979.

Vogler, Jon. *Work from Waste: Recycling Wastes to Create Employment.* Oxford: Intermediate Technology Publications Ltd. and OXFAM, 1981.

Werner, D. *Where There Is No Doctor: A Village Health Care Handbook.* Palo Alto, California: Hesperian Foundation, 1980.

## AGRICULTURE

Abrahams, P.J. *Understanding Soil Preparation.* Arlington, Virginia: Volunteers in Technical Assistance, 1984

Archer, Sellers G. *Soil Conservation.* Norman, Oklahoma: University of Oklahoma Press, 1969.

Attfield, Harlan. *Gardening With the Seasons.* Arlington, Virginia: Volunteers in Technical Assistance, 1979.

Bartholomew, W.V. *Soil Nitrogen--Supply Processes and Crop Requirements.* Technical Bullentin 6. Raleigh, North Carolina: North Carolina State University, 1972.

Bird, H.R. *Understanding Poultry Meat and Egg Production.* Arlington, Virginia: Volunteers in Technical Assistance, 1984.

Bradenburg, N.R. *Bibliography of Harvesting and Processing Forage Seed, 1949-1964.* U.S. Department of Agriculture, Agricultural Research Service, ARS 42-135, Washington: USDA, 1968.

Branch, Diana S. (ed.). *Tools for Homesteaders, Gardeners, and Small-Scale Farmers,* Emmaus, Pennsylvania, 1978.

Corven, James. *Basic Soil Improvement for Everyone.* Arlington, Virginia: Volunteers in Technical Assistance, 1983.

Ensminger, M.E., and Olentine, C.G., Jr. *Feeds and Nutrition.* Clovis, California: Ensminger Publishing Co., 1978.

Fitts, J.W., and Fitts, J.B. *Understanding Composting.* Arlington, Virginia: Volunteers in Technical Assistance, 1984

Freeman, John A. *Survival Gardening: Enough Nutrition from 1,000 Square Feet To Live On...Just in Case!* Rock Hill, South Carolina: John's Press, 1983.

Hughes, H.D. *Forages.* Ames, Iowa: Iowa State University Press, 1966.

Hunt, Marjorie, and Bartz, Brenda. *High Yield Gardening.* Emmaus, Pennsylvania: Rodale Press, Inc., 1986.

National Academy of Sciences. *Nutrient Requirements of Poultry.* Washington, D.C.: National Academy Press, 1977.

North, M.O. *Commercial Chicken Production Manual.* Second Edition. Westport, Connecticut: AVI Publishing Co., Inc., 1978.

Orr, H.L. *Duck and Goose Raising.* Publication 532. Ontario, Canada: Ministry of Agriculture and Food.

Piliang, W.G.; Bird, H.R.; Sunde, M.L.; and Pringle, D.J. "Rice Bran as the Major Energy Source for Laying Hens." *Poultry Science.* 61 (1982): 357.

Reddy, K.R.; Khaleel, R.; and Overcash, M.R. "Behavior and Transport of Microbial Pathogens and Indicator Organisms in Soils Treated with Organic Wastes." *Journal of Environmental Quality.* Madison, Wisconsin: American Society of Agronomy, 1981.

Rodale, J., ed. *The Complete Book of Composting.* Emmaus, Pennsylvania: Rodale Press, Inc., 1969.

Russel, F. W. *Soil Conditions and Plant Growth.* London, England: Logmans Green and Co., Ltd., 1961.

Stern, Peter. *Small Scale Irrigation.* London: Intermediate Technology Publications, 1979.

Young, J.A., Evans, R.A. & Budy, J.D. *Understanding Seed Collection and Handling.* Arlington, Virginia: Volunteers in Technical Assistance, 1986.

## FOOD PROCESSING AND PRESERVATION

Anderson, Jean. *The Green Thumb Preserving Guide.* New York: William Marrow & Company, Inc., 1976.

Barbour, Beverly. *The Complete Food Preservation Book.* New York: David McKay Company, Inc., 1978.

Burch, Joan, and Burch, Monte. *Home Canning and Preserving.* Reston, Virginia: Reston Publishing Company, Inc., 1977.

Carruthers, R.T. *Understanding Fish Preservation and Processing.* Arlington, Virginia: Volunteers in Technical Assistance, 1985.

Central Food Technological Research Institute. "Home-Scale Processing and Preservation of Fruits and Vegetables." Mysore, India: The Wesley Press, 1981.

Etchells, John L., and Jones, Ivan D. "Preservation of Vegetables by Salting or Brining." Farmers' Bulletin No. 1932. Washington, D.C.: U.S. Department of Agriculture, 1944.

Groppe, Christine C., and York, George K. "Pickles, Relishes,and Chutneys: Quick, Easy, and Safe Recipes." Leaflet No. 2275. Berkeley, California: University of California, Division of Agricultural Sciences, 1975.

Hertzberg, Ruth; Vaughan, Beatrice; and Greene, Janet. *Putting Food By.* Brattleboro, Vermont: The Stephen Greene Press.

Islam, Meherunnesa. *Food Preservation in Bangladesh.* Dacca, Bangladesh: Women's Development Programme, UNICEF/DACCA, 1977.

Kluger, Marilyn. *Preserving Summer's Bounty.* New York: M. Evans and Company, Inc., 1978.

Levinson, Leonard Louis. *The Complete Book of Pickles and Relishes.* New York: Hawthorn Books, Inc., 1965.

Lindblad, Carl, and Druben, Laurel. *Small Farm Grain Storage.* Arlington, Virginia: Volunteers in Technical Assistance, 1976.

Murry, Sue T. *Home Curing Fish.* Washington, D.C.: Agriculture and Rural Development Service, Agency for International Development, 1967.

Schuler, Stanley, and Schuler, Elizabeth Meriwether. *Preserving the Fruits of the Earth.* New York: The Dial Press, 1973.

Stiebeling, Jazel K. "Solar Food Preservation." Chicago, Illinois: Illinois Institute of Technology, 1981.

Stoner, Carol Hupping, Editor. *Stocking Up: How To Preserve the Foods You Grow, Naturally.* Emmaus, Pennsylvania: Rodale Press, 1977.

U.S. Department of Agriculture. Human Nutrition Research Division. "Home Canning of Fruits and Vegetables." Washington, D.C.: U.S. Department of Agriculture, 1965.

Weber, Fred, with Stoney, Carol. *Reforestation in Arid Lands.* Arlington, Virginia: Volunteers in Technical Assistance, 1986.

Worgan, J.T. "Canning and Bottling as Methods of Food Preservation in Developing Countries." *Appropriate Technology.* 4 (November 1977): 15-16.

## CONSTRUCTION

Action Peace Corps. *Handbook for Building Homes of Earth.* Washington, D.C.: Department of Housing and urban Develop ment, (undated).

Ahrens, C. *Manual for Supervising Self-Help Home Construction with Stabilized Earth Blocks Made in the CINVA-Ram Machine.* Kanawha County, West Virginia, 1965.

American Concrete Institute. *Handbook of Concrete Engineering.* ACI-82 Manual of Practice. Detroit, Michigan: American Concrete Institute, 1982.

Buchanan, W. *Hand Moulded Burnt Clay Bricks: Labour Intensive Production.* Malawi Ministry of Trade, Industry, and Tourism (United Nations Industrial Development Organization, Project DP/MLW/78/003), undated.

*Building with Adobe and Stabilized Earth Blocks.* Washington, D.C.: United States Department of Agriculture, 1972.

Bush, Alfred. *Understanding Stabilized Earth Construction.* Arlington, Virginia: Volunteers in Technical Assistance, 1984.

Groben, E. W. *Adobe Architecture: Its Design and Construction.* Seattle, Washington: The Shorey Book Store, 1975.

International Institute of Housing Technology. *The Manufacture of Asphalt Emulsion Stabilized Soil Bricks and Brick Maker's Manual.* Fresno, California: California State University, 1972.

Lunt, M.G. *Stabilized Soil Blocks for Building.* Garston, Watford, England: Building Research Establishment, 1980.

________. "Stabilized Soil Blocks for Building." Overseas Building Notes No. 184. Garston, England: Building Research Establishment, February 1980.

*Making Building Blocks with the CINVA-Ram Block Press.* Arlington, Virginia: Volunteers in Technical Assistance, 1975.

Metalibec Ltd. *CINVA-Ram Block Cement Soil in Large Scale Housing Construction in East Punjab.* Bombay, India: Government of India Press, 1948.

*Methods for Characterizing Adobe Building Materials.* Washington, D.C.: National Bureau of Standards, 1978.

Parry, J.P. *Brickmaking in Developing Countries.* Prepared for Overseas Division, Building Research Establishment, UK. Garston, Watford, United Kingdom: Building Research Establishment, 1979.

Salvadorean Foundation for Development and Low Cost Housing Research Unit. *Stabilized Adobe.* Washington, D.C.: Organization of American States, (undated)

Sidibe, B. *Understanding Adobe.* Arlington, Virginia: Volunteers in Technical Assistance (VITA), 1985.

U.S. Agency for International Development. *Handbook for Building Homes of Earth.* Action Pamphlet No. 4200.36. By Lyle A. Wolfskill, Wayne A. Dunlop, and Bob M. Callaway. Washington, D.C.: Peace Corps, December 1979.

U. S. Dept of the Army. *Concrete, Masonry and Brickwork: A Practical Handbook for the Home Owner and Small Builder.* New York: Dover Publications, Inc., 1975.

**HOME IMPROVEMENT**

Baldwin, S. *Biomass Stoves: Engineering Design, Development, and Dissemination.* Arlington, Virginia: Volunteers in Technical Assistance, 1986.

Bruyere, John. *Country Comforts: The New Homesteaders Handbook.* New York: Sterling Publishing Co., Inc., 1979.

Bramson, Ann. *Soap.* New York: Workman Publishing Co., 1975.

Clarke, R. (ed.). *Wood-Stove Dissemination: Proceedings of the Conference Held at Wolfheze, The Netherlands.* London: Intermediate Technology Publications, Ltd., 1985.

de Silva, D. "A Charcoal Stove From Sri Lanka," *Appropriate Technology,* Vol. 7, No. 4, 1981, pp. 22-24.

Donkor, Peter. *Small-Scale Soapmaking: A Handbook.* London: Intermediate Technology Publications, 1986.

Foley, G. and Moss, P. "Improved Cooking Stoves In Developing Countries." *Earthscan* Technical Report No. 2, 1983, 175 pp. Illus.

Hassrick, P. "Umeme: A Charcoal Stove from Kenya." *Appropriate Technology* Vol. 9, No. 1, 1982, pp. 6-7.

*Making Soap and Candles.* Pownal, Vermont: P. H. Storey Communications, Inc., 1973.

Tata Energy Research Institute. *Solid Fuel Cooking Stoves.* Bombay, India, 1980.

*Testing the Efficiency of Wood-Burning Cookstoves:* International Standards. Arlington, Virginia: Volunteers in Technical Assistance, 1985.

## CRAFTS AND VILLAGE INDUSTRY

Berold, Robert, and Caine, Collette (eds.). *People's Workbook.* Johannesburg, South Africa: Enrironmental and Development Agency, 1981.

Cardew, M. *Pioneer Pottery.* New York, New York: St. Martin's press, 1976.

Conrad, J.W. Ceramic Formulas: *The Complete Compendium (A Guide to Clay, Glazes, Enamel, Glass, and Their Colors).* New York, New York: MacMillan Publishing Co., 1975.

Cooper, E. *The Potter's Book of Glaze Recipes.* New York, New York: Charles Scribner's Sons, 1980.

Green, D. *Pottery Glazes.* New York: Watson Guptill Publishing, 1973.

Lawrence and West. *Ceramic Science for the Potter.* Radnor, Pennsylvania: Chilton Book Co.

Nelson, G. *Ceramics: A Potter's Handbook.* New York: Holt, Reinhart & Winston, 1984.

Norton, F.H. *Elements of Ceramics.* Redding, Massachusetts: Addison-Wesley Publishing Co., 1974.

__________. *Kilns: Design, Construction and Operation.* Philadelphia, Pennsylvania: Chilton Book Co., 1968.

Peter Starkey. *Salt Glaze,* London: Pitman Publishing Co., 1977.

Petersham, M. *Understanding the Small-Scale Clay Products Enterprise.* Arlington, Virginia: Volunteers in Technical Assistance, 1984.

Schurecht, H.G. "Salt Glazing and Ceramic Ware." *Bulletin of the American Ceramic Society,* Vol. 23, No. 2.

"Simple Methods of of Candle Manufacture," London: Intermediate Technology Publications, Ltd., 1985.

Small-Scale Papermaking. Technical Memorandum No. 8. Geneva: International Labor Office, 1985.

Troy, J. *Salt Glazed Ceramics.* New York: Watson Guptill Publications Co., 1977.

Troy, J. *Glazes for Special Effects.* New York: Watson Guptill Publications Co.

Vogler, Jon, and Sarjeant, Peter. *Understanding Small-Scale Papermaking.* Arlington, Virginia: Volunteers in Technical Assistance, 1986.

Weygers, A.G. *The Making of Tools.* New York: Van Nostrand Reinhold Company, 1973.

Young, Jean (ed.). *Woodstock Craftsman's Manual.* New York: Praeger Publishers, 1972.

**COMMUNICATIONS AND GENERAL REFERENCE**

Berold, Robert, and Caine, Collette (eds.). *People's Workbook.* Johannesburg, South Africa: Enrironmental and Development Agency, 1981.

Darrow, Ken, and Saxenian, Mike. *Appropriate Technology Sourcebook.* Stanford, California: Volunteers in Asia, 1986.

McLaren, 1. *The Sten-Screen: Making and Using a Low-Cost Printing Process.* London: Intermediate Technology Publications, Inc., 1983.

Seymour, John. *The Complete Book of Self Sufficiency.* London: Corgi Books div. Transworld Publishers, Ltd., 1981

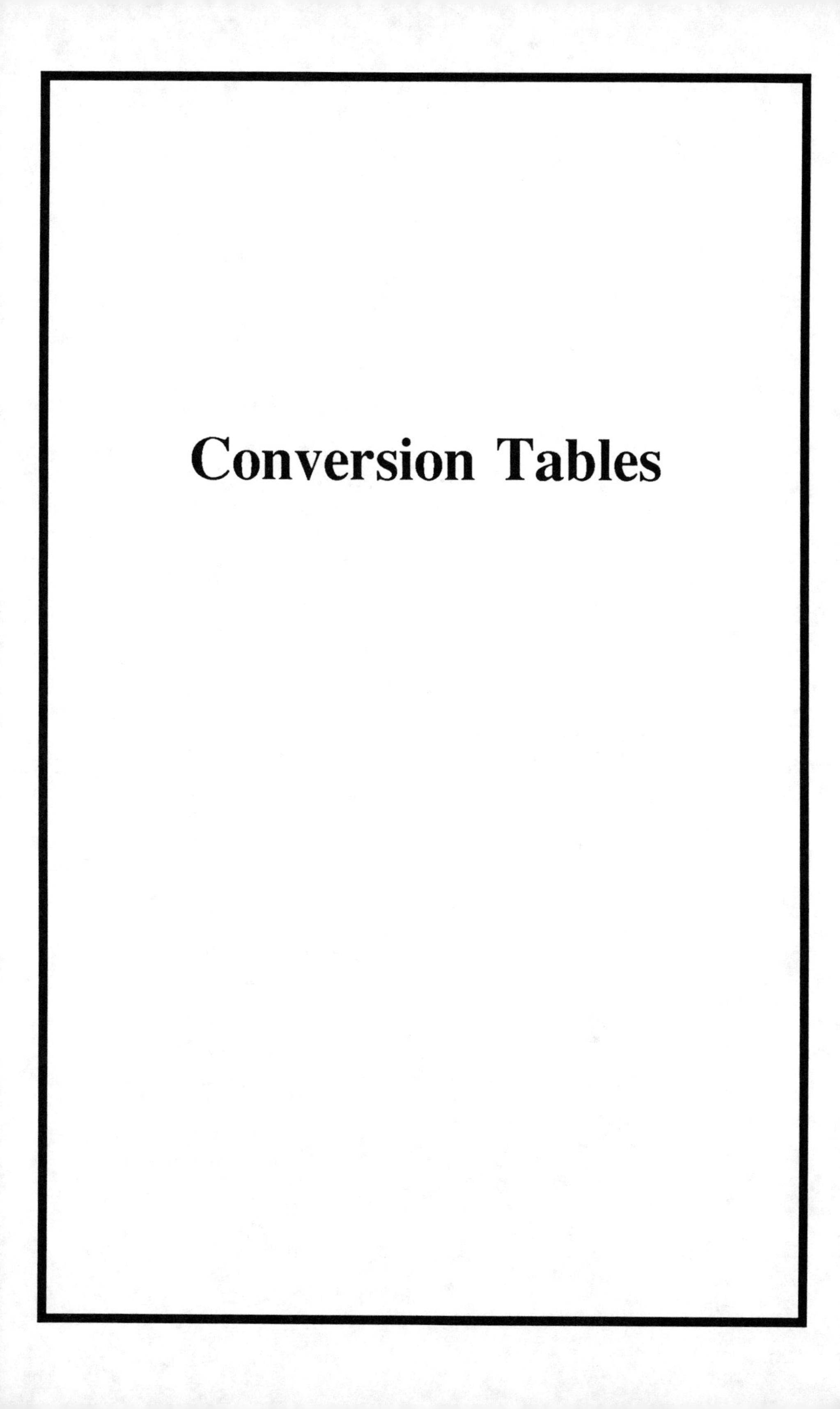

# Conversion Tables

## CONVERSION TABLES

| MULTIPLY | BY | TO OBTAIN |
|---|---|---|
| acres | 43,560 | square feet |
| acres | 4,047 | square meters |
| acres | $1.562 \times 10^{-3}$ | square miles |
| acres | 0.004047 | square kilometers |
| acres | 4840 | square yards |
| atmospheres | 76.0 | cms of mercury |
| atmospheres | 29.92 | inches of mercury |
| atmospheres | 10,333 | kgs/square meter |
| atmospheres | 14.70 | pounds/square inch |
| British thermal units | 0.2530 | kilogram-calories |
| B.t.u. | 777.5 | foot-pounds |
| B.t.u. | $3.927 \times 10^{-4}$ | horsepower-hours |
| B.t.u. | 1,054 | joules |
| B.t.u. | 107.5 | kilogram-meters |
| B.t.u. | $2.928 \times 10^{-4}$ | kilowatt-hours |
| B.t.u./min. | 0.02356 | horsepower |
| B.t.u./min. | 0.01757 | kilowatts |
| B.t.u./min. | 17.57 | watts |
| calories | 0.003968 | B.t.u. |
| calories | 3.08596 | foot-pounds |
| calories | $1.1622 \times 10^{-6}$ | kilowatt-hours |
| centimeters | 0.3937 | inches |
| centimeters | 0.01 | meters |
| centimeters of mercury | 0.1934 | pounds/square inch |
| centimeters/second | 1.969 | feet/minute |
| centimeters/second | 0.036 | kilometer/hour |
| centimeters/second | 0.6 | meters/minute |
| centimeters/second | 0.02237 | miles/hour |
| cubic centimeters | $10^{-6}$ | cubic meters |
| cubic centimeters | $6.102 \times 10^{-2}$ | cubic inches |
| cubic centimeters | $3.531 \times 10^{-5}$ | cubic feet |
| cubic centimeters | $1.308 \times 10^{-6}$ | cubic yards |
| cubic feet | 1,728 | cubic inches |
| cubic feet | 0.02832 | cubic meters |
| cubic feet | $2.832 \times 10^{4}$ | cubic centimeters |
| cubic feet | 7.481 | gallons |
| cubic feet | 28.32 | liters |
| cubic feet/minute | 472.0 | cubic cms/second |
| cubic feet/minute | 0.1247 | gallons/second |
| cubic feet/minute | 0.4720 | liters/second |
| cubic feet/minute | 62.4 | pounds water/min |
| cubic inches | $5.787 \times 10^{-4}$ | cubic feet |
| cubic inches | $1.639 \times 10^{-5}$ | cubic meters |
| cubic inches | $2.143 \times 10^{-5}$ | cubic yards |
| cubic meters | 35.31 | cubic feet |
| cubic meters | 264.2 | gallons |
| cubic meters | $10^{3}$ | liters |
| cubic yards | $7.646 \times 10^{5}$ | cubic centimeters |
| cubic yards | 27.0 | cubic feet |
| cubic yards | 46,656 | cubic inches |
| cubic yards | 0.7646 | cubic meters |
| cubic yards | 202.0 | gallons |
| cubic yards | 764.6 | liters |
| cubic yards/min. | 0.45 | cubic feet/second |

| MULTIPLY | BY | TO OBTAIN |
|---|---|---|
| cubic yards/min. | 3.367 | gallons/second |
| cubic yards/min. | 12.74 | liters/second |
| degrees (angle) | 60 | minutes |
| degrees (angle) | 0.01745 | radians |
| degrees (angle) | 3,600 | seconds |
| dynes | $1.020 \times 10^{-3}$ | grams |
| dynes | $2.248 \times 10^{-6}$ | pounds |
| ergs | $9.486 \times 10^{-11}$ | B.t.u. |
| ergs | 1 | dyne-centimeters |
| ergs | $7.376 \times 10^{-8}$ | foot-pounds |
| ergs | $10^{-7}$ | joules |
| ergs | $2.390 \times 10^{-11}$ | kilogram-calories |
| ergs | $1.020 \times 10^{-8}$ | kilogram-meters |
| ergs/second | $1.341 \times 10^{-10}$ | horsepower |
| ergs/second | $10^{-10}$ | kilowatts |
| feet | 30.48 | centimeters |
| feet | 0.3048 | meters |
| feet/second | 18.29 | meters/minute |
| foot-pounds | $1.286 \times 10^{-3}$ | B.t.u. |
| foot-pounds | $1.356 \times 10^{7}$ | ergs |
| foot-pounds | $5.050 \times 10^{-7}$ | horsepower-hours |
| foot-pounds | $3.241 \times 10^{-4}$ | kilogram-calories |
| foot-pounds | 0.1383 | kilogram-meters |
| foot-pounds | $3.766 \times 10^{-7}$ | kilowatt-hours |
| foot-pounds/minute | $1.286 \times 10^{-3}$ | B.t.u./minute |
| foot-pounds/minute | 0.01667 | foot-pounds/second |
| foot-pounds/minute | $3.241 \times 10^{-4}$ | kg-calories/min |
| foot-pounds/minute | $2.260 \times 10^{-5}$ | kilowatts |
| foot-pounds/second | $7.172 \times 10^{-2}$ | B.t.u./minute |
| foot-pounds/second | $1.818 \times 10^{-3}$ | horsepower |
| foot-pounds/second | $1.945 \times 10^{-2}$ | kg-calories/min |
| foot-pounds/second | $1.356 \times 10^{-3}$ | kilowatts |
| gallons | 0.1337 | cubic feet |
| gallons | 231 | cubic inches |
| gallons | $3.785 \times 10^{-3}$ | cubic meters |
| gallons | 3.785 | liters |
| gallons/minute | $2.228 \times 10^{-3}$ | cubic feet/second |
| gallons/minute | 0.06308 | liters/second |
| grams | $10^{-3}$ | kilograms |
| grams | $10^{3}$ | miligrams |
| grams | 0.03527 | ounces |
| grams | 0.03215 | troy ounces |
| grams/cubic centimeter | 62.43 | pounds/cubic feet |
| grams centimeters | $9.297 \times 10^{-8}$ | B.t.u. |
| horsepower | 42.44 | B.t.u./minute |
| horsepower | 33,000 | foot-pounds/minute |
| horsepower | 550 | foot-pounds/second |
| horsepower | 10.70 | kg-calories/min |
| horsepower | 0.7457 | kilowatts |
| horsepower | 745.7 | watts |
| horsepower | 1.014 | horsepower(metric) |
| horsepower-hours | 2547 | B.t.u. |
| horsepower-hours | $1.98 \times 10^{6}$ | foot-pounds |
| horsepower-hours | 641.7 | kilogram-calories |
| horsepower-hours | $2.737 \times 10^{5}$ | kilogram-meters |
| horsepower-hours | 0.7457 | kilowatt-hours |
| horsepower-hours | $2.684 \times 10^{6}$ | joules |
| inches | 2.540 | centimeters |
| inches | 254.0 | millimeters |

| MULTIPLY | BY | TO OBTAIN |
|---|---|---|
| inches of mercury | 0.03342 | atmospheres |
| inches of mercury | 1.133 | feet of water |
| inches of mercury | 345.3 | kgs/sq meter |
| inches of mercury | 70.73 | pounds/sq foot |
| inches of mercury | 0.4912 | pounds/sq inch |
| inches of water | 0.002458 | atmospheres |
| inches of water | 0.07355 | inches of mercury |
| inches of water | 25.40 | kgs/square meter |
| inches of water | 0.5781 | ounces/square inch |
| inches of water | 5.204 | pounds/square foot |
| inches of water | 0.03613 | pounds/square inch |
| joules | 0.0009458 | B.t.u. |
| joules | 0.73756 | foot-pounds |
| joules | 0.0002778 | watt-hours |
| joules | 1.0 | watt-seconds |
| kilograms | 980,665 | dynes |
| kilograms | $10^3$ | grams |
| kilograms | 2.2046 | pounds |
| kilograms | $1.102 \times 10^{-3}$ | short tons |
| kilogram-calories | 3.968 | B.t.u. |
| kilogram-calories | 3,086 | foot-pounds |
| kilogram-calories | $1.558 \times 10^{-3}$ | horsepower-hours |
| kilogram-calories | 4,183 | joules |
| kilogram-calories | 426.6 | kilogram-meters |
| kilogram-calories/min. | 51.43 | foot-pounds/second |
| kilogram-calories/min. | 0.09351 | horsepower |
| kilogram-calories/min. | 0.06972 | kilowatts |
| kilograms/hectare | .893 | pounds/acre |
| kilometers | $10^5$ | centimeters |
| kilometers | 0.6214 | miles |
| kilometers | 3,281 | feet |
| kilometers | 1,000 | meters |
| kilometers | 1093.6 | yards |
| kilometers/hour | 27.78 | centimeters/sec |
| kilometers/hour | 54.68 | feet/minute |
| kilometers/hour | 0.9113 | feet/second |
| kilometers/hour | 0.5396 | knots/hour |
| kilometers/hour | 16.67 | meters/hour |
| kilometers/hour | 0.6214 | miles/hour |
| kilowatts | 56.92 | B.t.u./minute |
| kilowatts | $4.425 \times 10^4$ | foot-pounds/minute |
| kilowatts | 737.6 | foot-pounds/second |
| kilowatts | 1.341 | horsepower |
| kilowatts | 14.34 | kg-calories/min |
| kilowatts | $10^3$ | watts |
| kilowatts-hours | 3,412 | B.t.u. |
| kilowatts-hours | $2.655 \times 10^6$ | foot-pounds |
| kilowatts-hours | 1.341 | horsepower-hours |
| kilowatts-hours | $3.6 \times 10^6$ | joules |
| kilowatts-hours | 860.5 | kilogram-calories |
| kilowatts-hours | $3.671 \times 10^5$ | kilogram-meters |
| meters | 100 | centimeters |
| meters | 3.2808 | feet |
| meters | 39.37 | inches |
| meters | $10^{-3}$ | kilometers |
| meters | $10^3$ | millimeters |
| meters | 1.0936 | yards |
| meter-kilograms | $9.807 \times 10^7$ | centimeter-dynes |

| MULTIPLY | BY | TO OBTAIN |
|---|---|---|
| meter-kilograms | $10^5$ | centimeter-grams |
| meter-kilograms | 7.233 | pound-feet |
| meters/minute | 1.667 | centimeters/second |
| meters/minute | 3.281 | feet/minute |
| meters/minute | 0.05468 | feet/second |
| meters/minute | 0.06 | kilometers/hour |
| meters/minute | 0.03728 | miles/hour |
| meters/second | 196.8 | feet/minute |
| meters/second | 3.281 | feet/second |
| meters/second | 3.6 | kilometers/hour |
| meters/second | 0.06 | kilometers/minute |
| meters/second | 2.237 | miles/hour |
| meters/second | 0.03728 | miles/minute |
| miles | $1.609 \times 10^5$ | centimeters |
| miles | 5,280 | feet |
| miles | 1.6093 | kilometers |
| miles | 1,760 | yards |
| miles/min | 88.0 | feet/second |
| miles/min | 1.6093 | kilometers/minute |
| miles/min | 0.8684 | knots/minute |
| ounces | 8.0 | drams |
| ounces | 437.5 | grains |
| ounces | 28.35 | grams |
| ounces | 0.625 | pounds |
| ounces/square inch | 0.0625 | pounds/square inch |
| pints (dry) | 33.60 | cubic inches |
| pints (liquid) | 28.87 | cubic inches |
| pounds | 444,823 | dynes |
| pounds | 7,000 | grains |
| pounds | 453.6 | grams |
| pounds | 0.45 | kilograms |
| pounds of water | 0.01602 | cubic feet |
| pounds of water | 27.68 | cubic inches |
| pounds of water | 0.1198 | gallons |
| pounds of water/min. | $2.669 \times 10^{-4}$ | cubic feet/second |
| pounds/cubic foot | 0.01602 | grams/cubic cms. |
| pounds/cubic foot | 16.02 | kgs/cubic meter |
| pounds/cubic foot | $5.787 \times 10^{-4}$ | pounds/cubic inch |
| pounds/square foot | 4.882 | kgs/sq meter |
| pounds/square foot | $6.944 \times 10^{-3}$ | pounds/square inch |
| pounds/square inch | 0.06304 | atmospheres |
| pounds/square inch | 703.1 | kgs/square meter |
| pounds/square inch | 144.0 | pounds/square foot |
| quarts (dry) | 67.20 | cubic inches |
| quarts (liquid) | 57.75 | cubic inches |
| quadrants (angle) | 90 | degrees |
| quadrants (angle) | 5,400 | minutes |
| quadrants (angle) | 1.571 | radians |
| radians | 57.30 | degrees |
| radians | 3,438 | minutes |
| radians/second | 57.30 | degrees/second |
| raidans/second | 0.1592 | revolutions/second |
| revolutions | 360.0 | degrees |
| revolutions | 4.0 | quadrants |
| revolutions | 6.283 | radians |
| revolutions/minute | 6.0 | degrees/second |
| square centimeters | $1.076 \times 10^{-3}$ | square feet |
| square centimeters | 0.1550 | square inches |
| square centimeters | $10^{-6}$ | square meters |

| MULTIPLY | BY | TO OBTAIN |
|---|---|---|
| square centimeters | 100 | square millimeters |
| square feet | $2.296 \times 10^{-5}$ | acres |
| square feet | 929.0 | square centimeters |
| square feet | 144.0 | square inches |
| square feet | 0.09290 | square meters |
| square feet | $3.587 \times 10^{-8}$ | square miles |
| square feet | 0.1111 | square yards |
| square inches | 6.452 | square centimeters |
| square inches | 645.2 | square millimeters |
| square meters | $2.471 \times 10^{-4}$ | acres |
| square meters | 10.764 | square feet |
| square meters | $3.861 \times 10^{-7}$ | square miles |
| square meters | 1.196 | square yards |
| square miles | 640.0 | acres |
| square miles | $2.7878 \times 10^{7}$ | square feet |
| square miles | 2.590 | square kilometers |
| square miles | $3.098 \times 10^{6}$ | square yards |
| square yards | $2.066 \times 10^{-4}$ | acres |
| square yards | 9.0 | square feet |
| square yards | 0.8361 | square meters |
| square yards | $3.228 \times 10^{-7}$ | square miles |
| temp (degs C) + 237 | 1.0 | abs temp (degs K) |
| temp (degs C) + 17.8 | 1.8 | temp (degs F) |
| temp (degs F) - 32 | 5/9 | temp (degs C) |
| tons (long) | 1,016 | kilograms |
| tons (long) | 2,240 | pounds |
| tons (metric) | $10^{3}$ | kilograms |
| tons (metric) | 2,205 | pounds |
| tons (short) | 907.2 | kilograms |
| tons (short) | 2,000 | pounds |
| tons (short)/sq. foot | 9,765 | kgs/square meter |
| tons (short)/sq. foot | 13.89 | pounds/square inch |
| tons (short)/sq. inch | $1.406 \times 10^{6}$ | kgs/square meter |
| tons (short)/sq. inch | 2,000 | pounds/square inch |
| yards | 0.9144 | meters |

FIGURE 1

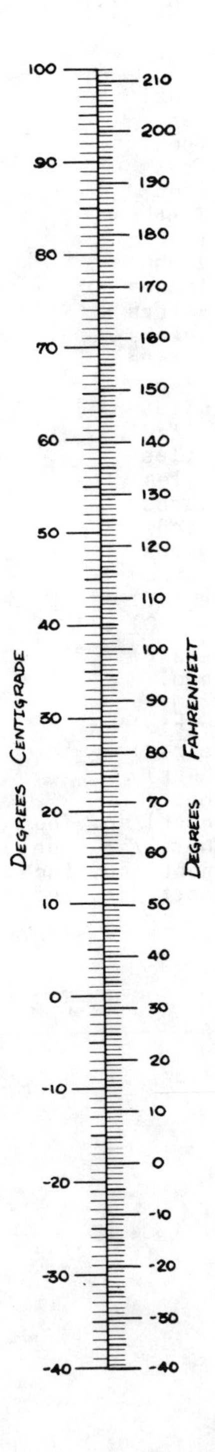


## TEMPERATURE CONVERSION

The chart in Figure 1 is useful for quick conversion from degrees Celsius (Centigrade) to degrees Fahrenheit and vice versa. Although the chart is fast and handy, you must use the equations below if your answer must be accurate to within one degree.

Equations:

Degrees Celsius = 5/9 x (Degrees Fahrenheit -32)

Degrees Fahrenheit = 1.8 x (Degrees Celsius) +32

Example:

This example may help to clarify the use of the equations; 72F equals how many degrees Celsius?

72F = 5/9 (Degrees F -32)

72F = 5/9 (72 -32)

72F = 5/9 (40)

72F = 22.2C

Notice that the chart reads 22C, an error of about 0.2C.

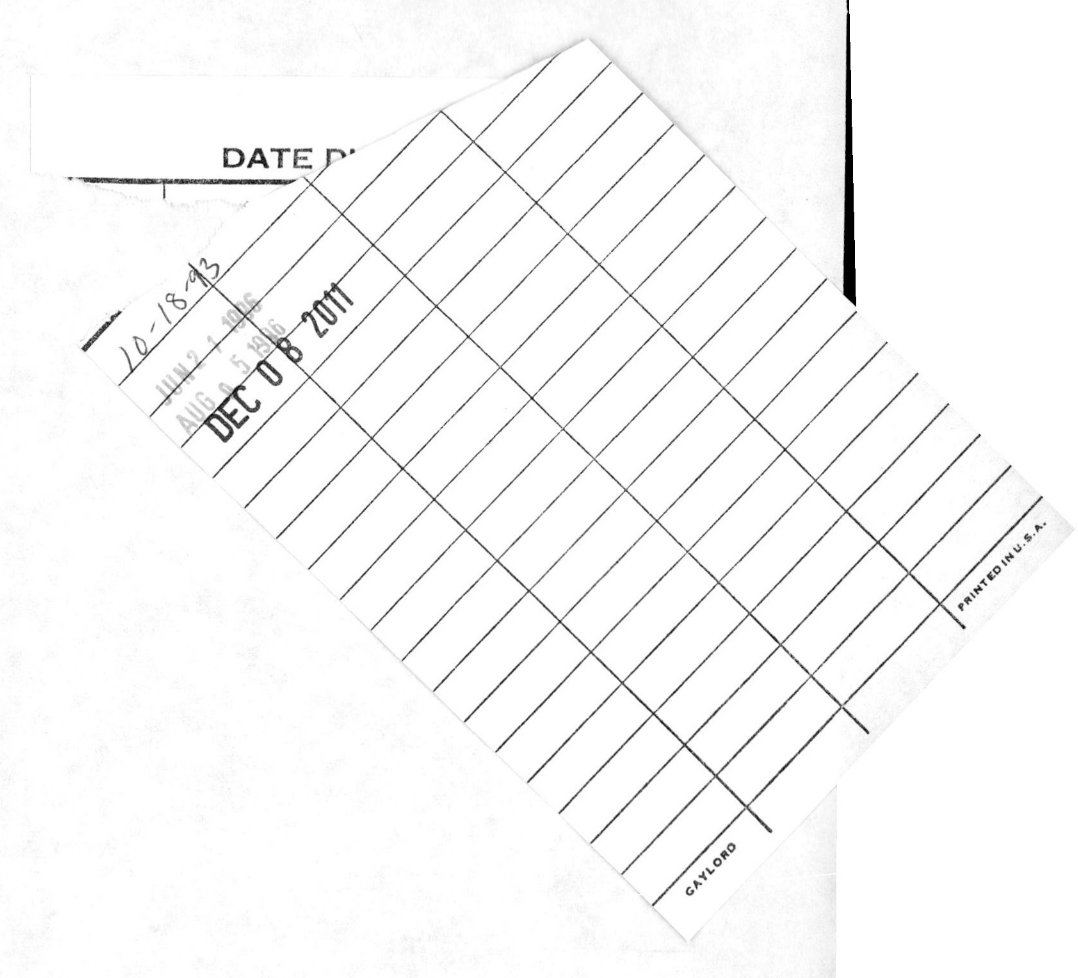
DATE D
10-18-93
JUN 2 1 1996
AUG 0 5 1996
DEC 0 8 2011
GAYLORD
PRINTED IN U.S.A.